Charles Seale-Hayne Library
University of Plymouth
(01752) 588 588
LibraryandITenquiries@plymouth.ac.uk

Oxford Lecture Series in
Mathematics and its Applications 3

Series editors

John Ball   Dominic Welsh

OXFORD LECTURE SERIES
IN MATHEMATICS AND ITS APPLICATIONS

Mathematical Topics in Fluid Mechanics

Volume 1
Incompressible Models

Pierre-Louis Lions

University Paris-Dauphine
and
Ecole Polytechnique

CLARENDON PRESS · OXFORD
1996

Oxford University Press, Walton Street, Oxford OX2 6DP

Oxford New York
Athens Auckland Bangkok Bombay
Calcutta Cape Town Dar es Salaam Delhi
Florence Hong Kong Istanbul Karachi
Kuala Lumpur Madras Madrid Melbourne
Mexico City Nairobi Paris Singapore
Taipei Tokyo Toronto
and associated companies in
Berlin Ibadan

Oxford is a trade mark of Oxford University Press

Published in the United States by
Oxford University Press Inc., New York

A catalogue record for this book is available from the British Library

Library of Congress Cataloging in Publication Data
(Data available)

ISBN 0 19 851487 5

Typeset by the author in T$_E$X
Printed in Great Britain by
Bookcraft (Bath) Ltd
Midsomer Norton, Avon

to Lila and Dorian

PREFACE

Our goal in this book is to present various mathematical results on fluid mechanics models such as, for instance, Navier–Stokes equations both in the incompressible case and in the compressible case. Most of these results are new even if some have been announced in various places. For each of these recent results, we shall give complete proofs and we try to present as much as possible self-contained proofs that do not assume from the reader really technical prerequisites other than a basic training in (nonlinear) partial differential equations. The book is divided into two volumes, the first one being essentially devoted to incompressible models while the second one is concerned with compressible models and asymptotic problems.

Before we briefly describe the topics covered here, we wish to mention that this book does not pretend to be a complete survey of the existing mathematical results even if we recall quite a few works on fluid mechanics equations. We selected what we consider to be the most significant results and in making such a biased selection we certainly omitted many relevant contributions to the field. We tried to compensate for these omissions by a rather extensive bibliography (even if some of the references included there are not quoted in the text). Let us also warn the reader that this book is concerned only with newtonian fluids and that many important subjects, such as the numerical approximation of the models we study, turbulence models, qualitative properties of solutions (bifurcation theories, attractors, inertial manifolds), reactive flows and combustion models, magnetohydrodynamics (MHD), multi-phase flows, and free boundary problems, are not even touched on here. Also, as we shall see, many basic open questions are left unanswered and we shall recall a large number of open problems. More than two centuries after the introduction by L. Euler (and later by Navier) of the fluid mechanics equations, much remains to be understood mathematically even if considerable progress has been (slowly) made. We only hope that these notes will be a small contribution to the formidable task of the mathematical understanding of fluid mechanics models.

Let us now go a little bit into the contents of this book. We begin in chapter 1 by recalling the fundamental equations modelling newtonian fluids together with the basic approximated and simplified models—more are to be found in the following chapters. Much more could be said on the derivation of these models and we strongly advise the reader to consult such classical references as G.K. Batchelor [25], and L. Landau and E. Lifschitz [281]. Many references may be found in the bibliography here.

The rest of the book is divided into three parts. The first one (which concludes volume 1) is concerned with *incompressible models* and is divided into three chapters (2, 3, 4). We begin in chapter 2 with a study of the so-called density-dependent Navier–Stokes equations. We present in section 2.1 the most general existence results for such problems, namely global existence results of weak solutions in arbitrary dimensions with possibly vanishing density (i.e. a vacuum is allowed in some regions). These results are due to R.J. DiPerna and the author—they were announced in [126], P.L. Lions [307]—and extend previous results due to various authors like S.N. Antontsev, A.V. Kazhikov and V.N. Monakhov [17], S.N. Antontsev and A.V. Kazhikov [16], J. Simon [436],[437]. Complete proofs are indicated in sections 2.3 and 2.4, while in section 2.2 we discuss regularity questions, stationary problems and mention various open questions like uniqueness, although we devote section 2.5 to partial "uniqueness results" showing that global weak solutions are equal to a strong one if it exists.

The following chapter (3) deals with the classical Navier–Stokes equations for homogeneous, incompressible fluids. In section 3.1 we deduce from the analysis performed in chapter 2 the celebrated results of J. Leray [283],[284],[285] concerning the global existence of weak solutions, and we recall various classical facts on Navier–Stokes equations that can be found in many existing books (see for instance P. Constantin and C. Foias [102], and the bibliography). Sections 3.2 and 3.3 are devoted to some recent "regularity" results in three dimensions consisting of variants or extensions of results shown by C. Guillopé, C. Foias, R. Temam [203], L. Tartar [470], P. Constantin [98], R. Coifman, P.L. Lions, Y. Meyer and S. Semmes [95]. These results aim to make precise the regularity of the global weak solutions. Finally, in section 3.4, we consider briefly the so-called Rayleigh–Benard equations and we indicate existence results of global weak solutions.

Chapter 4 is the last chapter of this part on incompressible models. The first four sections are devoted to the classical Euler equations: we first recall (4.1) the state of the art on Euler equations. Let us mention at this stage that more details can be found in A. Madja [316] and J.Y. Chemin [90]. We next (4.2) make a few remarks on the two-dimensional case, comparing the multiple notions of weak solutions and showing the existence and uniqueness of weak solutions for almost all initial conditions in L^2. We then briefly discuss in section 4.3 the fundamental open question of a priori estimates in three dimensions, and we give some details about the simple examples introduced in R.J. DiPerna and P.L. Lions [126] showing the lack of "intermediate" a priori estimates. Finally, we introduce in section 4.4 the notion of "ultra weak" solutions, which we call dissipative solutions, whose only merits are their global existence and the fact that they coincide with classical solutions as long as they exist. Despite their weakness, they will turn out to be an extremely useful tool for recovering Euler equations from some compressible models in part III. The last two sections are devoted to other incompressible models, namely density-dependent Euler equations in section 4.5, and the models obtained via the hydrostatic approximation in section 4.6.

We then present in Appendices A–E various "technical" results used in chapters 2–4.

The second volume will consist of parts II and III. The second part deals with compressible models. The first three chapters deal with the so-called (somewhat inappropriately) compressible isentropic Navier–Stokes equations and detail the results (and their proofs) announced by P.L. Lions [**303**],[**304**], [**305**]. The first chapter (5) deals with the compactness properties of sequences of solutions: we first explain (in section 5.1) difficulties encountered in such models which are due to the possible propagation of oscillations in densities (related to acoustic modes) and we state the available compactness results. Those results are shown in section 5.2, while the slightly more difficult case of Dirichlet boundary conditions is treated in section 5.3.

Chapter 6 deals with stationary (or time-discretized) problems associated with the compressible isentropic Navier–Stokes equations: we state our existence and regularity results in section 6.1 (in two and three space dimensions), we prove these results in section 6.2 and we treat the isothermal case in two dimensions in section 6.3. All these three sections are concerned with time-discretized problems. The final section of this chapter (6.4) is concerned with real stationary problems.

Next, chapter 7 is concerned with global existence results for the Cauchy problem. Our main results are stated in section 7.1 and two "different" proofs are respectively given in sections 7.2 and 7.3 which rely crucially on the results and methods introduced in chapter 5. Finally, we consider in section 7.4 the case of other (and more realistic from the applications point of view) boundary conditions.

Chapter 8 collects additional results and information on compressible models and related systems of equations. Section 8.1 is concerned with the exact compressible model and we present, in particular, some rather preliminary existence results. We come back in section 8.2 to isentropic and isothermal models and we present some global existence results of a different nature. Next, we investigate in sections 8.3 and 8.4 a shallow water model and we present some results essentially taken from P.L. Lions and P. Orenga [**309**]: section 8.3 is devoted to the global existence of weak solutions while section 8.4 is concerned with the global existence of smooth solutions. Then we discuss in section 8.5 the compressible isentropic Euler equations (i.e. the inviscid case) in one space dimension, reviewing the results due to R.J. DiPerna [124], G.Q. Chen [92], P.L. Lions, B. Perthame and E. Tadmor [311], and P.L. Lions, B. Perthame and P.E. Souganidis [310]. We also show in that section some new bounds (valid for all weak solutions and not only entropy solutions). Finally, we discuss in section 8.6 a tentative low Mach number model, directly inspired by A. Majda [315].

The final part of the book consists of a single chapter devoted to asymptotic limits. Sections 9.1, 9.2, and 9.3 are devoted to incompressible limits (small Mach number, density nearly constant) from the solutions of compressible, isentropic, Navier–Stokes equations built in chapter 7. In particular, we consider in section 9.1 the problem of recovering the global weak solutions (à la Leray) of the incompressible Navier–Stokes equations, and in section 9.2 we strengthen the convergence results in two space dimensions. Finally, in section 9.3, we obtain the solutions of the incompressible Euler equations in the smooth regime (globally in time in two space dimensions, and on the maximal time interval in three space dimensions). This analysis relies upon the notion of dissipative solutions of Euler equations introduced in section 4.4. Next, section 9.4 is devoted to the rigorous derivation of the linearized system (around a constant flow) and thus of a simple acoustics limit. Finally, section 9.5 is concerned with some asymptotic problems (of homogenization type) for the compressible, isentropic Navier–Stokes equations.

P.L.L.

Paris
December 1995

CONTENTS

Contents

INTENDED CONTENTS OF VOLUME 2

PRESENTATION OF THE MODELS

1.1 Fundamental equations for newtonian fluids

We recall here the standard derivation of the classical fluid dynamics equations in eulerian form. For the evolution of a fluid (or a gas) in N spatial dimensions $(N \geq 1)$, the description involves $(N + 2)$ fields, namely the mass density, the velocity field and the energy. We shall derive the corresponding $(N + 2)$ evolution equations in the case of a fluid filling the whole space, and we postpone the discussion of boundary conditions. Finally, let us mention that more details on the considerations which follow can be found in G.K. Batchelor [25], L. Landau and E. Lifschitz [281], and P. Germain [170] (see also the bibliography of this book).

We begin with the evolution equation for the (mass) density $\rho(= \rho(x,t))$. Applying the principle of *conservation of mass*, we simply observe that for any volume $\mathcal{O}$—say, a smooth bounded open set of $\mathbb{R}^N$—the variation of mass inside $\mathcal{O}$, i.e. $\int_{\mathcal{O}} \frac{\partial \rho}{\partial t} \, dx$, must be equal to the flux of mass on $\partial \mathcal{O}$. Since particles of fluids are moving along the integral curves of $\dot{X} = u(X,t)$, this flux is clearly equal to $-\int_{\partial \mathcal{O}} \rho \, u \cdot n \, dS$, where we denote by n the unit outward normal to $\partial \mathcal{O}$. Therefore,

$$\int_{\mathcal{O}} \frac{\partial \rho}{\partial t} \, dx = -\int_{\partial \mathcal{O}} \rho \, u \cdot n \, dS, \tag{1.1}$$

and, if we apply Stokes' formula, we deduce from (1.1)

$$\int_{\mathcal{O}} \left\{ \frac{\partial \rho}{\partial t} + \operatorname{div}(\rho u) \right\} = 0.$$

Since $\mathcal{O}$ is arbitrary, we obtain finally

$$\frac{\partial \rho}{\partial t} + \operatorname{div}(\rho u) = 0. \tag{1.2}$$

Another way (clearly equivalent) to derive (1.2) is to say that the transport of mass on a time interval $(t, t+h)$ and the conservation of mass yield

$$\rho(X(t+h), t+h)\, J(h) \ = \ \rho(x,t)$$

where $X(t) = x$, $\dot{X}(s) = u(X(s), s)$ and $J(h)$ is the jacobian of the transformation $(x \mapsto X(t+h, x))$. Standard considerations on ordinary differential equations give: $J(h) = 1 + h \operatorname{div} u(x,t) + o(h)$; therefore, we obtain

$$\rho(x,t) + h \left\{ \frac{\partial \rho}{\partial t} + u \cdot \nabla \rho + (\operatorname{div} u)\rho \right\} (x,t) + o(h) \ = \ \rho(x,t),$$

and we recover (1.2).

We now turn to the evolution of the velocity field $u = (u_1(x,t), \ldots, u_N(x,t))$ or equivalently of the momentum ρu. If we follow the first argument above and if we apply the principle of conservation of momentum, we find

$$\int_{\mathcal{O}} \frac{\partial}{\partial t} (\rho u) \, dx \ = \ -\int_{\partial \mathcal{O}} \rho u (u \cdot n) \, dS + \int_{\mathcal{O}} \rho f \, dx + \int_{\partial \mathcal{O}} F \cdot n \, dS. \qquad (1.3)$$

The two last terms of the right-hand side of (1.3) represent the forces acting on the fluid, namely volume forces corresponding to external forces (gravity, Coriolis, electromagnetic forces), and surface forces which, roughly speaking, are due to the fact that we are dealing with a continuum and which can be (somewhat improperly) thought of as contact forces with, for instance, the fluid particles lying outside $\mathcal{O}$. Clearly, F is a tensor and if we set $\sigma = -F$, the tensor σ is called the stress tensor. Classically, a fluid in motion is submitted to two kinds of stresses corresponding to compression effects and viscous effects, and one writes

$$\sigma \ = \ -p\mathbf{1} + \tau \qquad (1.4)$$

where p, a scalar function, is the pressure and τ is the viscous stress tensor. We denote by $\mathbf{1}$ the identity matrix (tensor), i.e. $\mathbf{1} = (\delta_{ij})_{ij}$.

Therefore, using (1.3) and (1.4) and the Stokes formula, we deduce finally

$$\frac{\partial}{\partial t} (\rho u) + \operatorname{div}(\rho u \otimes u) - \operatorname{div}(\tau) + \nabla p \ = \ \rho f \qquad (1.5)$$

or (in a given orthonormal basis)

$$\frac{\partial}{\partial t} (\rho u_i) + \partial_j (\rho u_i u_j - \tau_{ij} + p\, \delta_{ij}) \ = \ \rho f_i, \quad \text{for} \quad 1 \le i \le N. \qquad (1.6)$$

Here and below, we write equivalently $\partial_j = \frac{\partial}{\partial x_j}$ and systematically use the convention of implicit summation over repeated indices.

Let us observe also that if we expand the derivatives of ρu_i and $\rho u_i u_j$ in (1.6) and we use (1.2), we can also write (at least, if all functions are smooth)

$$\rho \frac{\partial u}{\partial t} + \rho(u \cdot \nabla)u - \operatorname{div}(\tau) + \nabla p = \rho f. \tag{1.7}$$

Another way (again equivalent) to derive equation (1.7) is to use Newton's equations, noticing that the acceleration of a fluid particle at (x,t) is $\frac{d}{d\tau}\{u(X(\tau),\tau)\}|_{\tau=t} = \left(\frac{\partial u}{\partial t} + (u \cdot \nabla)u\right)(x,t)$. We then obtain

$$\int_O \rho\left\{\frac{\partial u}{\partial t} + (u \cdot \nabla)u\right\} dx = \int_O \rho f\, dx + \int_{\partial O} F \cdot n\, dS$$

and (1.7) follows easily.

We have now to describe the scalar unknown p and the unknown tensor τ. Postponing the discussion of p, we recall a few facts concerning τ. First of all, the conservation of angular momentum—which we shall not describe here—leads to the following fact: τ is a symmetric tensor. Next, a classical fluid is a continuum where the constitutive law for τ is of the following form

$$\tau = \tau(Du, \rho, T) \tag{1.8}$$

where the temperature T will be discussed later on. Then, if we *postulate* that τ is a linear function of Du, is invariant under a change of reference frame (translation and rotation) and that the fluid is isotropic, we deduce from elementary algebraic manipulations that necessarily

$$\tau = \lambda \operatorname{div} u\, \mathbf{1} + 2\mu\, d \tag{1.9}$$

where d is the so-called deformation tensor

$$d = \frac{1}{2}(Du + {}^t Du) \tag{1.10}$$

and λ, μ are the so-called Lamé viscosity coefficients. Observe that in general λ and μ are functions of ρ and T in view of (1.8). The three assumptions mentioned above that lead to (1.9) correspond to the so-called *newtonian fluids*, the only case we shall consider in this book. In other words, we shall always assume here (1.9) and very often we shall consider λ and μ as fixed constants (i.e. independent of ρ and T). Furthermore, the kinetic theory of gases (monatomic gases) indicates that the Stokes relationship should hold, namely

$$\lambda = -\frac{2\mu}{3}. \tag{1.11}$$

This is the three-dimensional situation. For an "N-dimensional gas" we would obtain $\lambda = -\frac{2\mu}{N}$, while of course we still have $\lambda = -\frac{2\mu}{3}$ for the evolution of a "3-dimensional gas" depending only on one or two coordinates.

For most fluids and gases, experiments indicate that $\lambda + \frac{2\mu}{3}$ is very small and this is why it is often set to 0 for common fluids, an assumption we shall not need in the rest of this book. In practice, and this is also crucial for the mathematical analysis of these models, we have

$$\mu \geq 0, \quad \lambda + \frac{2}{N}\mu \geq 0. \tag{1.12}$$

Since $\tau = 0$ if $\lambda = \mu = 0$, this case is called the inviscid (non-viscous) case, while if $\mu > 0$, $\lambda + \mu > 0$, we have the viscous case.

We finally derive the last equation which corresponds to the conservation of energy, which expresses the first law of thermodynamics. Before we discuss the equation in more detail, let us make a physical comment: the derivation of the energy equation relies upon the assumption that, in a fluid in motion, the fluctuations around thermodynamic equilibria are sufficiently weak so that the classical thermodynamics results hold at every point and at all times. In particular, the thermodynamical state of the fluid is determined by the same state variables (thermodynamic pressure p, internal energy per unit mass e, thermodynamic temperature T, density ρ) as in classical thermodynamics, and these variables are determined by the same state equations.

Next, we observe that the total energy E is given by the sum of the kinetic energy $\rho|u|^2/2$ and of the internal energy ρe. Then the conservation of energy reads

$$\left.\begin{aligned}
\frac{d}{dt}\int_{\mathcal{O}} \rho\left(\frac{|u|^2}{2} + e\right) dx &= -\int_{\partial\mathcal{O}} \rho\left(\frac{|u|^2}{2} + e\right) u \cdot n\, dS \\
&+ \int_{\mathcal{O}} \rho f \cdot u\, dx + \int_{\partial\mathcal{O}} u \cdot (\sigma \cdot n)\, dS + Q.
\end{aligned}\right\} \tag{1.13}$$

The two last integral terms of the right-hand side of (1.13) correspond to the work done by the forces and, assuming there are no sources of heat (we are dealing with non-reacting fluids), Q is simply the amount of heat received (or lost) across the boundary $\partial\mathcal{O}$ characterized by a heat flux. In other words

$$Q = -\int_{\partial\mathcal{O}} q \cdot n\, dS$$

for some vector q to be determined. Using Stokes' formula, we deduce from (1.13) and (1.4)

$$\left.\begin{aligned}
\frac{\partial}{\partial t}\left\{\rho\left(\frac{|u|^2}{2} + e\right)\right\} &+ \operatorname{div}\left\{u\left[\rho\left(\frac{|u|^2}{2} + e\right) + p\right]\right\} \\
&= \operatorname{div}\{\tau \cdot u\} - \operatorname{div}(q) + \rho f \cdot u.
\end{aligned}\right\} \tag{1.14}$$

In order to close the system, it remains to describe p, e and q. If we choose as independent (thermodynamic) variables ρ and T (the thermodynamic temperature), then p and e are functions of ρ and T, i.e. obey some given state equations of the following form

$$p = p(\rho, T), \quad e = e(\rho, T). \tag{1.15}$$

We shall come back to this point later on. There remains to describe the heat conduction, or in other words to examine the dependence of q upon T (and ρ). If we suppose that the fluid is isotropic, we are led to

$$q = -k(\rho, T, |\nabla T|)\, \nabla T \tag{1.16}$$

for some scalar function k which, in most cases, is taken to be simply a function of ρ, T or even a constant called the thermal conduction coefficient.

We wish now to recall some basic facts from thermodynamics that will be useful later on. First of all, there exists a state variable called entropy (or entropy per unit mass), denoted by s, satisfying

$$ds = \frac{1}{T}\left\{ de + p\, d\!\left(\frac{1}{\rho}\right) \right\}, \tag{1.17}$$

and the basic assumption on thermodynamic equilibria yields the so-called Gibbs equation

$$\frac{ds}{dt} = \frac{1}{T}\left\{ \frac{de}{dt} + p\frac{d}{dt}\!\left(\frac{1}{\rho}\right) \right\} \tag{1.18}$$

where $\frac{d}{dt}$ denotes the total time derivative, i.e. along fluid particle paths, namely

$$\frac{d}{dt} = \frac{\partial}{\partial t} + u\cdot\nabla_x. \tag{1.19}$$

Then the entropy of a given volume $\mathcal{O}$ is given by $\int_{\mathcal{O}} \rho s\, dx$. Applying the second law of thermodynamics, we find

$$\int_{\mathcal{O}} \frac{\partial}{\partial t}(\rho s)\, dx \geq -\int_{\partial\mathcal{O}} \rho s\, u\cdot n\, dS - \int_{\partial\mathcal{O}} \frac{q}{T}\cdot n\, dS$$

or

$$\frac{\partial}{\partial t}(\rho s) + \operatorname{div}(u\rho s) \geq -\operatorname{div}\left(\frac{q}{T}\right). \tag{1.20}$$

On the other hand, we deduce from (1.18) and (1.2)

$$\rho\frac{ds}{dt} = \frac{1}{T}\left(\rho\frac{de}{dt} + p\operatorname{div} u \right). \tag{1.21}$$

In addition, multiplying (1.7) by u, we have

$$\rho \frac{\partial}{\partial t}\left(\frac{|u|^2}{2}\right) + \rho(u \cdot \nabla)\frac{|u|^2}{2} - \mathrm{div}\,(\tau u) + u \cdot \nabla p$$
$$= \rho f \cdot u - \tau \cdot Du = \rho f \cdot u - \tau \cdot d$$

or, in view of (1.2),

$$\frac{\partial}{\partial t}\left(\rho \frac{|u|^2}{2}\right) + \mathrm{div}\left\{u\left(\rho \frac{|u|^2}{2}\right)\right\} - \mathrm{div}\,(\tau u) + u \cdot \nabla p = \rho f \cdot u - \tau \cdot d. \quad (1.22)$$

Comparing (1.14) and (1.22), we find

$$\frac{\partial}{\partial t}(\rho e) + \mathrm{div}\,(u\rho e) + p\,\mathrm{div}\,u = -\mathrm{div}\,q + \tau \cdot d, \quad (1.23)$$

hence in view of (1.2)

$$\rho \frac{de}{dt} + p\,\mathrm{div}\,u = -\mathrm{div}\,q + \tau \cdot d,$$

and inserting this relation into (1.21), we find, using (1.2) once more

$$\frac{\partial}{\partial t}(\rho s) + \mathrm{div}\,(u\rho s) = -\frac{1}{T}\mathrm{div}\,q + \frac{1}{T}\tau \cdot d. \quad (1.24)$$

In particular, we find, comparing (1.20) and (1.24)

$$\tau \cdot d - \frac{1}{T}q \cdot \nabla T \geq 0, \quad (1.25)$$

an inequality which must thus be satisfied for all (ρ, u, T). In particular, choosing $u \equiv 0$, we deduce

$$-q \cdot \nabla T \geq 0,$$

an inequality which, combined with (1.16), yields $k \geq 0$, a fact which is consistent with experiments. Similarly, if we choose T to be constant, we deduce

$$\tau \cdot d \geq 0.$$

For a newtonian fluid, we deduce from (1.9) for all u

$$2\mu|d|^2 + \lambda(\mathrm{div}\,u)^2 \geq 0$$

and this is easily seen to be equivalent to (1.12), namely the "natural" constraint on μ and λ.

At this stage there only remains to discuss the "thermodynamical constraints" on the state equations (1.15); therefore, we choose ρ and T as independent variables for the thermodynamic relations. Then we deduce from (1.17)

$$\frac{\partial s}{\partial T} = \frac{1}{T}\frac{\partial e}{\partial T}, \qquad \frac{\partial s}{\partial \rho} = \frac{1}{T}\left\{\frac{\partial e}{\partial \rho} - \frac{p}{\rho^2}\right\}, \tag{1.26}$$

and, in particular, s can be deduced from the expressions (1.15). But (1.26) also implies the following compatibility equation: $\frac{\partial}{\partial \rho}\left(\frac{1}{T}\frac{\partial e}{\partial T}\right) = \frac{\partial}{\partial T}\left(\frac{1}{T}\left\{\frac{\partial e}{\partial \rho} - \frac{p}{\rho^2}\right\}\right)$, i.e.

$$\frac{1}{\rho^2}\left\{p - T\frac{\partial p}{\partial T}\right\} = \frac{\partial e}{\partial \rho}. \tag{1.27}$$

This relationship constrains possible laws for p and e.

Additional constraints can be derived from the second law of thermodynamics. Choosing now (e, τ) or (s, τ), where $\tau = \frac{1}{\rho}$, as independent variables, one deduces from the second law the following equivalent statements

$$s(e, \tau) \quad \text{is concave in} \quad (e, \tau) \tag{1.28}$$

or

$$e(s, \tau) \quad \text{is convex in} \quad (s, \tau). \tag{1.29}$$

These statements are obviously equivalent since (1.28) is equivalent to the convexity of the set $\{(s, e, \tau) \,/\, s \leq s(e, \tau)\}$ while (1.29) is equivalent to the convexity of the set $\{(s, e, \tau) \,/\, e(s, \tau) \leq e\}$, and those two sets are identical since (1.7) yields $\frac{\partial s}{\partial e} = \frac{1}{T} > 0$.

Let us mention two consequences of (1.28)–(1.29). First of all, ρ or τ being fixed, $\frac{\partial e}{\partial s} = T$ and thus $\frac{\partial^2 e}{\partial s^2} = \frac{\partial T}{\partial e}$. Hence, (1.29) yields

$$e(\rho, T) \quad \text{is increasing in} \quad T. \tag{1.30}$$

Similarly, s being fixed, $\frac{\partial e}{\partial \tau} = -p$, $\frac{\partial s}{\partial \tau} = \frac{p}{T}$ and thus (1.28), (1.29) imply

$$\left. \begin{array}{ll} p(\rho, T) & \text{is increasing in } \rho \text{ for } s \text{ fixed} \quad (s = s(\rho, T)) \\ p(\rho, T)T^{-1} & \text{is increasing in } \rho \text{ for } e \text{ fixed} \quad (e = e(\rho, T)). \end{array} \right\} \tag{1.31}$$

In conclusion, the restrictions on the state equations (1.15) are given by (1.27) and (1.29), while (1.26) indicates how to derive s once (1.15) is given. We wish now to conclude by giving two examples and we begin with the most common model, namely the case of an *ideal gas*. An ideal gas is a fluid which obeys the following laws: Mariotte's law, namely $p = \rho f(T)$, and Joule's law, namely $e = e(T)$ for some scalar functions f and e. Observe then that (1.27) implies that $f(T)$ is linear and we deduce

$$p = R\rho T, \quad e = e(T) \tag{1.32}$$

where, because of (1.31), $R > 0$, e is increasing in T. The constant R is called the ideal gas constant. Writing $C_v = e'(T)$ (≥ 0), $C_p = \frac{\partial h}{\partial T} = R + e'(T)$ where $h = e + \frac{p}{\rho}$ is the so-called enthalpy, we see that $C_p > C_v$ and we set $\gamma = \frac{C_p}{C_v}$.

Common fluids and gases in "normal" conditions (i.e. not with high densities or at high temperatures) are often well described by the ideal gas model, which assumes furthermore that C_p and C_v are *constants*, i.e. $e = C_v T$ with $C_v > 0$ and $\gamma = 1 + \frac{R}{C_v} > 1$. We then deduce from (1.26) that, in this case, $s = R \log \left(\frac{T^{1/(\gamma-1)}}{\rho}\right) = C_v \log \frac{T}{\rho^{\gamma-1}}$. It is possible to deduce from the kinetic theory of gases that $\gamma = \frac{N+2}{N}$ (for a monatomic gas) or $\gamma = \frac{5}{3}$ if $N = 3$. The most interesting region for physical applications is $\left(1, \frac{5}{3}\right]$.

As a mathematical exercise, it is worth looking at the case when we use only Joule's law, namely $e = e(T)$. Then, because of (1.27), we deduce

$$e = e(T), \quad p = p_1(\rho)T \qquad (1.33)$$

and because of (1.30) and (1.31), $e(T)$ and $p_1(\rho)$ have to be increasing (or at least non-decreasing). In fact, these monotonicity conditions on e and ρ are, in this case, equivalent to (1.28) since, up to some irrelevant constant, $s = h(e) = P(\tau)$ where $h'(e) = \frac{1}{T}$ $(e = e(T))$, $P'(\tau) = p_1(\frac{1}{\tau})$.

Let us conclude by summarizing the model we derived for newtonian fluids

$$
\left.
\begin{aligned}
&\frac{\partial \rho}{\partial t} + \operatorname{div}(\rho u) = 0 \\[1mm]
&\frac{\partial(\rho u_i)}{\partial t} + \operatorname{div}(\rho u u_i) - \partial_j\{\mu(\partial_j u_i + \partial_i u_j)\} \\[1mm]
&\qquad - \partial_i(\lambda \operatorname{div} u) + \partial_i p = \rho f_i, \quad \text{for } 1 \leq i \leq N, \\[1mm]
&\frac{\partial(\rho e)}{\partial t} + \operatorname{div}(\rho u e) + p \operatorname{div} u - \partial_j(k \partial_j T) \\[1mm]
&\qquad = \frac{\mu}{2}(\partial_i u_j + \partial_j u_i)^2 + \lambda(\operatorname{div} u)^2
\end{aligned}
\right\} \qquad \text{(C)}
$$

where λ, u, p, e are functions of ρ and T; k is a non-negative function of ρ, T and possibly $|\nabla T|$; λ, μ satisfy (1.12); and (1.27), (1.29) hold. The most common case being the one when λ, μ, k are *constants*, we have $p = R\rho T$ $(R > 0)$, $e = C_v T$ $(C_v > 0)$ (the ideal gas) and we set $\gamma = 1 + \frac{R}{C_v}$. Finally, s, defined (up to a constant) by (1.26), satisfies

$$
\left.
\begin{aligned}
&\frac{\partial(\rho s)}{\partial t} + \operatorname{div}(u \rho s) \\[1mm]
&\qquad = -\frac{1}{T} \operatorname{div}(k \nabla T) + \frac{\mu}{2}(\partial_i u_j + \partial_j u_i)^2 + \lambda(\operatorname{div} u)^2.
\end{aligned}
\right\} \qquad (1.34)
$$

When $\mu > 0$, $\lambda + \frac{2\mu}{N} > 0$, the system of equations (C) is called the "*compressible Navier–Stokes equations*".

1.2 Approximated and simplified models

In this section we wish to discuss successively *incompressible models*, *low Mach number expansions*, *ideal fluids* and some simplified mathematical models. Many more approximations or simplifications are possible (shallow waters, quasi-geostrophic approximations, etc.), some of which will be studied in the rest of this book, in which case we will say a few words on their derivation.

We thus begin with *incompressible models* or, more accurately, models for incompressible fluids, which are very important for applications since many common fluids (liquids) are incompressible or only very slightly compressible. Mathematically, the incompressibility means

$$\operatorname{div} u \ = \ 0. \tag{1.35}$$

Indeed, if at fixed time $t > 0$, we consider the volume of an open set $\mathcal{O}$ filled with the fluid, then, at time $t + h$ ($h > 0$), the corresponding fluid particles will fill the set

$$\mathcal{O}_h \ = \ \{x(t{+}h, y) \,/\, y \in \mathcal{O}\}$$

where $\dot{x}(\tau) = u(x(\tau), \tau)$, $x(t, y) = y$. The volume of $\mathcal{O}_h$ is given by

$$\int_{\mathcal{O}} J(h, y)\, dy$$

where $J(h, y)$ is the jacobian of $(y \mapsto x(t{+}h, y))$. Finally, it is well known that we have

$$\dot{J}(\tau) \ = \ \operatorname{div} u(x(\tau), \tau)\, J(\tau).$$

Therefore, the volume of $\mathcal{O}_h$ is equal to the volume of $\mathcal{O}$ (for all $\mathcal{O}$, $h \geq 0$) if and only if (1.35) holds.

Next, we may follow the same derivation of the equations as in the preceding section and we recover the conservation of mass (1.2) and the integral form (1.3) of the conservation of momentum. Notice that since $\operatorname{div} u = 0$, if ρ is constant initially it remains so, in which case we say the fluid is *homogeneous* ($\rho \equiv \bar{\rho} \in (0, \infty)$). Also, the stress tensor σ has to preserve the incompressibility and we deduce from the hypothesis of a newtonian flow

$$\sigma \ = \ 2\mu d - p\,\mathbf{1} \tag{1.36}$$

where $\mu \geq 0$. The *hydrostatic pressure*, p, is in fact a Lagrange multiplier associated to the incompressibility constraint (notice that $\tilde{\sigma}_{ij}\partial_i u_j = 0$ for all u such that $\operatorname{div} u = 0$ if and only if $\tilde{\sigma}_{ij} = p\,\delta_{ij}$ for some p). Notice finally that $\operatorname{div} d = \frac{1}{2}\partial_j(\partial_i u_j + \partial_j u_i) = \frac{1}{2}\Delta u_i$ since $\operatorname{div} u = 0$, and we obtain finally

$$\left.\begin{aligned}\frac{\partial}{\partial t}(\rho u_i) + \operatorname{div}(\rho u u_i) - \partial_j(\mu(\partial_i u_j + \partial_j u_i)) + \partial_i p = \rho f_i,\\ \text{for } 1 \leq i \leq N, \quad \operatorname{div} u = 0,\end{aligned}\right\}\quad(1.37)$$

or, if μ is constant,

$$\left.\begin{aligned}\frac{\partial}{\partial t}(\rho u_i) + \operatorname{div}(\rho u u_i) - \mu\Delta u_i + \partial_i p = \rho f_i,\\ \text{for } 1 \leq i \leq N, \ \operatorname{div} u = 0.\end{aligned}\right\}\quad(1.38)$$

In conclusion, we look for the (ρ, u, p) solution of the following system of equations

$$\left.\begin{aligned}\frac{\partial \rho}{\partial t} + \operatorname{div}(\rho u) = 0\\ \frac{\partial(\rho u_i)}{\partial t} + \operatorname{div}(\rho u u_i) - \partial_j(\mu(\partial_i u_j + \partial_j u_i)) + \partial_i p = \rho f_i\\ \text{for } 1 \leq i \leq N, \quad \operatorname{div} u = 0.\end{aligned}\right\}\quad(1.39)$$

If $\mu > 0$ (and is constant), i.e. if viscous effects are present, this system of equations is the so-called *non-homogeneous incompressible Navier–Stokes equations* (quite often, incompressible is omitted), and a particularly relevant case consists in choosing $\rho \equiv \bar{\rho} > 0$, in which case (1.39) reduces to

$$\left.\begin{aligned}\bar{\rho}\frac{\partial u_i}{\partial t} + \bar{\rho}\operatorname{div}(u u_i) - \mu\Delta u_i + \partial_i p = \bar{\rho} f_i,\\ \text{for } 1 \leq i \leq N, \ \operatorname{div} u = 0.\end{aligned}\right\}\quad(1.40)$$

This system is called the *homogeneous, incompressible Navier–Stokes equations*, often simply called the Navier–Stokes equations(!). Replacing μ by $\nu = \frac{\mu}{\rho}$, p by $\frac{p}{\rho}$, we see it is enough to study the case when $\bar{\rho} = 1$.

It remains to write down an equation for the internal energy $e\ (= e(\rho, T))$: following the procedure described in the preceding section and taking into account the incompressibility condition (1.35), we find

$$\frac{\partial(\rho e)}{\partial t} + \operatorname{div}(u\rho e) - \operatorname{div}(k\nabla T) = \frac{\mu}{2}(\partial_i u_j + \partial_j u_i)^2 \quad(1.41)$$

where $k \geq 0$ may depend on ρ, T. The most common case corresponds to $k > 0$, k constant (i.e. independent of T) and, if we are in the homogeneous situation, $e = e(\bar{\rho}, T)$ is only a function of T, for instance $e = C_v T$, for some $C_v > 0$.

We next discuss low *Mach number expansions*. We just want to present here some *formal* asymptotics, some of which allow us to go from the compressible models introduced in the preceding section to the incompressible models we stated above. First of all, let us recall the precise definition of the Mach number: $M = \frac{|u|}{\sqrt{p'(\rho)}}$. Therefore, letting M go to 0 means that, keeping ρ, T of a typical size 1, we consider u of order ε where $\varepsilon > 0$ is a small parameter (going to zero). We also rescale the time variable, considering finally

$$\left. \begin{aligned} \rho_\varepsilon(x,t) = \rho\left(x, \frac{t}{\varepsilon}\right), \quad u_\varepsilon(x,t) = \frac{1}{\varepsilon} u\left(x, \frac{t}{\varepsilon}\right), \\ T_\varepsilon(x,t) = T\left(x, \frac{t}{\varepsilon}\right) \end{aligned} \right\} \tag{1.42}$$

where (ρ, u, T) solve (C) and λ, μ, k depend upon the scaling parameter ε in a way to be determined.

We begin with the "traditional" derivation of the homogeneous, incompressible Navier–Stokes equations. We then take $\mu(\varepsilon) = \varepsilon\mu$, $\lambda(\varepsilon) = \varepsilon\lambda$, $k(\varepsilon) = \varepsilon k$, $f(\varepsilon) = \varepsilon^2 f$, where λ, μ, k may be functions of (ρ, T), and we deduce from (C)

$$\left. \begin{aligned} &\frac{\partial \rho_\varepsilon}{\partial t} + \operatorname{div}(u_\varepsilon \rho_\varepsilon) = 0 \\ &\frac{\partial (\rho_\varepsilon u_\varepsilon)}{\partial t} + \operatorname{div}(\rho_\varepsilon u_\varepsilon \otimes u_\varepsilon) - \operatorname{div}(2\mu d_\varepsilon + \lambda \operatorname{div} u_\varepsilon) \\ &\qquad\qquad + \nabla\left(\frac{1}{\varepsilon^2} p_\varepsilon\right) = \rho_\varepsilon f \\ &\frac{\partial (\rho_\varepsilon e_\varepsilon)}{\partial t} + \operatorname{div}(\rho_\varepsilon u_\varepsilon e_\varepsilon) - \operatorname{div}(k \nabla T_\varepsilon) + p_\varepsilon \operatorname{div} u_\varepsilon \\ &\qquad\qquad = \varepsilon^2 \left\{ \frac{\mu}{2} (\partial_i u_{\varepsilon,j} + \partial_j u_{\varepsilon,i})^2 + \lambda (\operatorname{div} u_\varepsilon)^2 \right\}. \end{aligned} \right\} \tag{1.43}$$

The equation corresponding to the conservation of momentum indicates that p_ε should behave like $P(t) + \varepsilon^2 p + o(\varepsilon^2)$ where $P(t)$ is independent of x. One way to ensure this fact consists in assuming that ρ_ε and T_ε are initially constant (up to terms of order ε) in order to avoid initial layers, and that p_ε is initially constant up to terms of order ε^2 (this is clearly the case if ρ_ε and T_ε are also constant up to terms of order ε^2). Then, at least formally, if we set

$$\rho_\varepsilon = \bar{\rho} + \varepsilon\rho_1 + \mathcal{O}(\varepsilon^2), \quad T_\varepsilon = \bar{T} + \varepsilon T_1 + \mathcal{O}(\varepsilon^2), \quad u_\varepsilon = u + \mathcal{O}(\varepsilon)$$

we deduce from (1.43)

$$p(\overline{\rho}, \overline{T}) = P(t) > 0 \quad \text{is independent of } x \tag{1.44}$$

$$\frac{\partial p}{\partial \rho}(\overline{\rho}, \overline{T})\rho_1 + \frac{\partial p}{\partial T}(\overline{\rho}, \overline{T})T_1 = P_1(T) \quad \text{is independent of } x \tag{1.45}$$

$$\frac{\partial \overline{\rho}}{\partial t} + \text{div}\,(u\overline{\rho}) = 0 \tag{1.46}$$

$$\frac{\partial(\overline{\rho}u)}{\partial t} + \text{div}\,(\overline{\rho}u \otimes u) - \text{div}\,(2\mu d + \lambda\,\text{div}\,u) + \nabla\pi = \overline{\rho}f \tag{1.47}$$

$$\frac{\partial(\overline{\rho}\,\overline{e})}{\partial t} + \text{div}\,(\overline{\rho}u\overline{e}) - \text{div}(k\nabla\overline{T}) + P(t)\,\text{div}\,u = 0 \tag{1.48}$$

where $\overline{e} = e(\overline{\rho}, \overline{T})$, $\mu = \mu(\overline{\rho}, \overline{T})$, $k = k(\overline{\rho}, \overline{T})$, $\pi = \lim_{\varepsilon \to 0_+} \frac{1}{\varepsilon^2}(p_\varepsilon - P - \varepsilon P_1)$.

From now on, we assume, to simplify the calculations, the ideal gas laws: $p = R\rho T$, $e = C_v T$. Then we deduce from (1.48) that $\overline{\rho}\,\overline{e} = (C_v/R)P(t) = \frac{1}{\gamma-1}P(t)$ is independent of t; this, in fact, requires some boundary conditions like, for instance, if we consider the evolution in the whole space $\mathbb{R}^n$, u and ∇T vanishing at infinity. Hence, $P(t) = P_0 > 0$. We then deduce from (1.48) and (1.46)

$$\frac{\partial}{\partial t}(\log \overline{T}) + u \cdot \nabla(\log \overline{T}) = -\left(\frac{\partial}{\partial t} + u \cdot \nabla\right)(\log \overline{\rho})$$

$$= \text{div}\,u = \frac{1}{\gamma P_0}\,\text{div}\,(k\nabla\overline{T}).$$

Hence, if $\overline{T}$ is constant initially, $\overline{T}$ is equal to that constant for all (x, t). Therefore, $\overline{\rho}$ and $\overline{T}$ are positive constants and $\text{div}\,u = 0$. In conclusion, (1.46)–(1.48) reduce to

$$\overline{\rho}\frac{\partial u}{\partial t} + \overline{\rho}u \cdot \nabla u - \overline{\mu}\Delta u + \nabla\pi = \overline{\rho}f, \quad \text{div}\,u = 0$$

where $\overline{\mu} = \mu(\overline{\rho}, \overline{T})$, $\overline{\rho} > 0$, $\overline{T} = \frac{P_0}{\rho}$, and we recover the homogeneous, incompressible Navier–Stokes equations.

We next present some variants of the above formal derivation. First of all, we consider the case when $\mu(\varepsilon) = \varepsilon\mu$, $\lambda(\varepsilon) = \varepsilon\lambda$, $k(\varepsilon) = \varepsilon k_\varepsilon$ and $k_\varepsilon \to \overline{k} \geq 0$ as $\varepsilon \to 0_+$, $f(\varepsilon) = \varepsilon^2 f$. We still assume the ideal gas laws ($p = R\rho T$, $e = C_v T$) and we now assume that p is initially constant up to terms of order ε^2 but we do not assume that ρ or T are initially constant (or equivalent to constants). With the same notation as above, we still recover (1.46)–(1.48) with k replaced by $\overline{k}$. Exactly as before, we deduce that $P \equiv P_0 > 0$ and

$$\gamma P_0\,\text{div}\,u = \text{div}\,(\overline{k}\nabla\overline{T}). \tag{1.49}$$

In particular, if $\overline{k} = 0$, i.e. $k(\varepsilon) = o(\varepsilon)$, $\operatorname{div} u = 0$ and (1.46)–(1.48) become

$$\left. \begin{array}{l} \dfrac{\partial \overline{\rho}}{\partial t} + \operatorname{div}(u\overline{\rho}) = 0, \quad \operatorname{div} u = 0, \\[2mm] \dfrac{\partial(\overline{\rho}u)}{\partial t} + \operatorname{div}(\overline{\rho}u \otimes u) - \operatorname{div}(2\overline{\mu}d) + \nabla\pi = \overline{\rho}f, \end{array} \right\}$$

i.e. the non-homogeneous, incompressible Navier–Stokes equations. Notice that $\overline{\mu} = \mu\left(\overline{\rho}, \frac{P_0}{\rho}\right)$ and that, if μ is independent of ρ and T, $\operatorname{div}(2\overline{\mu}d) = \mu\Delta u$.

Next, if $\overline{k} > 0$, we obtain the following new system of equations

$$\left. \begin{array}{l} \dfrac{\partial \overline{\rho}}{\partial t} + \operatorname{div}(u\overline{\rho}) = 0, \quad \gamma \operatorname{div} u = \operatorname{div}\left(\overline{k}\nabla\left(\frac{1}{\overline{\rho}}\right)\right) \\[2mm] \dfrac{\partial \overline{\rho}u}{\partial t} + \operatorname{div}(\overline{\rho}u \otimes u) - \operatorname{div}(2\overline{\mu}d + \overline{\lambda}\operatorname{div} u) + \nabla\pi = \overline{\rho}f. \end{array} \right\} \quad (1.50)$$

We would like to point out that we are not aware of any formal derivation of incompressible models including a temperature equation (or energy equation) like (1.41). One possible physical explanation for this apparent lack of consistency between compressible and incompressible models is the following assertion: compressible models are valid for gases and the incompressible limit yields particular incompressible models, while the general incompressible models aim to describe liquids where compression effects are neglected.

We next briefly make a few remarks on *perfect fluids* which correspond to the particular case when $\lambda \equiv \mu \equiv k \equiv 0$. The compressible models (also called *compressible Euler equations* in this case) take the following form

$$\left. \begin{array}{l} \dfrac{\partial \rho}{\partial t} + \operatorname{div}(\rho u) = 0 \\[2mm] \dfrac{\partial \rho u}{\partial t} + \operatorname{div}(\rho u \otimes u) + \nabla p(\rho, T) = \rho f \\[2mm] \dfrac{\partial}{\partial t}\left(\rho\left(\frac{|u|^2}{2} + e\right)\right) + \operatorname{div}\left\{u\left\{\rho\frac{|u|^2}{2} + +\rho e + p\right\}\right\} = \rho f \cdot u \end{array} \right\} \quad (1.51)$$

and in the ideal gas case $p(\rho, T) = R\rho T$, $e = C_v T$, $\gamma = 1 + \frac{R}{C_v}$. Furthermore, we might expect that the entropy equation (1.34) simply becomes

$$\frac{\partial \rho s}{\partial t} + \operatorname{div}(u\rho s) = 0 \quad \text{or} \quad \rho\frac{\partial s}{\partial t} + \rho u \cdot \nabla s = 0.$$

However, this is the case only on regions where ρ, u, T are smooth. In fact, it is well known that shocks (discontinuities) appear in finite time and that the entropy equation does not hold "on shocks". Instead, only the following inequality, which is deduced from (1.34), remains true:

$$\frac{\partial \rho s}{\partial t} + \operatorname{div}(\rho u s) \geq 0. \tag{1.52}$$

As long as shocks do not form, the entropy equation is valid and if initially s is constant, it remains constant: $s \equiv s_0$. In the ideal gas case, this leads to $p = (Re^{s_0/C_0})\rho^\gamma = c_0 \rho^\gamma$ $(c_0 > 0)$ and we obtain the so-called *isentropic gas dynamics* system

$$\left.\begin{array}{l} \dfrac{\partial \rho}{\partial t} + \operatorname{div}(\rho u) = 0 \\[2mm] \dfrac{\partial (\rho u)}{\partial t} + \operatorname{div}(\rho u \otimes u) + c_0 \nabla \rho^\gamma = 0. \end{array}\right\} \tag{1.53}$$

This system has clearly a somewhat limited physical validity in view of what we have just recalled but has recently received a lot of (mathematical) attention. For a general pressure law, $s = s_0$ means $p = \bar{p}(\rho)$ $(= p(\rho, T)$ with T determined by $s = s_0$) where $\bar{p}$ is increasing in ρ (see (1.31)).

For a perfect fluid the incompressible models take the following form: in the *homogeneous* case, we obtain the classical Euler equations

$$\frac{\partial u}{\partial t} + \operatorname{div}(u \otimes u) + \nabla p = 0, \quad \operatorname{div} u = 0 \tag{1.54}$$

(the constant density $\rho \equiv \bar{\rho}$ is scaled out by considering $p/\bar{\rho}$ instead of p).

In the inhomogeneous case, we obtain the density-dependent Euler equations

$$\left.\begin{array}{l} \dfrac{\partial \rho}{\partial t} + \operatorname{div}(\rho u) = 0, \quad \operatorname{div} u = 0 \\[2mm] \dfrac{\partial (\rho u)}{\partial t} + \operatorname{div}(\rho u \otimes u) + \nabla p = 0. \end{array}\right\} \tag{1.55}$$

We now turn to *some models* obtained by further simplifications or different (singular) asymptotic limits. We first consider the compressible system (C) with μ, λ, k constants ($\mu, \lambda > 0$, $k \geq 0$) and we neglect the heating due to viscous dissipation, an approximation which is reasonable except for hypersonic gases. We then obtain for the temperature (or energy) equation

$$\frac{\partial \rho e}{\partial t} + \operatorname{div}(\rho u e) + \operatorname{div} u p - k \Delta T = 0. \tag{1.56}$$

We also assume Joule's law: $e = e(T)$, $p = \bar{p}(\rho)T$ (for instance the ideal gas law $e = C_v T$, $p = R\rho T$) where $\bar{p}$ is increasing. Next, we take $k = 0$ (we do not claim this is a relevant physical assumption!) obtaining an equation which is easily seen to be equivalent to the following entropy equation

$$\rho \frac{\partial s}{\partial t} + \rho u \cdot \nabla s = 0. \tag{1.57}$$

In particular, if s is constant initially, we now expect to deduce $s \equiv s_0$ (we do not expect shocks since $\lambda, \mu > 0$), and we obtain the *compressible isentropic Navier–Stokes equations*

$$\left. \begin{aligned} &\frac{\partial \rho}{\partial t} + \operatorname{div}(\rho u) = 0 \\ &\frac{\partial(\rho u)}{\partial t} + \operatorname{div}(\rho u \otimes u) - \mu \Delta u - (\lambda + \mu)\nabla \operatorname{div} u + \nabla p(\rho) = \rho f \end{aligned} \right\} \tag{1.58}$$

where $p(\rho) = \bar{p}(\rho T)$ (and $s(\rho, T) = s_0$) is increasing in ρ because of (1.31). In particular, in the ideal gas case, $p(\rho) = c_0 \rho^\gamma$ with $c_0 = Re^{s_0/C_v}$.

Let us make at this stage a general remark, which is related to isentropic or barotropic pressure laws. Precise state equations like $p = p(\rho, T)$ are not an easy matter nor are they fully understood. In particular, many semi-empirical laws exist (depending on the physical phenomena to be studied). For instance, one can find the following pressure law for water in R. Courant and K.O. Friedrichs [108] (p. 8): $p = A(\rho/\rho_0)^\gamma - B$ with the proposed values $\gamma = 7$, $A = 3001$, $B = 3000$ (atm.) and ρ_0 is the density at 0°C.

We finally describe some (mathematical) formal asymptotics. First of all, we take $p = p(\rho)T$, $e = C_v T$ and we let C_v go to $+\infty$ (!). The temperature equation then yields

$$\rho \frac{\partial T}{\partial t} + \rho u \cdot \nabla T = 0$$

or

$$\rho \frac{\partial T}{\partial t} + \rho u \cdot \nabla T - \kappa \Delta T = 0$$

if $\frac{k}{C_v} \to \kappa > 0$. A particular solution is clearly $T \equiv T_0 > 0$. We then obtain the *compressible, isothermal Navier–Stokes equations*

$$\left. \begin{aligned} &\frac{\partial \rho}{\partial t} + \operatorname{div}(\rho u) = 0 \\ &\frac{\partial(\rho u)}{\partial t} + \operatorname{div}(\rho u \otimes u) - \mu \Delta u - (\lambda + \mu)\nabla \operatorname{div} u + T_0 \nabla p(\rho) = 0 \end{aligned} \right\} \tag{1.59}$$

and $p(\rho) = R\rho$ in the ideal gas case ($p(\rho)$ is increasing, as usual).

The other limit consists in letting k go to $+\infty$. We deduce $T \equiv T(t)$ (independent of x), and in the ideal gas case, the function T is determined by the conservation of energy

$$\frac{d}{dt} \left\{ \int \rho \frac{|u|^2}{2} + TM \right\} = \int \rho f \cdot u$$

where $M = \int \rho$ (independent of t), or equivalently by the entropy

$$R \frac{d}{dt} \int \rho \log \frac{T^{1/(\gamma-1)}}{\rho} = \int \frac{1}{T} \left\{ \frac{\mu}{2} |\partial_i u_j + \partial_j u_i|^2 + \lambda (\operatorname{div} u)^2 \right\}.$$

We now conclude this section (and the chapter) with a brief discussion of *initial* and *boundary conditions*. Our main goal in this book is to study the Cauchy problems for the models derived and described above. In other words, we study the systems of equations for $t \geq 0$ prescribing the values of $\rho, \rho u, \rho e$ at $t = 0$. Of course, for the models introduced in this section in which one does not write an energy equation or one takes $\rho \equiv \overline{\rho}$ constant, we only prescribe ρ and ρu, or u at $t = 0$.

The question of boundary conditions is much more delicate and would require a detailed discussion. Our ambition in this book is somewhat limited since we shall consider problems set in a domain Ω, with standard (mathematical) boundary conditions on $\partial\Omega$ for (possibly) ρ, u, T. We shall essentially always restrict ourselves to three cases:

(i) Ω is a smooth, bounded, connected, open set in $\mathbb{R}^N$, and we impose Dirichlet-type boundary conditions on $\partial\Omega$ on ρ and u while we impose on T Dirichlet, Neumann or mixed boundary conditions. A typical example would be the case of a homogeneous Dirichlet condition, i.e. $u \cdot n = 0$ on $\partial\Omega$ or $u = 0$ on $\partial\Omega$ if viscous terms are included in the model, in which case no conditions on ρ on $\partial\Omega$ are imposed. If we incorporate a temperature equation, a rare event unfortunately in this book, we can impose for instance homogeneous Neumann conditions: $\frac{\partial T}{\partial n} = 0$ on $\partial\Omega$. Here and above, n denotes the unit outward normal to $\partial\Omega$.

(ii) $\Omega = \prod_{i=1}^{N}(0, L_i)$ with $L_i > 0$ ($\forall\, 1 \leq i \leq N$) and all functions are periodic in each x_i of period L_i, for all $1 \leq i \leq N$.

(iii) $\Omega = \mathbb{R}^N$ and (ρ, u, T) "vanish at infinity" or are "constant at infinity".

PART I

INCOMPRESSIBLE MODELS

2

DENSITY-DEPENDENT
NAVIER–STOKES EQUATIONS

This chapter is devoted to the so-called density-dependent Navier–Stokes equations or inhomogeneous, incompressible Navier–Stokes equations, namely

$$\frac{\partial \rho}{\partial t} + \mathrm{div}\,(\rho u) = 0, \quad \rho \geq 0, \quad \text{in } \Omega \times (0, +\infty) \tag{2.1}$$

$$\left.\begin{array}{c} \dfrac{\partial \rho u}{\partial t} + \mathrm{div}\,(\rho u \otimes u) - \mathrm{div}\,(2\mu d) + \nabla p = \rho f, \\[2mm] \mathrm{div}\, u = 0, \quad \text{in } \Omega \times (0, +\infty) \end{array}\right\} \tag{2.2}$$

where $d = \frac{1}{2}\left(\partial_i u_j + \partial_j u_i\right)$, f is given on $\Omega \times (0, +\infty)$ and $\mu = \mu(\rho)$ is a continuous, positive function on $[0, +\infty)$. Of course, (2.1), (2.2) are to be complemented with boundary and initial conditions.

The above model was derived in chapter 1 but we wish to add to that derivation the fact that it can be also seen, due to the possible dependence of μ on ρ, as a model for the evolution of a multi-phase flow consisting of several immiscible, incompressible fluids with constant densities and various viscosity coefficients.

2.1 Existence results

Let us first describe the boundary and initial conditions for (2.1)–(2.2) which we shall consider here. We study only three model cases (recall that $N \geq 2$):

(i) (Dirichlet case) Ω is a smooth, bounded, connected open subset of $\mathbb{R}^N$ and

$$u = 0 \quad \text{on } \partial\Omega. \tag{2.3}$$

(ii) (Periodic case) $\Omega = \prod_{i=1}^{N}(0, L_i)$ ($L_i > 0$, $\forall\, 1 \leq i \leq N$) and (ρ, u) are periodic so that we can consider that (2.1)–(2.2) hold on $\mathbb{R}^N \times (0, +\infty)$ with (ρ, u) periodic in each x_i of period L_i, for all $i \in \{1, \ldots, N\}$.

(iii) ($\mathbb{R}^N$ case) $\Omega = \mathbb{R}^N$ and we want ρ to be bounded (for example) while u satisfies, in a sense to be made precise,

$$u \to u_\infty \qquad \text{as } |x| \to +\infty, \quad \text{for all } t \geq 0 \qquad (2.4)$$

where u_∞ is fixed in $\mathbb{R}^N$.

We now discuss initial conditions. In view of (2.1)–(2.2), we need to impose conditions on ρ and ρu at $t = 0$. Observe that we cannot directly impose initial conditions on u in case ρ vanishes on some part of Ω, i.e. if there is some vacuum. We then consider, if Ω is bounded,

$$\left.\begin{array}{l} \rho_0 \geq 0 \quad \text{a.e. in } \Omega, \quad \rho_0 \in L^\infty(\Omega), \\[4pt] m_0 \in L^2(\Omega)^N, \quad m_0 = 0 \quad \text{a.e. on } \{\rho_0 = 0\}, \\[4pt] |m_0|^2/\rho_0 \in L^1(\Omega) \end{array}\right\} \qquad (2.5)$$

where we agree that $|m_0|^2/\rho_0 = 0$ a.e. on $\{\rho_0 = 0\}$. We also want to impose

$$\rho|_{t=0} = \rho_0 \quad \text{on } \Omega, \quad \rho u|_{t=0} = m_0 \quad \text{on } \Omega. \qquad (2.6)$$

Of course, in the periodic case, we may extend ρ_0 and m_0 periodically on $\mathbb{R}^N$.

If $\Omega = \mathbb{R}^N$ (case (iii) above), (2.5) is replaced (if $u_\infty \not\equiv 0$) by

$$\left.\begin{array}{l} \rho_0 \geq 0 \quad \text{a.e. in } \mathbb{R}^N, \quad \rho_0 \in L^\infty(\mathbb{R}^N), \\[4pt] m_0 - \rho_0 u_\infty \in L^2(\mathbb{R}^N), \quad m_0 = 0 \quad \text{a.e. on } \{\rho_0 = 0\}, \\[4pt] |m_0 - \rho_0 u_\infty|^2/\rho_0 \in L^1(\mathbb{R}^N). \end{array}\right\} \qquad (2.7)$$

For technical reasons, we need to assume, in the case when $\Omega = \mathbb{R}^N$, in addition to (2.7), one of the following three conditions

$$(1/\rho_0)\, 1_{(\rho_0 < \delta_0)} \in L^1(\mathbb{R}^N), \quad \text{for some } \delta_0 > 0, \qquad (2.8)$$

or

$$(\underline{\rho} - \rho_0)^+ \in L^p(\mathbb{R}^N), \quad \text{for some } \underline{\rho} \in (0, \infty),\ p \in \left(\frac{N}{2}, \infty\right), \qquad (2.9)$$

or

$$\left.\begin{array}{l} \text{if } N = 2, \quad \displaystyle\int_{\mathbb{R}^2} \rho_0^p <x>^{2(p-1)} (\log <x>)^r\, dx < \infty \\[6pt] \qquad\qquad \text{for some } p \in (1, +\infty],\ \text{with } r > 2p - 1 \\[6pt] \text{if } N \geq 3, \quad \rho_0 \in L^{\frac{N}{2}, \infty}(\mathbb{R}^N) \end{array}\right\} \qquad (2.10)$$

where we define $<x> = (1 + |x|^2)^{1/2}$.

Finally, we assume that the force f satisfies

$$f \in L^2(\Omega \times (0,T))^N, \quad \text{for all} \quad T \in (0,\infty). \tag{2.11}$$

Since we shall always make these assumptions in the rest of the chapter—unless explicitly mentioned—we shall not recall them.

Our main existence result will state the global existence of weak solutions that could be called "solutions à la Leray" by analogy with the classical global existence results for the homogeneous, incompressible Navier–Stokes equations obtained by J. Leray [283],[284],[285], and we wish now to define precisely what we mean by weak solutions. We look for solutions satisfying, for all $T \in (0,\infty)$, $R \in (0,\infty)$: $\rho \in L^\infty(\Omega \times (0,\infty))$; $u \in L^2(0,T;H_0^1(\Omega))^N$ (Dirichlet case (i)), $u \in L^2(0,T;H_{\text{per}}^1)^N$ (periodic case (ii)), $u \in L^2(0,T;H^1(B_R))^N$ and $u \in L^2(0,T;L^{\frac{2N}{N-2}}(\mathbb{R}^N))^N$ if $N \geq 3$ (whole space case (iii)) ; $\rho|u|^2 \in L^\infty(0,T;L^1(\Omega))$ if Ω is bounded, $\rho|u - u^\infty|^2 \in L^\infty(0,\infty;L^1(\mathbb{R}^N))$ if $\Omega = \mathbb{R}^N$; $\nabla u \in L^2(\Omega \times (0,T))$; $\rho \in C([0,\infty);L^p(\Omega))$ if Ω is bounded, $\rho \in C([0,\infty);L^p(B_R))$ if $\Omega = \mathbb{R}^N$ for all $1 \leq p < \infty$. Finally, if (2.8) or (2.9) hold, we require that $u \in L^2(\mathbb{R}^N \times (0,T))$ for all $T \in (0,\infty)$.

Here and everywhere below, $B_R = \{y \in \mathbb{R}^N \ / \ |y| < R\}$, $H^1(\mathcal{O}) = \{f \in L^2(\mathcal{O}) \ / \ \partial_i f \in L^2(\mathcal{O})\}$, $H_0^1(\Omega)$ is the subspace of $H^1(\Omega)$ consisting of functions whose trace on $\partial\Omega$ vanishes, $H_{\text{per}}^1 = \{f \in H^1(B_R)$ for all $R < \infty$, f is periodic in x_i of period L_i for all $1 \leq i \leq N\}$. Furthermore, (2.1) holds in the sense of distributions (for example in $\Omega \times (0,\infty)$ (case (i)) or in $\mathbb{R}^N \times (0,\infty)$ (cases (ii) and (iii)) and the initial condition on ρ (contained in (2.6)) is meaningful since we require ρ to be continuous in time (with values in L^p or L_{loc}^p). Of course, we require that div u vanishes as a distribution (on $\Omega \times (0,\infty)$ or $\mathbb{R}^N \times (0,\infty)$ in the periodic case or if $\Omega = \mathbb{R}^N$). In the periodic case (ii), ρ is periodic in x_i of period L_i for all $1 \leq i \leq N$, for all $t \in (0,\infty)$. It only remains to explain the meaning of (2.2) and of the initial condition on ρu (contained in (2.6)).

We shall use a weak formulation based upon a class of smooth test functions, namely the class Φ of $\phi \in C^\infty(\mathbb{R}^N \times [0,\infty))^N$, ϕ has compact support in $\Omega \times [0,\infty)$ in the Dirichlet case or if $\Omega = \mathbb{R}^N$, ϕ is periodic in x_i of period L_i for all $1 \leq i \leq N$ in the periodic case, div $\phi = 0$ on $\mathbb{R}^N \times [0,\infty)$. We want u to satisfy, for all $\phi \in \Phi$,

$$\left.\begin{aligned} -\int_\Omega m_0 \cdot \phi(x,0)\,dx + \iint_{\Omega \times (0,\infty)} \Big\{ -\rho u \cdot \frac{\partial\phi}{\partial t} - \rho u_i u_j\, \partial_i \phi_j \\ + \frac{1}{2}\mu(\partial_i u_j + \partial_j u_i)(\partial_i \phi_j + \partial_j \phi_i) - \rho f \cdot \phi \Big\}\,dx\,dt = 0. \end{aligned}\right\} \tag{2.12}$$

Let us also emphasize the fact that, in view of classical density results—see, for instance, R. Temam [472], R. Dautray and J.L. Lions [115]—we could equivalently take test functions ϕ, say in the Dirichlet case, such that $\operatorname{div}\phi = 0$, $\phi \in L^2(0,T; H^1_0(\Omega))^N$ and $\frac{\partial\phi}{\partial t} \in L^2(\Omega \times (0,T))^N$ for all $T \in (0,\infty)$, ϕ vanishes on Ω for t large, $\nabla\phi \in L^2(0,T; L^p(\Omega))$ (for instance) where $p > 2$ if $N = 2$, $p = N$ if $N \geq 3$. This last integrability requirement follows from the fact that, by Sobolev embeddings, we have, for all $T \in (0,\infty)$, if $N \geq 3$ (for example)

$$\rho|u|^2 \in L^\infty(0,T; L^1(\Omega)), \quad \rho|u|^2 \in L^1\left(0,T; L^{\frac{N}{N-2}}(\Omega)\right)$$

hence $\rho u_i u_j \in L^2\left(0,T; L^{\frac{N}{N-1}}(\Omega)\right)$ ($\forall\, 1 \leq i,j \leq N$).

At least formally, we expect solutions of (2.1)–(2.2) to satisfy the energy identities:

$$\frac{d}{dt}\int_\Omega \rho|u|^2\,dx + \int_\Omega \mu(\partial_i u_j + \partial_j u_i)^2\,dx = \int_\Omega 2\rho f \cdot u$$

if Ω is bounded, or

$$\frac{d}{dt}\int_\Omega \rho|u-u_\infty|^2\,dx + \int_\Omega \mu(\partial_i u_j + \partial_j u_i)^2\,dx = \int_\Omega 2\rho f \cdot (u-u_\infty)$$

if $\Omega = \mathbb{R}^N$. These identities are obtained upon multiplying (2.2) by u (or by $u - u_\infty$), observing that, because of (2.1), $\left(\frac{\partial}{\partial t}(\rho u) + \operatorname{div}(\rho u \otimes u)\right) \cdot u = \left((\rho\frac{\partial}{\partial t} + \rho u \cdot \nabla)u\right) \cdot u = \left(\rho\frac{\partial}{\partial t} + \rho u \cdot \nabla\right)\frac{|u|^2}{2} = \left\{\frac{\partial}{\partial t}\left(\rho\frac{|u|^2}{2}\right) + \operatorname{div}\left(\rho u \frac{|u|^2}{2}\right)\right\}$ and integrating by parts over Ω.

As it is often the case when dealing with global weak solutions of nonlinear partial differential equations, the global weak solutions we obtain satisfy the following *energy inequalities*

$$\frac{d}{dt}\int_\Omega \rho|u|^2\,dx + \int_\Omega \mu(\partial_i u_j + \partial_j u_i)^2\,dx \leq 2\int_\Omega \rho f \cdot u \quad \text{in } \mathcal{D}'(0,\infty), \quad (2.13)$$

$$\left. \begin{aligned} &\int_\Omega \rho|u|^2\,dx + \int_0^t\!\!\int_\Omega \mu(\partial_i u_j + \partial_j u_i)^2\,dx\,ds \\ &\quad \leq \int_\Omega \frac{|m_0|^2}{\rho_0}\,dx + 2\int_0^t\!\!\int_\Omega \rho f \cdot u\,dx\,ds \quad \text{a.e. } t \in (0,\infty), \end{aligned} \right\} \quad (2.14)$$

if Ω is bounded, and if $\Omega = \mathbb{R}^N$

$$\left. \begin{aligned} &\frac{d}{dt}\int_{\mathbb{R}^N} \rho|u-u_\infty|^2\,dx + \int_{\mathbb{R}^N} \mu(\partial_i u_j + \partial_j u_i)^2\,dx \\ &\quad \leq 2\int_{\mathbb{R}^N} \rho f \cdot (u-u_\infty) \quad \text{in } \mathcal{D}'(0,\infty), \end{aligned} \right\} \quad (2.15)$$

$$\left.\begin{array}{l} \displaystyle\int_{\mathbb{R}^N} \rho|u-u_\infty|^2 \, dx + \int_0^t \int_{\mathbb{R}^N} \mu(\partial_i u_j + \partial_j u_i)^2 \, dx \, ds \\[3mm] \displaystyle \leq \int_{\mathbb{R}^N} \frac{|m_0 - \rho_0 u_\infty|^2}{\rho_0} \, dx + 2 \int_0^t \int_\Omega \rho f \cdot (u - u_\infty) \, dx \, ds \\[3mm] \text{a.e. } t \in (0, \infty). \end{array}\right\} \quad (2.16)$$

We may now state our main existence result.

Theorem 2.1. *There exists a global weak solution (ρ, u) of (2.1)–(2.2) satisfying the boundary and initial conditions described above and the energy inequalities stated above. Furthermore, we have for all $0 \leq \alpha \leq \beta < \infty$*

$$\left.\begin{array}{l} \text{meas}\,\{x \in \mathbb{R}^N \,/\, \alpha \leq \rho(x,t) \leq \beta\} \quad (\in [0, +\infty]) \\[2mm] \text{is independent of } t \geq 0. \end{array}\right\} \quad (2.17)$$

In particular, if $\Omega = \mathbb{R}^N$ and $\rho_0 - \rho^\infty \in L^p(\mathbb{R}^N)$ for some $1 \leq p < \infty$, $\rho^\infty \in [0, \infty)$, $\rho - \rho^\infty \in C([0, \infty); L^p(\mathbb{R}^N))$ and $\|\rho - \rho^\infty\|_{L^p(\mathbb{R}^N)}$ is independent of $t \geq 0$. Also, if $\rho_0 \equiv \bar\rho$ on Ω for some $\bar\rho \in [0, \infty)$, then $\rho \equiv \bar\rho$ on $\Omega \times [0, \infty)$.

Remarks 2.1. 1) Global existence results for non-homogeneous, incompressible Navier–Stokes equations were first obtained by A.V. Kazhikov [254]—see also A.V. Kazhikov and S.H. Smagulov [260], S.N. Antontsev and A.V. Kazhikov [16], S.N. Antontsev, A.V. Kazhikov and V.N. Monakhov [17]—in the case when μ is independent of ρ, ρ_0 is bounded away from 0. These results were extended by various authors and in particular by J. Simon [435],[436],[437] allowing ρ_0 to vanish but still with a constant μ. The methods of proofs do not adapt to the situation treated here, namely μ depending on ρ. Furthermore, the continuity in t of ρ (with values in L^p) or properties like (2.17) were not known with the above generality. Particular cases of results similar to Theorem 2.1 were obtained by R.J. DiPerna and the author relying on general results on "transport equations" shown in [128] and were announced by R.J. DiPerna and P.L. Lions in [126]. This is the approach we detail (and extend) here.

2) Many questions are left open; we shall come back to these in the next section. Let us only mention a rather technical matter: the weak formulation (2.12) yields the existence of a pressure field p (defined up to a "distribution in t only") which is essentially a distribution (one can analyse its singularity in terms of negative Sobolev spaces, or sums of negative Sobolev spaces)—see J. Simon [437] for more details; however, we do not know if p (appropriately normalized) $\in L^1_{\text{loc}}(\Omega \times (0, \infty))$.

3) The final statement of Theorem 2.1 is a trivial consequence of (2.17). Obviously, if $\rho \equiv \bar\rho > 0$, u is a weak solution of the homogeneous, incompressible Navier–Stokes equations (1.40) (the classical Navier–Stokes equations!). We shall come back to this particular case in chapter 4.

4) Let us also explain how (2.17) yields the statements on $\rho - \rho^\infty$: indeed, (2.14) implies that the distribution function of $\rho - \rho^\infty$ is independent of $t \geq 0$, therefore $\|\rho - \rho^\infty\|_{L^p(\mathbb{R}^N)}$ is independent of t. Since we already know, by definition of weak solutions, that $\rho \in C([0, \infty); L^p_{\text{loc}}(\mathbb{R}^N))$, we deduce the fact stated in Theorem 2.1, namely $\rho \in C([0, \infty); L^p(\mathbb{R}^N))$, from the following classical observation from measure theory. Let $\omega_n \in L^p(\mathbb{R}^N)$ be such that $\omega_n \underset{n}{\rightarrow} \omega$ in $L^p_{\text{loc}}(\mathbb{R}^N)$, and ω_n and ω are equimeasurable, i.e. have the same distribution function. Then $\omega_n \underset{n}{\rightarrow} \omega$ in $L^p(\mathbb{R}^N)$. Working with $\sqrt{|\omega_n|}\,\text{sign}\,\omega_n$ instead of ω_n if $p = 1$, we see that it is enough to consider the case when $1 < p < \infty$. Then, obviously ω_n converges weakly in $L^p(\mathbb{R}^N)$ to ω while $\|\omega_n\|_{L^p} = \|\omega\|_{L^p}$. The strong convergence in $L^p(\mathbb{R}^N)$ follows.

5) At the end of this chapter (in section 2.4), we will briefly explain how the proof of Theorem 2.1 yields even more general results where we are allowed to take $\rho^0 \in L^p(\Omega)$ if Ω is bounded, $\rho^0 \in L^p_{\text{loc}}(\mathbb{R}^N)$ if $\Omega = \mathbb{R}^N$ and $p \geq \frac{N}{2}$ if $N \geq 3$ while $p > 1$ if $N = 2$. In this case, we need to assume that $(t \mapsto \mu(t))$ is bounded and bounded away from 0 on $[0, \infty)$ ($0 < \underline{\mu} \leq \mu(t) \leq \overline{\mu} < \infty$ for all $t \in [0, \infty)$).

6) We will show in section 2.3 that, for any weak solution (ρ, u) as in Theorem 2.1, the following property holds: if $\tilde{\rho} \in L^\infty(\mathbb{R}^N \times (0, \infty))$ solves (2.1), $\tilde{\rho}|_{t=0} \equiv \rho_0$ and $\tilde{\rho}|u|^2 \in L^1(\mathbb{R}^N \times (0, T))$ for all $T \in (0, \infty)$ if $\Omega = \mathbb{R}^N$, then $\tilde{\rho} \equiv \rho$. In other words, u being fixed, ρ is the unique solution of (2.1) with the initial condition given by ρ_0.

7) We wish to point out that the galilean invariance of the fluid mechanics models allows us to restrict our attention, without loss of generality, to the case when $u_\infty = 0$. Indeed, if $\Omega = \mathbb{R}^N$ and (ρ, u) solves (2.1)–(2.2), then $(\rho(x + V(t), t), (u(x + V(t), t) - v(t)))$ also solves (2.1)–(2.2) where $v \in L^1(0, T; \mathbb{R}^N)$ ($\forall\, T \in (0, \infty)$) and $V(t) = \int_0^t v(s)\, ds$. In particular, if we choose $v(t) \equiv u_\infty$, we only have to work with $(\rho(x + u_\infty t, t), u(x + u_\infty t, t) - u_\infty)$ and thus we only need to study the case $u_\infty = 0$.

8) Let us finally mention another consequence of (2.17) in the whole space case ($\Omega = \mathbb{R}^N$). We claim that if (2.8) or (2.9) hold then $u \in L^2(\mathbb{R}^N \times (0, T))$ for all $T \in (0, \infty)$. Indeed, if (2.9) holds, (2.17) implies that (2.9) holds with $\rho(t)$ replacing ρ_0. We then deduce from the Sobolev and Hölder inequalities, for all $t \geq 0$, defining $a = (\overline{\rho} - \rho)^T$, that

$$\underline{\rho}\|u\|^2_{L^2(\mathbb{R}^N)} \leq \|a\|_{L^p(\mathbb{R}^N)} \|u\|^{2\theta}_{L^2(\mathbb{R}^N)} \|\nabla u\|^{2(1-\theta)}_{L^2(\mathbb{R}^N)}$$

where $\frac{\theta}{2} + \frac{(1-\theta)N-2}{2N} = \frac{p-1}{p}$. Observe that $\theta \in (0, 1)$. Our claim follows easily since $\nabla u \in L^2(\mathbb{R}^N \times (0, T))$ for all $T \in (0, \infty)$.

If (2.8) holds, we observe that (2.17) also implies the equality

$$\|(1/\rho(t))\,1_{(\rho(t) < \delta_0)}\|_{L^1(\mathbb{R}^N)} = \|(1/\rho_0)\,1_{(\rho_0 < \delta_0)}\|_{L^1(\mathbb{R}^N)}.$$

Then, $u = u\,1_{(\rho<\delta_0)} + u\,1_{(\rho\geq\delta_0)}$ and $u\,1_{(\rho\geq\delta_0)} \in L^\infty(0,T;L^2(\mathbb{R}^N))$ since $|u|\,1_{(\rho\geq\delta_0)} \leq \frac{1}{\sqrt{\delta_0}}\sqrt{\rho}\,|u|$. In addition, we have

$$|u|\,1_{(\rho<\delta_0)} = (\sqrt{\rho}\,|u|)\frac{1}{\sqrt{\rho}})1_{(\rho<\delta_0)} \in L^\infty(0,T;L^1(\mathbb{R}^N)),$$

$$\text{for all } T \in (0,\infty).$$

By Sobolev embeddings, we also have, if $N \geq 3$: $u \in L^2(0,T;L^{2N/(N-2)}(\mathbb{R}^N))$ ($\forall\, T \in (0,\infty)$) and thus $|u|\,1_{(\rho<\delta_0)} \in L^2(\mathbb{R}^N \times (0,T))$ ($\forall\, T \in (0,\infty)$), and even $L^{2\frac{N+2}{N}}(0,T;L^2(\mathbb{R}^N))$. If $N = 2$, we obtain the L^2 bound using the results shown in Appendix B. $\square$

Finally, we would like to make a general comment on the initial condition imposed on ρu and on the weak formulation (2.12). Indeed, we did not require any time continuity (with values in an arbitrarily negative Sobolev space) upon ρu and, in fact, there is no reason why, in the sense of distributions, ρu should converge to m_0 at t goes to 0_+. The only information we can obtain is that, roughly speaking, ρu converges to m_0 up to a "gradient-like" distribution. A convincing argument which shows that we cannot expect more is to consider the case when $\rho_0 \geq \underline{\rho} > 0$. Then, by Theorem 2.1, $\rho \geq \underline{\rho}$ on $\Omega \times [0,\infty)$. If $(\rho u)(t) \to m_0$ in $L^1_{\text{loc}}(\Omega)$ as $t \to 0_+$, then, since $\rho \in C([0,\infty);L^1_{\text{loc}})$, $u(t) \to u_0 = \frac{m_0}{\rho_0}$ in $L^1_{\text{loc}}(\Omega)$ as $t \to 0_+$. In particular, we must have

$$\text{div}\,(u_0) = 0 \quad \text{in} \quad \mathcal{D}'(\Omega). \tag{2.18}$$

We did not impose such a condition on u_0 since, in general, u_0 does not exist when ρ_0 vanishes.

This is clearly a delicate point that we wish to clarify—in particular because we shall encounter similar difficulties in the study of incompressible limits (chapter 9). In order to keep ideas clear and avoid unpleasant technicalities, we only state and prove the following result in the case when $\Omega = \mathbb{R}^N$ and $u^\infty = 0$ even if similar results can be obtained in the case of Dirichlet boundary conditions—the periodic case or the case $\Omega = \mathbb{R}^N$, $u^\infty \neq 0$ are easily adapted from the case we treat. In order to avoid (too) negative Sobolev spaces, we introduce some notation: we denote by $R_i = (-\Delta)^{-1/2}\frac{\partial}{\partial x_i}$ ($1 \leq i \leq N$) the usual Riesz transform and we denote respectively by $R\wedge$, R and $R\cdot$ the operators $(-\Delta)^{-1/2}\,\text{rot}$, $(-\Delta)^{-1/2}\nabla$ and $(-\Delta)^{-1/2}\,\text{div}$. We shall also need the following elementary variant of Hodge–de Rham decomposition:

Lemma 2.1. *Let* $N \geq 2$, $\rho \in L^\infty(\mathbb{R}^N)$ *such that* $\rho \geq \underline{\rho} > 0$ *a.e. on* $\mathbb{R}^N$ *for some* $\underline{\rho} \in (0,\infty)$. *Then there exist two bounded operators* P_δ, Q_δ *on*

$L^2(\mathbb{R}^N)^N$ (whose norms depend only on ρ and $\|\rho\|_{L^\infty(\mathbb{R}^N)}$) such that for all $m \in L^2(\mathbb{R}^N)^N$, $(p,q) = (P_\rho m, Q_\rho m)$ is the unique solution in $L^2(\mathbb{R}^N)$ of

$$m = p + q, \quad R \cdot \left(\frac{1}{\rho} p \right) = 0, \quad R \wedge q = 0. \tag{2.19}$$

Furthermore, if $\rho_n \in L^\infty(\mathbb{R}^N)$, $\underline{\rho} \leq \rho_n \leq \overline{\rho}$ a.e. on $\mathbb{R}^N$ for some $0 < \underline{\rho} \leq \overline{\rho} < \infty$ and ρ_n converges a.e. to ρ, then $(P_{\rho_n} m_n, Q_{\rho_n} m_n)$ converges weakly in $L^2(\mathbb{R}^N)$ to $(P_\rho m, Q_\rho m)$ whenever m_n converges weakly to m.

Remarks 2.2. 1) Notice that we always have $\int_{\mathbb{R}^N} \frac{1}{\rho} (P_\rho m) \cdot (Q_\rho m) \, dx = 0$ and thus $P_\rho^* = \frac{1}{\rho} P_\rho(\rho \cdot)$, $Q_\rho^* = \frac{1}{\rho} Q_\rho(\rho \cdot)$. In other words, the above decomposition simply corresponds to a change of metric induced by $\frac{1}{\rho}$ in the usual decompositions.

2) If m_n converges strongly in L^2 to m, then $P_{\rho_n} m_n$, $Q_{\rho_n} \rho_n$ also converge strongly. Indeed, we have in view of the weak convergence stated above (recall that $\rho_n f_n \xrightarrow[n]{} \rho f$, $\frac{1}{\rho_n} f_n \xrightarrow[n]{} \frac{1}{\rho} f$ if $f_n \xrightarrow[n]{} f$ weakly in L^2)

$$\int_{\mathbb{R}^N} |P_{\rho_n} m_n|^2 \, dx = \int_{\mathbb{R}^N} \left\{ \frac{1}{\rho_n} P_{\rho_n} (\rho_n P_{\rho_n} m_n) \right\} \cdot m_n \, dx$$

$$\xrightarrow[n]{} \int_{\mathbb{R}^N} \left\{ \frac{1}{\rho} P_\rho(\rho P_\rho m) \right\} \cdot m \, dx = \int_{\mathbb{R}^N} |P_\rho m|^2 \, dx. \quad \square$$

Proof of Lemma 2.1. We first recall the usual "div–curl" decomposition, namely the Hodge decomposition

$$m = p_1 + q_1, \quad R \cdot p_1 = 0, \quad R \wedge q_1 = 0. \tag{2.20}$$

Recall that $p_1 = P_1 m$, $q_1 = Q_1 m$ define bounded operators on L^2. Next, (2.19) is equivalent to finding a unique $v \in L^2(\mathbb{R}^N)$ such that

$$p = p_1 + Rv, \quad R \cdot \left(\frac{1}{\rho} p \right) = 0$$

or equivalently to finding a unique $v \in L^2(\mathbb{R}^N)$ such that

$$R \cdot \frac{1}{\rho} (Rv + p_1) = 0 \quad \text{in} \quad \mathbb{R}^N. \tag{2.21}$$

The unknowns p and q are then given by $p = p_1 + Rv$, $q = q_1 - Rv$.

The equation (2.21) is an unusual way of writing the elliptic equation

$$\operatorname{div} \left(\frac{1}{\rho} (\nabla V + p_1) \right) = 0 \quad \text{in} \quad \mathbb{R}^N$$

where $V = (-\Delta)^{-1/2}v$. The only advantage of (2.21) is that $N = 2$ does not play any role there. In particular, the solution of (2.21) is the unique minimum over $L^2(\mathbb{R}^N)$ of

$$\int_{\mathbb{R}^N} \frac{1}{2\rho} |Rv|^2 + \frac{p_1}{\rho} \cdot Rv \, dx.$$

The bounds follow trivially since we have

$$\int_{\mathbb{R}^N} \frac{1}{\rho} |Rv|^2 \, dx = -\int_{\mathbb{R}^N} \frac{p_1}{\rho} Rv \, dx$$

hence

$$\|Rv\|_{L^2(\mathbb{R}^N)} \|\rho\|_{L^\infty(\mathbb{R}^N)}^{-1/2} \leq \left(\int_{\mathbb{R}^N} \frac{|p_1|^2}{\rho} \, dx \right)^{1/2}$$

$$\leq (\underline{\rho})^{-1/2} \|p_1\|_{L^2(\mathbb{R}^N)} \leq (\underline{\rho})^{-1/2} \|m\|_{L^2(\mathbb{R}^N)}.$$

In order to prove that the weak limit of $P_{\rho_n} m_n$ is $P_\rho m$, we use a standard "hilbertian strategy". We first observe that P_{ρ_n} is bounded on $L^2(\mathbb{R}^N)$ uniformly in n in view of the preceding bounds. Next, for any $\psi \in L^2(\mathbb{R}^N)$, $\rho \frac{\psi}{\rho_n} \xrightarrow{n} \psi$ in $L^2(\mathbb{R}^N)$ (by Lebesgue's theorem) and thus it is enough to check that

$$\int_{\mathbb{R}^N} (P_{\rho_n} m_n) \cdot \rho \frac{\psi}{\rho_n} \xrightarrow{n} \int_{\mathbb{R}^N} (P_\rho m) \cdot \psi \, dx.$$

But, we have, in view of Remarks 2.2 (1),

$$\int_{\mathbb{R}^N} \frac{1}{\rho_n} (P_{\rho_n} m_n) \cdot \rho \psi = \int_{\mathbb{R}^N} m_n \cdot \frac{1}{\rho_n} P_{\rho_n} (\rho \psi) \, dx$$

and it is enough to show that $P_{\rho_n} \varphi$ converges (strongly) in $L^2(\mathbb{R}^N)^N$ to $P_\rho \varphi$ for all $\varphi \in L^2(\mathbb{R}^N)^N$. In fact, it is enough to show that $P_{\rho_n} \varphi$ converges weakly in $L^2(\mathbb{R}^N)^N$ to $P_\rho \varphi$ since we have then, using again Remarks 2.2 (1),

$$\int_{\mathbb{R}^N} \frac{1}{\rho_n} |P_{\rho_n} \varphi|^2 \, dx = \int_{\mathbb{R}^N} \frac{1}{\rho_n} (P_{\rho_n} \varphi) \cdot \varphi \, dx = \int_{\mathbb{R}^N} (P_{\rho_n} \varphi) \cdot \left(\frac{\varphi}{\rho_n} \right) dx$$

$$\xrightarrow{n} \int_{\mathbb{R}^N} (P_\rho \varphi) \cdot \left(\frac{\varphi}{\rho} \right) dx = \int_{\mathbb{R}^N} \frac{1}{\rho} |P_\rho \varphi|^2 \, dx$$

while $\frac{1}{\sqrt{\rho_n}} (P_{\rho_n} \varphi)$ converges weakly in $L^2(\mathbb{R}^N)^N$ to $\frac{1}{\sqrt{\rho}} (P_\rho \varphi)$.

Without loss of generality, we may assume that $P_{\rho_n} \varphi, Q_{\rho_n} \varphi$ (extracting subsequences if necessary) converge weakly in $L^2(\mathbb{R}^N)$ to some $p, q \in$

$L^2(\mathbb{R}^N)$. Therefore, we have obviously: $\varphi = p + q$, $R \wedge q = 0$. Next, $\frac{1}{\rho_n}(P_{\rho_n}\varphi)$ converges weakly to $\frac{1}{\rho}p$ and we thus obtain $R \cdot (\frac{1}{\rho}p) = 0$. The uniqueness of the decomposition (2.19) completes the proof of Lemma 2.1. □

We may now go back to the issue of the "time-continuity" of ρu in Theorem 2.1. We shall use the following notation: $f \in C([0,T]; L^2_w(\mathbb{R}^N))$ if $f \in L^\infty(0,T; L^2(\mathbb{R}^N))$ and if f is continuous in t with values in $L^2(\mathbb{R}^N)$ endowed with the weak topology (picking up a metric for this topology on a large ball of $L^2(\mathbb{R}^N)$ containing the values of $f(t)$ for $t \in [0,T]$). Recall, as explained above, that in the following theorem we consider the case $\Omega = \mathbb{R}^N$ and $u_\infty = 0$.

Theorem 2.2. *Let (ρ, u) be a weak solution of (2.1)–(2.2) as in Theorem 2.1. Then, for all $T \in (0, \infty)$; $R \wedge (\rho u) \in C([0,T]; L^2_w)$ with $\{R \wedge (\rho u)\}(0) = R \wedge m_0$ and $\frac{\partial}{\partial t}\{R \wedge (\rho u)\} \in L^2(0,T; H^{-1}(\mathbb{R}^N)) + X$ where $X = L^p(0,T; W^{-1,q}(\mathbb{R}^N))$ with $1 \le p < \infty$, $q = \frac{Np}{Np-2}$ if $N \ge 3$, $1 < p < \infty$, $q = \frac{p}{p-1}$ if $N = 2$ and ρ_0 is bounded away from 0, and $X = L^p(0,T; W^{-1,q}_{loc}(\mathbb{R}^N))$ with $1 < p < \infty$, $1 \le q < \frac{p}{p-1}$ if $N = 2$; we can also take $X = L^\infty(0,T; W^{-1-\varepsilon,1}(\mathbb{R}^N))$ for any $\varepsilon > 0$ if $N = 2$. In addition, if we assume that ρ_0 is bounded way from 0, then we have*

1) *$\rho u, \sqrt{\rho}u, u \in C([0,T]; L^2_w)^N$ for all $T \in (0,\infty)$.*

2) *ρu (respectively $\sqrt{\rho}\,u, u$) converge weakly in $L^2(\mathbb{R}^N)$, as t goes to 0_+, to $P_{\rho_0}m_0$ (respectively $\frac{1}{\sqrt{\rho_0}}(P_{\rho_0}m_0), \frac{1}{\rho_0}(P_{\rho_0}m_0)$). Furthermore, if $\operatorname{div} u_0 = 0$ in $\mathcal{D}'(\mathbb{R}^N)$ where $u_0 = \frac{m_0}{\rho_0}$, ρu (respectively $\sqrt{\rho}u, u$) converge strongly in $L^2(\mathbb{R}^N)$, as t goes to 0_+, to m_0 (respectively $\sqrt{\rho_0}u_0, u_0$).*

3) *The energy inequality (2.11) holds for all $t \ge 0$.* □

We are going to prove Theorem 2.2, admitting temporarily Theorem 2.1 whose proof is given in the following sections.

Proof of Theorem 2.2. First of all, taking $\phi = R \wedge \psi$ ($\psi \in C_0^\infty(\mathbb{R}^N \times (0,\infty))$) in the weak formulation (2.12), we deduce easily that the following holds in the sense of distributions

$$\frac{\partial}{\partial t}\{R \wedge (\rho u)\} + \partial_j\{R \wedge (\rho u_j u)\} - \partial_j\{R \wedge (\mu(\nabla u_j + \partial_j u))\} = R \wedge (\rho f). \quad (2.22)$$

Recalling that R_k is bounded on $L^p(\mathbb{R}^N)$ for all $1 < p < \infty$, $1 \le k \le N$, we remark that for all $T \in (0, \infty)$

$$R \wedge (\rho f) \in L^2(\mathbb{R}^N \times (0,T)), \quad R \wedge (\mu(\nabla u_j + \partial_j u)) \in L^2(\mathbb{R}^N \times (0,T))$$

since $\rho, \mu \in L^\infty(\mathbb{R}^N \times (0,T))$, $f \in L^2(\mathbb{R}^N \times (0,T))$, $\nabla u \in L^2(\mathbb{R}^N \times (0,T))$.

Next, we observe that $\rho u_j u \in L^\infty(0,T; L^1(\mathbb{R}^N))$ ($\forall T$) and by Sobolev embeddings $u \in L^2(0,T; L^{\frac{2N}{N-2}}(\mathbb{R}^N))$ if $N \geq 3$, hence $\rho u_j u \in L^1(0,T; L^{\frac{N}{N-2}}(\mathbb{R}^N))$ for all $T \in (0,\infty)$. Therefore, if $N \geq 3$, $\rho u_j u \in L^p(0,T; L^q(\mathbb{R}^N))$ ($\forall T$) for $1 \leq p \leq \infty$, $q = \frac{Np}{Np-2}$. If $N = 2$, by the definition of a weak solution, $u \in L^2(0,T; H^1_{\mathrm{loc}}(\mathbb{R}^2))$, hence, by Sobolev embeddings, $u \in L^2(0,T; L^r_{\mathrm{loc}}(\mathbb{R}^2))$ for all $2 \leq r < \infty$ and $\rho u_j u \in L^\infty(0,T; L^1(\mathbb{R}^N)) \cap L^p(0,T; L^q_{\mathrm{loc}}(\mathbb{R}^N))$ for $1 \leq p < \infty$, $1 \leq q < \frac{p}{p-1}$, and for all $T \in (0,\infty)$. Finally, if $N = 2$ and ρ_0 is bounded away from 0, say $\rho_0 \geq \underline{\rho} > 0$ a.e. on $\mathbb{R}^N$, then (2.17) yields $\rho \geq \underline{\rho}$ a.e. on $\mathbb{R}^2 \times (0,\infty)$. Therefore, in this case, $u \in L^\infty(0,T; L^2(\mathbb{R}^2)) \cap L^2(0,T; H^1(\mathbb{R}^2))$ ($\forall T$) and, by the Gagliardo–Nirenberg inequalities, $u \in L^p(0,T; L^q(\mathbb{R}^2))$ for $1 \leq p \leq \infty$, $q = 2\frac{p}{p-1}$ ($\forall T$). Collecting all these bounds, we deduce from (2.21) the regularity of $\frac{\partial}{\partial t}\{R \wedge (\rho u)\}$ stated in Theorem 2.2.

This yields of course the continuity in t of $R \wedge (\rho u)$ with values, say, in $W^{-1,1}_{\mathrm{loc}}(\mathbb{R}^N)$ but, by definition, $R \wedge (\rho u)$ is bounded in $L^2(\mathbb{R}^N)$ on each interval $(0,T)$ (for all $T \in (0,\infty)$). Therefore, $R \wedge (\rho u) \in C([0,T]; L^2_w)$ ($\forall T$) and $R \wedge (\rho u)|_{t=0} = m_0$ because of (2.12).

We now turn to the proofs of claims 1)–3) in Theorem 2.2 in the case when $\rho_0 \geq \underline{\rho}$ a.e. in $\mathbb{R}^N$ for some $\underline{\rho} > 0$. As we just saw, this implies $\rho \geq \underline{\rho} > 0$ a.e. on $\mathbb{R}^N \times (0,\infty)$ and thus $u \in L^\infty(0,T; L^2(\mathbb{R}^N))$ ($\forall T$).

We next prove that $\rho u \in C([0,T]; L^2_w)$ ($\forall T$). With the notation of Lemma 2.1, setting $m = \rho u$, we see that $R \wedge (P_1 m) = R \wedge m \in C([0,T]; L^2_w)$ ($\forall T$) while, by definition, $R \cdot (P_1 m) = 0$. By classical properties of such div–curl decompositions (ellipticity!), we deduce that $P_1 m \in C([0,T]; L^2_w))$ ($\forall T$). Next, we claim that we have

$$m = P_\rho(P_1 m). \tag{2.23}$$

If it is the case, we deduce the fact that $m \in C([0,T]; L^2_w)$ from Lemma 2.1. Now, (2.23) follows from the obvious properties

$$R \wedge (P_1 m - m) = R \cdot Q_1 m = 0 \qquad \text{by definition of } P_1, Q_1,$$
$$R \cdot \left(\tfrac{1}{\rho}m\right) = R \cdot u = 0 \qquad \text{since} \quad \mathrm{div}\, u = 0.$$

The rest of claim 1) is an easy consequence of, on the one hand, the fact we have just proved, namely $\rho u \in C([0,T]; L^2_w))$ ($\forall T$) and, on the other hand, the properties of ρ, namely $\rho \in C([0,\infty); L^p(B_R))$ ($\forall R$, $\forall 1 \leq p < \infty$). Indeed observe that if $t_n \underset{n}{\to} t \geq 0$, $\frac{1}{\sqrt{\rho(t_n)}}$, $\frac{1}{\rho(t_n)} \underset{n}{\to} \frac{1}{\sqrt{\rho(t)}}$, $\frac{1}{\rho(t)}$ in $L^p(B_R)$ ($\forall R$, $\forall 1 \leq p < \infty$) and remain bounded in $L^\infty(\mathbb{R}^N)$ uniformly in n, and 1) follows by observing that $g_n f_n \underset{n}{\rightharpoonup} gf$ weakly in L^2, if $f_n \rightharpoonup f$ weakly in L^2, $g_n \underset{n}{\to} g$ in L^1_{loc} and g_n is bounded in L^∞ uniformly in n.

Claim 3) follows from 1): indeed, for all $t > 0$, we deduce from (2.14) the existence of $t_n \in (0, t)$, $t_n \underset{n}{\to} t$ such that

$$\left(\int_{\mathbb{R}^N} \rho |u|^2 \, dx \right)(t_n) + \int_0^{t_n} \int_{\mathbb{R}^N} \mu(\partial_i u_j + \partial_j u_i)^2 \, dx \, ds$$
$$\leq \int_{\mathbb{R}^N} \frac{|m_0|^2}{\rho_0} \, dx + \int_0^{t_n} \int_{\mathbb{R}^N} \rho f \cdot u \, dx \, ds.$$

We only have to let n go to $+\infty$ observing that

$$\lim_n \left(\int_{\mathbb{R}^N} \rho |u|^2 \, dx \right)(t_n) \geq \left(\int_{\mathbb{R}^N} \rho |u|^2 \, dx \right)(t)$$

since $\sqrt{\rho} u \in C([0, T]; L_w^2)$ ($\forall T$).

It remains to prove claim 2). First of all, in view of the properties of ρ, it is clearly enough to show that ρu converges weakly in $L^2(\mathbb{R}^N)$, as t goes to 0_+, to $P_{\rho_0}(m_0)$ while, if $\mathrm{div}\left(\frac{m_0}{\rho_0}\right) = 0$ in $\mathcal{D}'(\mathbb{R}^N)$ (or $R \cdot \left(\frac{m_0}{\rho_0}\right) = 0$), $\sqrt{\rho} u$ converges strongly in $L^2(\mathbb{R}^N)$, as t goes to 0_+, to $\frac{m_0}{\sqrt{\rho_0}}$. Since $m = \rho u \in C([0, 1]; L_w^2)$, we only have to show that $m(0) = P_{\rho_0} m_0$ in order to prove the weak convergence. But, in view of what we have already shown

$$R \wedge (m_0 - m(0)) = 0, \quad R \cdot \left(\frac{m(0)}{\rho_0}\right) = R \cdot u(0) = 0$$

since $u \in C([0, 1]; L_w^2)$. Hence, $m(0) = P_{\rho_0}(m_0)$.

Finally, if $\mathrm{div}\left(\frac{m_0}{\rho_0}\right) = 0$, we see that $m(0) = m_0$. Hence, $\sqrt{\rho} u$ converges weakly in $L^2(\mathbb{R}^N)$, as t goes to 0_+, to $\frac{m_0}{\sqrt{\rho_0}}$. But, we also have, since (2.4) holds for all $t \geq 0$ (claim 3) proved above),

$$\|\sqrt{\rho} u\|_{L^2(\mathbb{R}^N)}^2(t) \leq \left\| \frac{m_0}{\sqrt{\rho_0}} \right\|_{L^2(\mathbb{R}^N)}^2,$$

and the strong convergence in $L^2(\mathbb{R}^N)$ is proven. $\square$

We conclude this section by mentioning that the first statement in Theorem 2.2 can be expressed, in fact, in terms of $P_1(\rho u)$, i.e. of the projection, say in $L^2(\mathbb{R}^N)$, on the subspace of divergence-free vector fields—we used this fact in the proof of Theorem 2.2. In fact, the orthogonal decomposition (P_1, Q_1) is also possible in the case of Dirichlet boundary conditions (and in this case P_1 is the projection onto the subspace of divergence-free vector fields u with $u \cdot n = 0$ on $\partial\Omega$) or in the periodic case. If $\Omega = \mathbb{R}^N$ or in the periodic case, we obtain immediately

Theorem 2.3. *Let (ρ, u) be a weak solution of (2.1)–(2.2) as in Theorem 2.1. Then, for all $T \in (0, \infty)$, $P_1(\rho u) \in C([0, T]; L_w^2(\Omega))$ and $\frac{\partial}{\partial t}\{P_1(\rho u)\} \in L^2(0, T; H^{-1}(\Omega)) + X$ where $X = L^p(0, T; W^{-1,q}(\mathbb{R}^N))$ with $1 \le p < \infty$, $q = \frac{Np}{Np-2}$ if $N \ge 3$, $1 < p < \infty$, $q = \frac{p}{p-1}$ if $N = 2$ and ρ_0 is bounded away from 0, and $X = L^p(0, T; W_{\text{loc}}^{-1,q}(\mathbb{R}^N))$ with $1 < p < \infty$, $1 \le q < \frac{p}{p-1}$ if $N = 2$. If $N = 2$, we can also take $X = L^\infty(0, T; W^{-1-\varepsilon,1}(\mathbb{R}^N))$ for any $\varepsilon > 0$.*

In the case of Dirichlet boundary conditions, one must replace $W^{-1,q}(\Omega)$ by $V^{-1,q}(\Omega) = V_0^{1,q}(\Omega)'$ where $V_0^{1,q}(\Omega)$ is the closure for the $W^{1,q}$ norm of the space of functions φ in $C_0^\infty(\Omega)$ such that div $\varphi = 0$.

2.2 Regularity results and open problems

We begin this section by discussing some open problems on (2.1)–(2.2). First of all, the uniqueness of weak solutions is completely open in all dimensions. Of course, we expect this to be the case if $N \ge 3$ since, in view of Theorem 2.1 (and Remark 2.1 (3)), a particular case of our weak solution is $\rho \equiv \overline{\rho} \in (0, \infty)$ and, in that case, u is a weak solution "à la Leray" of the homogeneous, incompressible Navier–Stokes equations (the classical Navier–Stokes equations), and, for this particular case, the uniqueness of weak solutions is still an open question. But, even in two dimensions, the uniqueness of weak solutions is not known for (2.1)–(2.2). We show in section 2.5 some partial "uniqueness" results indicating that any weak solution is equal to a strong one if the latter exists.

Of course, the uniqueness of solutions is closely related to the regularity of solutions: "smooth enough" solutions are indeed unique—this is not difficult to check and results in that direction can be found in [17] for example. For the same reason as for the uniqueness, we cannot expect full regularity results (like $\rho_0, u_0 \in C^\infty$ yield C^∞ solutions for all $t \ge 0$) to be known since they would imply regularity results for the homogeneous Navier–Stokes equations. However, as we shall see and recall in chapter 4, regularity in the preceding sense holds for the homogeneous Navier–Stokes equations if $N = 2$ and if $N = 3$ various regularity results are available. If $N = 3$, we do not have any further regularity information on u different from what we stated in section 2.1 (or easy consequences of what we stated). In particular, as we already mentioned in Remark 2.1 (2), very little is known on the pressure field.

But, even when $N = 2$, regularity does not seem to be available at least when μ depends on ρ. However, in the very particular case when $N = 2$, μ is independent of ρ, i.e. μ is a positive constant and ρ_0 is bounded away from

0, it was shown by A.V. Kazhikov [254] (see also S.N. Antontsev and A.V. Kazhikov [16], S.N. Antontsev, A.V. Kazhikov and V.N. Monakhov [17]) that one can obtain more regularity and therefore also uniqueness results. We do not want to treat this case in great detail and we refer the interested reader to [17] for complete proofs. But we wish to explain the main idea of the proof of one regularity result, from which further regularity can be deduced by classical arguments (one way is to differentiate the equation and apply again the argument we are going to present).

More precisely, we wish to explain why $u \in L^2(0, T; H^2) \cap C([0, T]; H^1)$, $\frac{\partial u}{\partial t} \in L^2(0, T; L^2)$ ($\forall\, T$) if $\rho_0 \geq \underline{\rho} > 0$ a.e. for some $\underline{\rho} \in (0, \infty)$, $u_0 = \frac{m_0}{\rho_0} \in H^1$ and div $u_0 = 0$, $N = 2$, $\mu \in (0, \infty)$. We simply want to obtain, at least formally, a priori estimates on u and $\frac{\partial u}{\partial t}$ corresponding to the claimed regularity, and we shall do so only in the case when $\Omega = \mathbb{R}^2$, $u_\infty = 0$ to simplify the presentation. First of all, as we mentioned several times before, $\rho \geq \underline{\rho} > 0$ a.e. on $\mathbb{R}^2 \times (0, \infty)$. Next, using (2.1), we write (2.2) as

$$\left.\begin{aligned} &\rho \frac{\partial u}{\partial t} + \rho(u \cdot \nabla)u - \mu \Delta u + \nabla p = \rho f, \\ &\text{div } u = 0, \quad \text{in } \mathbb{R}^2 \times (0, \infty). \end{aligned}\right\} \tag{2.24}$$

We then multiply (2.24) by $\frac{\partial u}{\partial t}$, integrate (by parts) over $\mathbb{R}^2$ and find for all $t \geq 0$

$$\underline{\rho} \int_{\mathbb{R}^2} \left|\frac{\partial u}{\partial t}\right|^2 + \mu \frac{d}{dt}\left\{\frac{1}{2}\int_{\mathbb{R}^2} |\nabla u|^2 \, dx\right\} \leq \|\rho_0\|_{L^\infty(\mathbb{R}^2)} \left\|\frac{\partial u}{\partial t}\right\|_{L^2(\mathbb{R}^2)}$$

$$\cdot \| |u|\, |\nabla u| \|_{L^2(\mathbb{R}^2)} + \|\rho_0\|_{L^\infty(\mathbb{R}^2)} \|f\|_{L^2(\mathbb{R}^2)} \left\|\frac{\partial u}{\partial t}\right\|_{L^2(\mathbb{R}^2)}.$$

Hence, using the Cauchy–Schwarz inequality repeatedly, we deduce

$$\left\|\frac{\partial u}{\partial t}\right\|_{L^2(\mathbb{R}^2)}^2 + \frac{d}{dt}\|\nabla u\|_{L^2(\mathbb{R}^2)}^2 \leq C\left\{\|u\|_{L^4(\mathbb{R}^2)}^2 \|\nabla u\|_{L^4(\mathbb{R}^2)}^2 + \|f\|_{L^2(\mathbb{R}^2)}^2\right\}$$

where C denotes various constants independent of u, t. Next, we recall that we already have a bound (deduced from (2.14)) on $u \in L^\infty(0, T; L^2)$, $\nabla u \in L^2(0, T; L^2)$ and that the following inequality holds for all $v \in H^1(\mathbb{R}^2)$ (a particular case of the Gagliardo–Nirenberg inequalities)

$$\|v\|_{L^4(\mathbb{R}^2)} \leq C\|v\|_{L^2(\mathbb{R}^2)}^{1/2} \|\nabla v\|_{L^2(\mathbb{R}^2)}^{1/2}.$$

Therefore, $\int_0^T \int_{\mathbb{R}^2} |u|^4 \, dx \, dt \leq C\|u\|_{L^\infty(0,T;L^2(\mathbb{R}^2))}^2 \|\nabla u\|_{L^2(0,T;L^2(\mathbb{R}^2))}^2 \leq C$. Using (2.24) once more, we finally deduce that we have for all $\varepsilon > 0$

$$\left.\begin{aligned} &\left\|\frac{\partial u}{\partial t}\right\|_{L^2(\mathbb{R}^2)}^2 + \frac{d}{dt}|\nabla u|_{L^2(\mathbb{R}^2)}^2 \\ &\qquad\leq \frac{C_0(t)}{\varepsilon}\left(1 + \|\nabla u\|_{L^2(\mathbb{R}^2)}^2\right) + \varepsilon\|D^2 u\|_{L^2(\mathbb{R}^2)}^2 \end{aligned}\right\} \tag{2.25}$$

where $C_0 \geq 0$, $\int_0^T C_0(t)\,dt \leq C$ (for all $T \in (0,\infty)$).

Next, we observe that we have for all $t \geq 0$ in view of (2.24):

$$\| -\mu \Delta u + \nabla p \|_{L^2(\mathbb{R}^2)} \leq C \| f \|_{L^2(\mathbb{R}^2)} + C \left\| \frac{\partial u}{\partial t} \right\|_{L^2(\mathbb{R}^2)}$$
$$+ C \| |u| \, |\nabla u| \, \|_{L^2(\mathbb{R}^2)}.$$

Since div $u = 0$, we can use classical regularity results on (linear) Stokes equations—see for example R. Temam [**472**]—to deduce

$$\| u \|_{H^2(\mathbb{R}^2)} \leq C \Big\{ \| u \|_{L^2(\mathbb{R}^2)} + \| f \|_{L^2(\mathbb{R}^2)}$$
$$+ \left\| \frac{\partial u}{\partial t} \right\|_{L^2(\mathbb{R}^2)} + \| |u| \, |\nabla u| \, \|_{L^2(\mathbb{R}^2)} \Big\}.$$

Exactly as above, this yields for all $\varepsilon' > 0$

$$\| u \|_{H^2(\mathbb{R}^2)} \leq \frac{1}{\varepsilon'} C_1(t) + C \left\| \frac{\partial u}{\partial t} \right\|_{L^2(\mathbb{R}^2)} + \varepsilon' \| u \|_{H^2(\mathbb{R}^2)} \qquad (2.26)$$

where $C_1 \geq 0$, $\int_0^T C_1^2(t)\,dt \leq C$ (for all $T \in (0,\infty)$). Hence, choosing $\varepsilon' = 1/2$,

$$\| u \|_{H^2(\mathbb{R}^2)}^2 \leq C_2(t) + C \left\| \frac{\partial u}{\partial t} \right\|_{L^2(\mathbb{R}^2)}^2 \qquad (2.27)$$

where $C_2 \geq 0$, $\int_0^T C_2(t)\,dt \leq C$ (for all $T \in (0,\infty)$). Inserting (2.27) in (2.25) and choosing $\varepsilon = \frac{1}{2C}$, we deduce finally for all $t \geq 0$

$$\left\| \frac{\partial u}{\partial t} \right\|_{L^2(\mathbb{R}^2)}^2 + \frac{d}{dt} \| \nabla u \|_{L^2(\mathbb{R}^2)}^2 \leq C_3(t) \Big(1 + \| \nabla u \|_{L^2(\mathbb{R}^2)}^2 \Big)$$

where $C_3 \geq 0$, $\int_0^T C_3(t)\,dt \leq C$ (for all $T \in (0,\infty)$).

The desired a priori estimates on $\frac{\partial u}{\partial t}$ in $L^2(\mathbb{R}^2 \times (0,T))$, u in $L^\infty(0,T; H^1(\mathbb{R}^2))$ and thus u in $L^2(0,T; H^2(\mathbb{R}^2))$ ($\forall\, T$) follow using Grönwall's inequality. $\square$

Remark 2.3. Further regularity on $u(\rho, p)$ can be deduced from the regularity we just obtained. One way is to differentiate and apply similar arguments. Another way is to observe that, since $u \in L^2(0,T; H^2)$, u satisfies: $|u(x_1, t) - u(x_2, t)| \leq C(t)|x_1 - x_2| \, \big| \log \{ \min |x_1 - x_2|, \frac{1}{2} \} \big|$ for some $C \in L^2(0,T)$. This implies that, for each T, ρ is Hölder continuous in (x,t) on $[0,T]$ and this implies that $D_x^2 u$ is Hölder continuous in (x,t). $\square$

We would like to mention another interesting open question: suppose that $\rho_0 = 1_D$ for a smooth domain D ($\subset \Omega$), i.e. a patch of a homogeneous incompressible fluid "surrounded" by the vacuum (or a bubble of vacuum embedded in the fluid). Then, Theorem 2.1 yields at least one global weak solution and (2.17) implies that, for all $t \geq 0$, $\rho(t) = 1_{D(t)}$ for some set such that $\mathrm{vol}(D(t)) = \mathrm{vol}(D)$. In this case, (2.1)–(2.2) can be reformulated as a somewhat complicated free boundary problem. It is also very natural to ask whether the regularity of D is preserved by the time evolution.

Finally, we conclude this section with a few remarks on stationary problems associated with (2.1)–(2.2), namely

$$\left. \begin{array}{l} \rho \geq 0, \quad \mathrm{div}\,(\rho u) = 0, \quad \mathrm{div}\,u = 0, \\[2mm] \rho(u \cdot \nabla)u - \mu \Delta u + \nabla p = \rho f \text{ in } \Omega\,, \ u \in H_0^1(\Omega)^N, \ \rho \in L^\infty(\Omega) \end{array} \right\} \quad (2.28)$$

looking, for example, at the case of Dirichlet boundary conditions, and $\mu \in (0, \infty)$ independent of ρ. Choosing for instance $f \in L^2(\Omega)^N$, we claim that in general (2.28) has a "huge number of solutions". First of all, some "trivial" solutions are obtained by setting $\rho \equiv \lambda \in [0, \infty)$ and solving the stationary, homogeneous, incompressible Navier–Stokes equations:

$$\left. \begin{array}{l} \lambda(u \cdot \nabla)u - \mu \Delta u + \nabla p = \lambda f \quad \text{in} \quad \Omega, \\[2mm] u \in H_0^1(\Omega)^N, \quad \mathrm{div}\,u = 0 \quad \text{in } \Omega, \end{array} \right\} \quad (2.29)$$

and we know (see for example R. Temam [472]) that, for each $\lambda \geq 0$, there exists a solution (at least one) $u \in H^2(\Omega)$ ($p \in H^2(\Omega)$) at least if $N = 2$ or 3. In addition, uniqueness holds for instance if $N = 2$ and $\lambda \|f\|_{L^2}$ is small enough (μ and Ω being fixed).

In fact, there are many more (stationary) solutions of (2.28) than the preceding ones. Indeed, take for example $N = 2$, $\Omega = B_1$, $f_1 = x_2 g(r)$, $f_2 = -x_1 g(r)$ where $r = (x_1^2 + x_2^2)^{1/2}$, $g \in L^2(B_1)$ (i.e. $\int_0^1 g^2(s)\,ds < \infty$). Then, we look for solutions of (2.28) having the following forms : $\rho = \rho(r) \geq 0$, $u_1 = x_2 \psi(r)$, $u_2 = -x_1 \psi(r)$. Obviously, $\mathrm{div}\,u = \mathrm{div}\,(\rho u) = 0$. It is easy to check that $\rho(u \cdot \nabla)u = \nabla(r\rho\psi^2)$ and $-\Delta u_1 = -\frac{x_2}{r}((r\psi') + \psi)' = -x_2(\psi'' + \frac{3}{r}\psi')$, $-\Delta u_2 = x_1(\psi'' + \frac{3}{r}\psi')$. Therefore, if $\rho \geq 0$, $\rho \in L^\infty$ is given, solving (2.28) amounts to solving $-\psi'' - \frac{3}{r}\psi' = \rho g$, i.e. ψ is determined by

$$\psi(1) = 0, \quad \psi'(r) = -\frac{1}{r^3}\int_0^r s^3 \rho(s)\,g(s)\,ds, \quad (2.30)$$

and thus for each ρ, we obtain one stationary solution (smooth if ρ and g are smooth)!

A similar example can be built in the periodic case: take $f_1 = g\left(x_2 + \frac{L_2}{2}\right)$ where g is odd, periodic of period L_2, $\rho = \rho\left(x_2 + \frac{L_2}{2}\right)$ where $\rho \geq 0$, ρ is even, periodic of period $L_2/2$ ($g \in L^\infty$) and solve $-\mu u'' = \rho g$ on $\mathbb{R}$, u periodic of period L_2. Then, ρ, $u = (u(x_2), 0)$ solve (2.28). This indicates that the right way to formulate the stationary problem (2.28) might be to constrain (2.28) with an additional requirement on the distribution function of ρ, a direction that needs to be investigated in more detail.

Let us finally mention that the regularity analysis of (2.28) follows closely the known results on the steady-state homogeneous Navier–Stokes equations: in particular, if $N \leq 4$, $u \in H^2(\Omega)$ and $u \in W^{2,p}(\Omega)$ if $f \in L^p(\Omega)$ for any $2 \leq p < \infty$. In fact, since the case $N = 4$ does not seem to be well known, we shall come back to this point in chapter 4. However, even if $f \in C^\infty(\overline{\Omega})$, we cannot expect more regularity on u in view of the preceding examples: this is due to the fact that ρ may not even be continuous.

2.3 A priori estimates and compactness results

Let us first explain the organization of this section. We shall work mainly in the periodic case and after each proof we shall explain how to modify the preceding proofs in the Dirichlet case or in the case when $\Omega = \mathbb{R}^N$. Next, we begin with a priori (formal) estimates and then we state and prove some general compactness results on sequences of solutions. These compactness results will play a fundamental role in the existence proofs since they will allow us to deduce the existence of the global weak solutions upon passing to the limit in conveniently approximated problems and using the compactness results shown in this section.

We thus begin with a priori estimates. We first remark that (2.1) and the incompressibility condition ($\operatorname{div} u = 0$) immediately imply that the distribution function of $\rho(t)$—considered as a function of x—is independent of t. In other words, (2.17) holds. This is in fact nothing but the celebrated Liouville's theorem. A direct formal proof consists in observing that if $\beta \in C^1([0, \infty); \mathbb{R})$, $\beta(\rho)$ satisfies

$$\frac{\partial \beta(\rho)}{\partial t} + \operatorname{div}(u\beta(\rho)) = \left(\frac{\partial}{\partial t} + u \cdot \nabla\right)\beta(\rho) = \beta'(\rho)\left\{\frac{\partial \rho}{\partial t} + u \cdot \nabla \rho\right\} = 0.$$

Therefore, integrating over Ω (periodic case or Dirichlet case), we find, using the boundary conditions, that

$$\frac{d}{dt} \int_\Omega \beta(\rho) \, dx = 0$$

or equivalently that $(\int_\Omega \beta(\rho) \, dx)$ is independent of t.

In particular, choosing $\beta = g_n \in C^1([0, \infty[, \mathbb{R})$, $1 \geq g_n \geq 0$ such that $g_n(t) = 0$ if $t \notin [\alpha, \beta]$ (where $0 \leq \alpha < \beta < \infty$ are fixed) and $g_n(t) = 1$ if

$t \in [\alpha + \frac{1}{n}, \beta - \frac{1}{n}]$ (take $n \geq \frac{2}{\beta - \alpha}$), we deduce (2.17) from the preceding fact upon letting n go to $+\infty$. In particular (2.17) yields the following L^∞ a priori estimate

$$0 \leq \rho(x,t) \leq \|\rho_0\|_{L^\infty} \qquad \text{a.e.} \tag{2.31}$$

(in fact $\|\rho(t)\|_{L^\infty} = \|\rho_0\|_{L^\infty}$ for all $t \geq 0$!).

The other a priori estimate that we can obtain simply follows from the energy identity: indeed, we expect, at least formally, in view of (2.1), that (2.2) implies, multiplying by u and integrating by parts, that

$$\frac{d}{dt} \int_\Omega \rho \frac{|u|^2}{2} \, dx + \int_\Omega \mu(\partial_i u_j + \partial_j u_i) \partial_j u_i \, dx = \int_\Omega \rho f \cdot u \, dx$$

or

$$\frac{1}{2} \frac{d}{dt} \int_\Omega \rho |u|^2 \, dx + \frac{1}{2} \int_\Omega \mu(\partial_i u_j + \partial_j u_i)^2 \, dx = \int_\Omega \rho f \cdot u \, dx. \tag{2.32}$$

Next, the right-hand side of (2.32) is bounded, in view of (2.31), by

$$\left(\int_\Omega \rho |f|^2 \, dx \right)^{1/2} \left(\int_\Omega \rho |u|^2 \, dx \right)^{1/2} \leq \|\rho_0\|_{L^\infty}^{1/2} \|f\|_{L^2} \|\sqrt{\rho} u\|_{L^2},$$

and, because of (2.31), $\mu = \mu(\rho(x,t)) \geq \underline{\mu} = \inf \{ \mu(\lambda) \ / \ 0 \leq \lambda \leq \|\rho_0\|_{L^\infty} \} > 0$, and thus

$$\frac{1}{2} \int_\Omega \mu(\partial_i u_j + \partial_j u_i)^2 \, dx \geq \frac{\underline{\mu}}{2} \int_\Omega (\partial_i u_j + \partial_j u_i)^2 \, dx$$

$$= \underline{\mu} \left(\int_\Omega |\nabla u|^2 + 2 \partial_i u_j \, \partial_j u_i \, dx \right).$$

In addition, we find, integrating by parts,

$$\int_\Omega \partial_i u_j \, \partial_j u_i \, dx = \int_\Omega (\partial_i u_i)^2 \, dx = 0.$$

In conclusion, we obtain for all $T \in (0, \infty)$

$$\left. \begin{array}{l} \left(\frac{1}{2} \int_\Omega \rho |u|^2 \, dx \right)(t) + \underline{\mu} \int_0^t \int_\Omega |\nabla u|^2 \, dx \, ds \\[2mm] \leq \|\rho_0\|_{L^\infty}^{1/2} \int_0^t \|f\|_{L^2} \|\sqrt{\rho} u\|_{L^2} \, ds + \frac{1}{2} \int_\Omega \frac{|m_0|^2}{\rho_0} \, dx, \quad \forall \, t \in [0, T]. \end{array} \right\}$$
$$\tag{2.33}$$

In the case of Dirichlet boundary conditions, using the Cauchy–Schwarz inequality, we thus deduce

$$\|\nabla u\|_{L^2(\Omega \times (0,T))} \leq C \tag{2.34}$$

$$\sup_{0 \leq t \leq T} \||\rho|u|^2\|_{L^1(\Omega)} \leq C \tag{2.35}$$

where C denotes various constants which depend only on T, Ω and bounds on $\|\rho_0\|_{L^\infty}$, $\|f\|_{L^2(\Omega \times (0,T))}$, $\||\rho_0|u_0|^2\|_{L^1(\Omega)} = \left\|\frac{|m_0|^2}{\rho_0}\right\|_{L^1(\Omega)}$.

In the case of Dirichlet boundary conditions, we then deduce from Poincaré's inequality

$$\|u\|_{L^2(0,T;H^1(\Omega))} \leq C. \tag{2.36}$$

In the periodic case, we claim that (2.35) also holds. In order to see this, we introduce

$$<f> = \fint_\Omega f \, dx \qquad \text{where} \qquad \fint_\Omega = \frac{1}{\text{meas}(\Omega)} \int_\Omega$$

and we deduce from (2.35)

$$\left(\int_\Omega \rho \, dx\right) |<u>|^2 \leq 2 \int_\Omega \rho|u|^2 \, dx + 2 \int_\Omega \rho|u- <u>|^2 \, dx$$
$$\leq C + 2\|\rho_0\|_{L^\infty} \|u- <u>\|_{L^2}^2$$
$$\leq C + C\|\nabla u\|_{L^2}^2$$

for all $t \in (0,T)$. Then, we can assume without loss of generality that $\rho_0 \not\equiv 0$ (otherwise the problem is trivial: $\rho \equiv 0$), in which case we deduce from the argument made above on ρ that we have for all $t \geq 0$

$$\left(\int_\Omega \rho \, dx\right)(t) = \int_\Omega \rho_0 \, dx = M_0 > 0.$$

Hence, we have

$$\int_0^T |<u>|^2 \, dt \leq C \tag{2.37}$$

which, combined with (2.34), yields (2.36).

We conclude this brief discussion of a priori estimates by explaining the modifications of the preceding arguments needed to treat the whole space case ($\Omega = \mathbb{R}^N$). First of all, the derivation of (2.17) is simply identical. Next, instead of multiplying by u, we now multiply by $u - u_\infty$ and find in a similar way

$$\frac{1}{2}\frac{d}{dt}\int_{\mathbb{R}^N} \rho|u-u_\infty|^2 + \mu \int_{\mathbb{R}^N} |\nabla u|^2 \, dx \leq \|\rho_0\|_{L^\infty}^{1/2} \|f\|_{L^2} \|\sqrt{\rho}(u-u_\infty)\|_{L^2},$$

and we still obtain (2.34), while (2.35) is now replaced by

$$\sup_{0 \le t \le T} \| \rho |u - u_\infty|^2 \|_{L^1(\mathbb{R}^N)} \le C. \tag{2.38}$$

In particular, if $\rho_0 \ge \underline{\rho}$ a.e. on $\mathbb{R}^N$ for some $\underline{\rho} > 0$, then, by (2.37), we deduce $\rho(x, t) \ge \underline{\rho} > 0$ a.e. on $\mathbb{R}^N \times (0, \infty)$. Therefore, (2.38) yields

$$\| u - u_\infty \|_{L^2(0,T;H^1(\mathbb{R}^N))} \le C, \tag{2.39}$$

and we deduce from Sobolev embeddings if $N \ge 3$

$$\| u - u_\infty \|_{L^2\left(0,T;L^{\frac{2N}{N-2}}(\mathbb{R}^N)\right)} \le C. \tag{2.40}$$

At least formally, we in fact deduce (2.40) from (2.34) if $N \ge 3$ in all cases even without assuming that ρ_0 is bounded away from 0. In particular, if $N \ge 3$, we obtain

$$\| u \|_{L^2(0,T;H^1(B_R))} \le C, \tag{2.41}$$

for all $R \in (0, \infty)$ (C depends now on R).

Finally, we claim that, even if $N = 2$, (2.41) holds. In order to prove this claim, we first observe that, assuming again that $\rho_0 \not\equiv 0$, we can find for all $T \in (0, \infty)$ fixed, some $R_0 \in (0, \infty)$ such that

$$m_0 = \inf_{0 \le t \le T} \int_{B_{R_0}} \rho(x, t)\, dx > 0. \tag{2.42}$$

In fact, as we shall see, m_0 and R_0 depend only on ρ_0 and on "a modulus of continuity in t of ρ in L^1_{loc}". Indeed, arguing by contradiction, if such an R_0 does not exist, we find that for each $n \ge 1$, there exists $t_n \in [0, T]$ such that $\rho(t_n) = 0$ a.e. on B_n. Extracting a subsequence if necessary, we may assume that $t_n \underset{n}{\to} \bar{t} \in [0, T]$. Then, since $\rho \in C([0, T]; L^1_{\text{loc}})$, we see that $\rho(\bar{t}) \equiv 0$ a.e. on $\mathbb{R}^N$. Then, by general uniqueness results shown later on in this section, this implies that $\rho \equiv 0$ on $\mathbb{R}^N \times [0, \infty)$ and we reach a contradiction.

This proof, however, does not yield uniform bounds, i.e. bounds independent of the solutions, and we re-prove (2.42) below by a different and more efficient argument. But, before we do so, we wish to explain how (2.42) yields (2.41). Indeed, we just have to copy the argument that led to (2.37) in order to obtain

$$\int_0^T | <u>_R |^2\, dt \le C, \quad \text{for} \quad R \ge R_0, \tag{2.43}$$

where $<u>_R = \fint_{B_R} u \, dx$, and (2.43) combined with (2.34) yields (2.41).

We finally give another proof of (2.42) that yields uniform bounds. Since the difficulty encountered here will be encountered many times in this book, we state and prove a general lemma which is more general than what we really need here.

Lemma 2.2. *Let* $T \in (0, \infty)$, $\rho^n \in C([0, T]; L^1(B_R))$ *for all* $R \in (0, \infty)$. *We assume that* $\rho^n \geq 0$ *a.e. and that* $\rho_0^n = \rho^n(0)$ *satisfies, for some* $R_0 \in (0, \infty)$ *and for some* $\nu > 0$ *independent of* n,

$$\int_{B_{R_0}} \rho_0^n \, dx \geq \nu > 0. \tag{2.44}$$

We also assume that ρ^n *satisfies*

$$\frac{\partial \rho^n}{\partial t} + \operatorname{div}(m^n) = 0 \qquad in \quad \mathcal{D}'(\mathbb{R}^2 \times (0, \infty)) \tag{2.45}$$

where $m^n = m_1^n + m_2^n$, *and we have, for some* $C > 0$ *independent of* n,

$$\|m_1^n\|_{L^1(\mathbb{R}^2 \times (0,T))} \leq C, \quad \|m_2^n\|_{L^2(\mathbb{R}^2 \times (0,T))} \leq C. \tag{2.46}$$

Then there exist $R \geq R_0$ *and* $n_0 \geq 1$ *such that for* $n \geq n_0$ *we have*

$$\inf_{0 \leq t \leq T} \int_{B_R} \rho^n \, dx \geq \frac{\nu}{2}. \tag{2.47}$$

Before we prove this result, let us explain how we use this lemma in the above context: assume that (ρ^n, u^n) is a sequence of solutions of (2.1)–(2.2) with the bounds already shown, which we assume to be uniform in n. As we have seen in Remark 2.1 (7), it is enough to treat the case when $u_\infty = 0$, hence we have for all $T \in (0, \infty)$

$$\sup_{0 \leq t \leq T} \int_{\mathbb{R}^2} \rho^n |u^n|^2 \, dx \leq C, \quad \|\rho^n\|_{L^\infty(\mathbb{R}^2 \times (0,\infty))} \leq C.$$

We claim that if $\rho_0^n = \rho^n(0)$ satisfies (2.44), then (2.47) holds. Indeed, we just have to check that (2.48) holds with $m_2^n = m^n = \rho^n u^n$, and this is obvious since

$$\int_{\mathbb{R}^2} |m^n|^2 \, dx \leq \left(\int_{\mathbb{R}^2} \rho^n |u^n|^2 \, dx \right) \|\rho^n\|_{L^\infty(\mathbb{R}^2)}.$$

Proof of Lemma 2.2. Without loss of generality, extracting subsequences if necessary, we may assume that $|m_1^n|$ and $|m_2^n|^2$ converge weakly in the sense of measures to some bounded, non-negative measures on $\mathbb{R}^2 \times [0, T]$ denoted respectively by μ_1 and μ_2. Next, we choose $\varphi \in C_0^\infty(\mathbb{R}^2)$ such that $\varphi \equiv 1$ on B_1, $0 \leq \varphi \leq 1$ on $\mathbb{R}^2$, $\varphi(x) = 0$ if $|x| \geq 2$ and we define $C_1 = \sup_{\mathbb{R}^2} |\nabla\varphi|$, $C_2 = \left(\int_{\mathbb{R}^2} |\nabla\varphi|^2 \, dx\right)^{1/2}$.

We then consider $\alpha > 0$ such that $C_1\alpha + C_2\alpha^{1/2}T^{1/2} \leq \nu/2$. Then, we observe that there exists $R \geq \max(R_0, 1)$ such that for $i = 1, 2$

$$\int_{\mathbb{R}^2 \times [0,T]} 1_{(R-1 \leq |x| \leq 2R+1)} \, d\mu_i \leq \alpha/2.$$

Then, for n large, we also have for $i = 1, 2$

$$\int_0^T dt \int_{\mathbb{R}^2} dx \, 1_{(R \leq |x| \leq 2R)} \mu_i^n \leq \alpha$$

where $\mu_1^n = |m_1^n|$, $\mu_2^n = |m_2^n|^2$.

We next multiply (2.45) by $\varphi_R(x) = \varphi(\frac{x}{R})$ and integrate over $\mathbb{R}^2 \times [0, t]$ (for all $t \in [0, T]$) to find in view of (2.44)

$$\left(\int_{\mathbb{R}^2} 1_{|x| \leq 2R} \rho^n \, dx\right)(t)$$

$$\geq \nu - \int_0^T dt \int_{\mathbb{R}^2} dx \, 1_{R \leq |x| \leq 2R} \left(|m_1^n| + |m_2^n|\right) |\nabla\varphi_R|$$

$$\geq \nu - \frac{C_1}{R}\alpha - C_2 \int_0^T dt \left(\int_{\mathbb{R}^2} 1_{R \leq |x| \leq 2R} |m_2^n|^2 \, dx\right)^{1/2}$$

$$\geq \nu - C_1\alpha - C_2 T^{1/2}\alpha^{1/2} \geq \frac{\nu}{2}$$

in view of the choices of α and R. The proof of the lemma is then complete. $\square$

We finally briefly explain some bounds *in the case when* $\Omega = \mathbb{R}^N$. Let us first observe that (2.17) obviously implies that (2.8) and (2.9) hold uniformly in $t \geq 0$. Next, if $N = 3$ and (2.10) holds, $\rho(t) \in L^{\frac{N}{2}, \infty}(\mathbb{R}^N)$ for all $t \geq 0$ and $\|\rho(t)\|_{L^{\frac{N}{2}, \infty}(\mathbb{R}^N)} = \|\rho_0\|_{L^{\frac{N}{2}, \infty}(\mathbb{R}^N)}$. Furthermore, as explained in Remark 2.1 (8), if (2.8) or (2.9) hold, we obtain a priori estimates on u in $L^2(\mathbb{R}^N \times (0, T))$ and thus in $L^2(0, T; H^1(\mathbb{R}^N))$ for all $T \in (0, \infty)$. We now consider the case when (2.10) holds and $N = 2$: recall that we wish to show that this bound propagates (in t). In order to prove this claim, we wish to multiply (2.1) by $\varphi(x) = <x>^{2(p-1)} (\log <x>)^r$. Of course, we need to justify the integration by parts to be performed but this point can be checked easily in view of the bounds that follow. We then obtain, since ρ^p also satisfies (2.1), for some $m \in [2, \infty)$ to be determined later on,

$$\frac{d}{dt} \int_{\mathbb{R}^2} \rho^p \varphi \, dx$$

$$\leq \int_{\mathbb{R}^2} \rho^p |u| \, |\nabla \varphi| \, dx \leq C \int_{\mathbb{R}^2} \rho^p |u| <x>^{-1} \varphi \, dx$$

$$\leq C \left(\int_{\mathbb{R}^2} \frac{|u|^m}{<x>^2} (\log <x>)^{-\alpha} \, dx \right)^{1/m}$$

$$\cdot \left(\int_{\mathbb{R}^2} \rho^{p \frac{m}{m-1}} \varphi^{\frac{m}{m-1}} <x>^{\frac{2-m}{m-1}} (\log <x>)^{\frac{\alpha}{m-1}} \, dx \right)^{\frac{m-1}{m}}$$

$$\leq C X_m \|\rho\|_{L^\infty}^{p/m} \left(\int_{\mathbb{R}^2} \rho^p \varphi \left\{ \varphi^{\frac{1}{m-1}} <x>^{\frac{2-m}{m-1}} (\log <x>)^{\frac{\alpha}{m-1}} \right\} dx \right)^{\frac{m-1}{m}}$$

where α is chosen in $(\frac{m}{2}+1, \infty)$ and $X_m = \left(\int_{\mathbb{R}^2} \frac{|u|^m}{<x>^2} (\log <x>)^{-\alpha} \, dx \right)^{1/m}$. In view of Appendix B, X_m is bounded in $L^2(0,T)$ ($\forall \, T \in (0, \infty)$), and our claim follows if we choose m large enough so that $m > 2p$ and thus $\varphi^{\frac{1}{m-1}} <x>^{\frac{1}{m-1}} (\log <x>)^{\frac{\alpha}{m-1}}$ is bounded on $\mathbb{R}^2$.

We now turn to the fundamental compactness results that we need in the existence proofs presented in the next section. We consider, for the reasons explained above, the periodic case and we suppose that two sequences ρ^n, u^n are given satisfying: $\rho^n \in C([0,T]; L^1(B_R))$ ($\forall \, R \in (0, \infty)$), $\rho^n \geq 0$ is periodic in x_i of period L_i ($\forall \, 1 \leq i \leq N$), $u^n \in L^2(0,T; H^1_{\text{per}})^N$ where $T \in (0, \infty)$ is fixed. We define $\rho_0^n = \rho^n(0)$ and we assume

$$0 \leq \rho^n \leq C \qquad \text{a.e. on } \Omega \times (0,T) \tag{2.48}$$

$$\text{div } u^n = 0 \quad \text{a.e. on } \Omega \times (0,T), \quad \|u^n\|_{L^2(0,T;H^1(\Omega))} \leq C \tag{2.49}$$

$$\frac{\partial \rho^n}{\partial t} + \text{div } (\rho^n u^n) = 0 \qquad \text{in } \mathcal{D}'(\mathbb{R}^N \times (0,T)) \tag{2.50}$$

$$\rho_0^n \underset{n}{\to} \rho_0 \quad \text{in } L^1(\Omega), \quad u^n \underset{n}{\rightharpoonup} u \quad \text{weakly in } L^2(0,T; H^1_{\text{per}}), \tag{2.51}$$

for some ρ_0 which thus satisfies $0 \leq \rho_0 \leq C$ a.e., and where C denotes various positive constants independent of n. Notice that, because of the bound (2.48), the convergence of ρ_0^n to ρ_0 also holds in $L^p(\Omega)$ for all $1 \leq p < \infty$ and that, because of (2.49), div $u = 0$ a.e. on $\Omega \times (0,T)$.

Theorem 2.4. *1) With the above assumptions, ρ^n converges in $C([0,T]; L^p(\Omega))$ for all $1 \leq p < \infty$ to the unique periodic solution ρ, bounded on $\Omega \times (0,T)$, of*

$$\left. \begin{array}{l} \dfrac{\partial \rho}{\partial t} + \text{div } (\rho u) = 0 \qquad \text{in } \mathcal{D}'(\mathbb{R}^N \times (0,T)), \\[2mm] \rho \in C([0,T]; L^1(\Omega)), \quad \rho(0) = \rho_0 \quad \text{a.e. in } \Omega. \end{array} \right\} \tag{2.52}$$

2) We assume in addition that $\rho^n |u^n|^2$ is bounded in $L^\infty(0, T; L^1(\Omega))$ and that we have for some $q \in (1, \infty)$, $m \geq 1$

$$\left| < \frac{\partial}{\partial t} (\rho^n, u^n), \varphi > \right| \leq C \|\varphi\|_{L^q(0,T;W^{m,q}(\Omega))} \qquad (2.53)$$

for all $\varphi \in L^q(0, T; W^{m,q}(\Omega))$ periodic such that $\operatorname{div} \varphi = 0$ on $\mathbb{R}^N \times (0, T)$. Then, for all $1 \leq i \leq N$, $\sqrt{\rho^n} u_i^n$ converges to $\sqrt{\rho} u_i$ in $L^p(0, T; L^r(\Omega))$ for $2 < p < \infty$, $1 \leq r < \frac{2Np}{Np-4}$, and u_i^n converges to u_i in $L^\theta(0, T; L^{\frac{N\theta}{N-2}}(\Omega))$ for $1 \leq \theta < 2$ on the set $\{\rho > 0\}$ (if $N = 2$, $\frac{N\theta}{N-2}$ is replaced by an arbitrary r in $[1, \infty)$).

Remarks 2.4. 1) Part 1 is essentially contained in R.J. DiPerna and P.L. Lions [**128**] and we re-prove it for the reader's convenience.

2) It is possible to weaken the bounds on ρ^n and u^n. For instance, if we keep (2.49), it is enough to assume instead of (2.48) that ρ^n is bounded, uniformly in $t \in [0, T]$, in $L^p(\Omega)$ where $p > \frac{2N}{N+2}$—one can even treat the case when $p = \frac{2N}{N+2}$ if $N \geq 3$. In fact, if we consider renormalized solutions instead of solutions in the sense of distributions, the above result holds with $p = 1$! This is shown in R.J. DiPerna and P.L. Lions [**128**].

3) The same result holds with some obvious adaptations in the case of Dirichlet boundary conditions replacing $L^2(0, T; H^1_{\text{per}})$ by $L^2(0, T; H^1_0(\Omega))$, and assuming that (2.50) holds in $\Omega \times (0, T)$ and that $\varphi \in C_0^\infty(\Omega \times (0, T))^N$ with $\operatorname{div} \varphi = 0$ in $\Omega \times (0, T)$ in (2.53). $\quad\square$

Proof of part 1 of Theorem 2.4. The proof is divided into several steps. Without loss of generality, we may assume, extracting a subsequence if necessary, that ρ_n converges weakly to some ρ in $L^p(\Omega \times (0, T))$ for all $1 < p < \infty$ where ρ satisfies (2.52), ρ is periodic. In addition, since $\rho^n u^n$ is bounded in $L^2(0, T; L^q(\Omega))$ with $1 \leq q \leq \frac{2N}{N-2}$ ($q < \infty$ if $N = 2$), we deduce easily from (2.50) that ρ^n converges to ρ in $C([0, T], W^{-m,p}(\Omega))$ for $1 \leq p < \infty$, $m > 0$; see for instance J.L. Lions [**293**] for very general compactness results of that sort. If we equip $L^p(\Omega)$ ($1 < p < \infty$) with the weak topology and an associated distance over a large ball containing all values $\rho^n(t)$ ($n \geq 1$, $t \in [0, T]$), we also deduce easily that ρ^n converges to ρ in $C([0, T]; L^p_w(\Omega))$, and, in particular, $\rho(0) = \rho_0$ a.e. in Ω.

Then, we first prove (step 1) that ρ uniquely solves (2.52). Next, we give a general regularization procedure for solutions of transport equations like (2.50) (step 2). In step 3, we complete the proof of part 1.

Step 1. In order to check that ρ solves (2.52), that is

$$\frac{\partial \rho}{\partial t} + \operatorname{div}(\rho u) = 0 \quad \text{in} \quad \mathcal{D}'(\mathbb{R}^N \times (0, T)),$$

we have only to show that $\rho^n u^n$ converges to ρu in $\mathcal{D}'(\mathbb{R}^N \times (0,T))$. This is in fact rather straightforward since ρ^n converges to ρ in $L^2(0,T;H^{-1}(B_R))$ for all $R < \infty$, while $u^n \varphi$ converges weakly to $u\varphi$ in $L^2(0,T;H_0^1(B_R))$ for all $\varphi \in C_0^\infty(\mathbb{R}^N \times (0,T))$ supported, say, in $B_R \times (0,T)$. Hence,

$$\int_0^T \int_{\mathbb{R}^N} \rho^n u^n \, \varphi \, dt \, dx = <\rho^n, u^n \varphi> \underset{n}{\to} <\rho, u\varphi> = \int_0^T \int_{\mathbb{R}^N} \rho u \, \varphi \, dt \, dx,$$

and our claim is shown.

We next explain why ρ uniquely solves (2.52). More generally, if $g \in L^\infty(\mathbb{R}^N \times (0,T))$, g periodic, $g \in C([0,T];L_w^p(\Omega))$ $(1 < p < \infty)$ satisfies: $g(0) = 0$ a.e. on $\mathbb{R}^N$,

$$\frac{\partial g}{\partial t} + \operatorname{div}(ug) = 0 \qquad \text{in} \quad \mathcal{D}'(\mathbb{R}^N \times (0,T)),$$

then $g \equiv 0$. Indeed, we deduce from the regularization property proved below (step 2) that $|g|$ also solves the same equation. Then, we simply integrate the equation in x using the periodicity to find

$$\frac{d}{dt} \int_\Omega |g| \, dx = 0 \qquad \text{in} \quad \mathcal{D}'(0,T)$$

and $\int_\Omega |g| \, dx = m(t) \in C([0,T])$ satisfies $m(0) = 0$. Therefore, $m \equiv 0$ and $g \equiv 0$.

Step 2. A general regularization for solutions of transport equations. This regularization is based upon the following classical lemma that we re-prove for the reader's convenience. We denote by $\omega_\varepsilon = \frac{1}{\varepsilon^N} \omega(\frac{\cdot}{\varepsilon})$ a smoothing sequence, i.e. $\omega \in C_0^\infty(\mathbb{R}^N)$, $\int_{\mathbb{R}^N} \omega \, dx = 1$, Support$(\omega) \subset B_1$, $\omega \geq 0$ and $\varepsilon \in (0,1]$.

Lemma 2.3. *Let* $v \in W^{1,\alpha}(\mathbb{R}^N)$, $g \in L^\beta(\mathbb{R}^N)$ *with* $1 \leq \alpha, \beta \leq \infty$, $\frac{1}{\alpha} + \frac{1}{\beta} \leq 1$. *Then, we have*

$$\left. \begin{array}{l} \|\operatorname{div}(vg) * \omega_\varepsilon - \operatorname{div}(v(g * \omega_\varepsilon))\|_{L^\gamma(\mathbb{R}^N)} \\[2mm] \qquad \leq C\|v\|_{W^{1,\alpha}(\mathbb{R}^N)} \|g\|_{L^\beta(\mathbb{R}^N)} \end{array} \right\} \qquad (2.54)$$

for some $C \geq 0$ *independent of* ε, v *and* g *and* γ *is determined by* $\frac{1}{\gamma} = \frac{1}{\alpha} + \frac{1}{\beta}$. *In addition,* $\operatorname{div}(vg) * \omega_\varepsilon - \operatorname{div}\{v(g * \omega_\varepsilon)\}$ *converges to 0 in* $L^\gamma(\mathbb{R}^N)$ *as* ε *goes to 0 if* $\gamma < \infty$.

Proof of Lemma 2.3. Once (2.54) is proven, the rest of Lemma 2.3 is clear using the density of $C_0^\infty(\mathbb{R}^N)$ in $W^{1,\alpha}(\mathbb{R}^N)$ (if $\alpha < \infty$) or in $L^\beta(\mathbb{R}^N)$ (if $\beta < \infty$). Next, in order to prove (2.54), we define $C_\varepsilon =$

div $(vg) * \omega_\varepsilon -$ div $(v(g * \omega_\varepsilon))$, which is nothing but a commutator, and we write $C_\varepsilon = r_\varepsilon - (\text{div } v)(g * \omega_\varepsilon)$ where

$$r_\varepsilon = \int_{\mathbb{R}^N} \frac{1}{\varepsilon}\, (v(y) - v(x)) \cdot \nabla \omega\left(\frac{x-y}{\varepsilon}\right) \frac{1}{\varepsilon^N}\, g(y)\, dy.$$

Obviously, we have

$$|(\text{div } v)(g * \omega_\varepsilon)| \leq \sqrt{N}\, |Dv|\, |g * \omega_\varepsilon|.$$

On the other hand, we have, using Hölder's inequality,

$$|r_\varepsilon| \leq C\left[\fint_{B(x,\varepsilon)} \left\{\frac{1}{\varepsilon}\, |v(y) - v(x)|\right\}^s\right]^{1/s} \cdot \left[\fint_{B(x,\varepsilon)} |g|^t\right]^{1/t}$$

where $1 \leq s, t \leq \infty$, $\frac{1}{s} + \frac{1}{t} = 1$, $1 \leq t \leq \beta$, $1 \leq s \leq \alpha$ and C denotes various positive constants independent of ε, v and g.

Next, we write

$$\left|\frac{1}{\varepsilon}\, (v(y) - v(x))\right|^s = \left|\frac{1}{\varepsilon} \int_0^1 \nabla v(x + \lambda(y-x)) \cdot (y-x)\, d\lambda\right|^s$$

$$\leq \int_0^1 |\nabla v(x + \lambda(y-x))|^s \left|\frac{y-x}{\varepsilon}\right|^s d\lambda.$$

Therefore

$$\fint_{B(x,\varepsilon)} \left|\frac{1}{\varepsilon}(v(y) - v(x))\right|^s ds \leq \int_0^1 d\lambda \int_{B_1} |\nabla v(x + \lambda \varepsilon w)|^s\, |w|^s\, dw$$

$$\leq \int_0^1 d\lambda \int_{B_1} |\nabla v(x + \lambda \varepsilon w)|^s\, dw = |\nabla v|^s * \overline{\chi}_\varepsilon$$

where $\overline{\chi}_\varepsilon(z) = \int_0^1 d\lambda\, \frac{1}{(\varepsilon\lambda)^N}\, 1_{B_{\lambda\varepsilon}}(z) = \frac{1}{N-1}\left(\left(\frac{\varepsilon}{|z|}\right)^{N-1} - 1\right) 1_{B_\varepsilon}\, \varepsilon^{-N}$ (and $\int_{\mathbb{R}^N} \overline{\chi}_\varepsilon = \text{meas}(B_1)$). Thus we obtain, defining $\chi_\varepsilon = (\text{meas}(B_\varepsilon))^{-1} 1_{B_\varepsilon}$

$$|C_\varepsilon| \leq C\{|Dv|\, |g * \omega_\varepsilon| + (|Dv|^s * \overline{\chi}_\varepsilon)^{1/s}(|g|^t * \chi_\varepsilon)^{1/t} \quad \text{a.e. on } \mathbb{R}^N, \quad (2.55)$$

and we conclude easily since we have, by classical properties of convolutions, for all $\varepsilon \geq 0$

$$\|g * \omega_\varepsilon\|_{L^\beta} \leq \|g\|_{L^\beta}, \quad \left\|(|Dv|^s * \overline{\chi}_\varepsilon)^{1/s}\right\|_{L^\alpha} \leq \|Dv\|_{L^\alpha} \|\overline{\chi}_\varepsilon\|_{L^1}^{1/s},$$

$$\left\|(|g|^t * \chi_\varepsilon)^{1/t}\right\|_{L^\beta} \leq \|g\|_{L^\beta},$$

and $\|\overline{\chi}_\varepsilon\|_{L^1} = \text{meas}(B_1)$.

In fact, the manipulations leading to (2.55) need to be justified and one way is to argue by density on (2.55). □

In particular, we deduce from Lemma 2.3 the following fact: if $v \in L^2(0,T; H^1(\mathbb{R}^N))$, $g \in L^\infty(\mathbb{R}^N \times (0,T))$, v and g are periodic, $\text{div}\, v = 0$ a.e. and

$$\frac{\partial g}{\partial t} + \text{div}\,(vg) = 0 \qquad \text{on} \quad \mathcal{D}'(\mathbb{R}^N \times (0,T)), \qquad (2.56)$$

then, for any $\beta \in C(\mathbb{R};\mathbb{R})$, $\beta(g)$ also solves (2.56). Indeed, in view of Lemma 2.3, we have

$$\frac{\partial g_\varepsilon}{\partial t} + v \cdot \nabla g_\varepsilon = r_\varepsilon \qquad \text{in} \quad \mathbb{R}^N \times (0,T),$$

where $g_\varepsilon = g * \omega_\varepsilon$, $r_\varepsilon \underset{\varepsilon}{\to} 0$ in $L^2(B_R \times (0,T))$ ($\forall\, R < \infty$). Hence, if $\beta \in C^1(\mathbb{R};\mathbb{R})$, $\beta(g_\varepsilon)$ satisfies

$$\frac{\partial \beta(g_\varepsilon)}{\partial t} + \text{div}(v\beta(g_\varepsilon)) = \left(\frac{\partial}{\partial t} + v \cdot \nabla\right)\beta(g_\varepsilon) = \beta'(g_\varepsilon)v_\varepsilon \quad \text{in } \mathbb{R}^N \times (0,T).$$

Then, our claim follows upon letting ε go to 0, at least when β is C^1. If β is merely continuous, we simply approximate it (uniformly on $[-\|g\|_{L^\infty}, \|g\|_{L^\infty}]$) by C^1 functions and pass to the limit in the sense of distributions.

Let us point out that we already used this fact (with $\beta(t) = |t|$) in step 1 above. Let us also remark that this regularization allows us to show that $g \in C([0,T]; L^p(\Omega))$ for all $1 \le p < \infty$: indeed, we observe that, for all $\eta, \varepsilon > 0$,

$$\frac{d}{dt} \int_\Omega |g_\varepsilon - g_\eta|^p\, dx \le \left(\int_\Omega |r_\varepsilon - r_\eta|^p\, dx\right)^{1/p}\left(\int_\Omega |g_\varepsilon - g_\eta|^p\, dx\right)^{1/p'}.$$

Hence,

$$\sup_{[0,T]} \|g_\varepsilon - g_\eta\|_{L^p} \le \|g(0)*\omega_\varepsilon - g(0)*\omega_\eta\|_{L^p} + \int_0^T \|r_\varepsilon - r_\eta\|_{L^p(\Omega)}\, dt.$$

Therefore, g_ε converges to g in $C([0,T]; L^p(\Omega))$.

Step 3. We have only to show that ρ^n converges to ρ, say, in $C([0,T]; L^2(\Omega))$ (to deduce the convergence in $C([0,T]; L^p(\Omega))$, for all $1 \le p < \infty$). We already know from step 2 that $\rho \in C([0,T]; L^2(\Omega))$ and from the argument given before step 1 that ρ^n converges to ρ in $C([0,T]; L^2_w(\Omega))$. Therefore, we have only to show that $\rho^n(t_n)$ converges in $L^2(\Omega)$ to $\rho(t)$ if

t_n ($\in [0, T]$) converges to t, while we already know that $\rho^n(t_n)$ converges weakly in $L^2(\Omega)$ to $\rho(t)$.

Hence, the proof of part 1 is complete if we show that we have for all $t \geq 0$:

$$\int_\Omega (\rho^n(t))^2 \, dx = \int_\Omega (\rho_0^n)^2 \, dx \underset{n}{\rightarrow} \int_\Omega \rho_0^2 \, dx = \int_\Omega \rho(t)^2 \, dx. \qquad (2.57)$$

In fact, the convergence is obvious in view of (2.48) and thus we have only to check the fact that $\rho^n(t)$ (resp. $\rho(t)$) has the same L^2 norm as ρ_0^n (resp. ρ_0). Next, in view of step 2, $(\rho^n)^2$ (resp. ρ^2) also solves (2.50) (resp. (2.52)) and the claimed conservations simply follow upon integrating these equations and using the periodicity of all the functions considered. □

Remark 2.5. As mentioned in Remark 2.4 (3), the proof of part 1 is easily modified if we replace the periodicity requirement on ρ^n, u^n by Dirichlet boundary conditions, namely $u^n = u = 0$ on $\partial\Omega$, or in other words $u^n, u \in L^2(0, T; H_0^1(\Omega))$. Of course, in that case, all equations are set in $\Omega \times (0, T)$.

The only argument which needs some explanation is the "integration over Ω of div (ρu)" where $\rho \in L^\infty(\Omega \times (0, T))$, $u \in L^2(0, T; H_0^1(\Omega))$. This is done by observing that, by classical Hardy-type inequalities, $\frac{u}{d} \in L^2(\Omega \times (0, T))$ where $d = \text{dist}\,(x, \partial\Omega)$. Then, we consider, for ε small enough, $\varphi_\varepsilon \in C_0^\infty(\Omega)$ such that

$$\left. \begin{array}{l} 0 \leq \varphi_\varepsilon \leq 1 \quad \text{in } \Omega, \quad \varphi_\varepsilon(x) = 1 \quad \text{if} \quad d(x) \geq \varepsilon, \\[2mm] \varphi_\varepsilon(x) = 0 \quad \text{if} \quad d(x) \leq \frac{\varepsilon}{2}, \quad |\nabla\varphi_\varepsilon| \leq \frac{C}{\varepsilon} \quad \text{in } \Omega \end{array} \right\}$$

for some $C \geq 0$ independent of ε. Then we have

$$|<\text{div}(\rho u), \varphi_\varepsilon>| = \left| \int_\Omega \rho u \cdot \nabla\varphi_\varepsilon \right|$$

$$\leq C \int_\Omega |u| \frac{1}{\varepsilon} 1_{(d \leq \varepsilon)} \, dx \leq C \int \frac{|u|}{d} \, dx$$

and $\left(\int_{(d \leq \varepsilon)} \frac{|u|}{d} \, dx \right) \to 0$ in $L^2(0, T)$ as $\varepsilon \to 0_+$. □

Proof of part 2 of Theorem 2.4. We first prove that we have

$$\int_0^T dt \int_\Omega dx \, \rho^n |u^n|^2 \underset{n}{\rightarrow} \int_0^T dt \int_\Omega dx \, \rho |u|^2. \qquad (2.58)$$

Indeed the condition (2.53) shows that $\frac{\partial}{\partial t} \{P_1(\rho^n u^n)\}$ is bounded in $L^q(0, T; W^{-m,q}(\Omega))^N$ while, by assumption, $\rho^n u^n$ and thus $P_1(\rho^n u^n)$ are bounded in $L^\infty(0,T;L^2(\Omega))^N$. Hence, by classical compactness theorems (see for instance J.L. Lions [**293**], R. Temam [**472**]), $P_1(\rho^n u^n)$ is compact in $L^2(0,T;H^{-1}(\Omega))^N$. In particular, since $\rho^n u^n$ converges weakly to ρu (step 1 of the proof of part 1 in $L^\infty(0,T;L^2(\Omega))^N$ for the weak-$*$ topology, $P_1(\rho^n u^n)$ converges to $P_1(\rho u)$ in $L^2(0,T;H^{-1}(\Omega))^N$. Hence, we have

$$\int_0^T dt \int_\Omega dx\, \rho^n |u^n|^2 = \int_0^T dt\, (\rho^n u^n, u^n)_{L^2(\Omega)}$$

$$= \int_0^T dt\, (P_1(\rho^n u^n), u^n)_{L^2(\Omega)} = \int_0^T dt <P_1(\rho^n u^n), u^n>_{H^{-1}\times H^1}$$

$$\xrightarrow[n]{} \int_0^T dt <P_1(\rho u), u>_{H^{-1}\times H^1} = \int_0^T dt\, (P_1(\rho u), u)_{L^2(\Omega)}$$

$$= \int_0^T dt\, (\rho u, u)_{L^2(\Omega)} = \int_0^T dt \int_\Omega dx\, \rho |u|^2$$

where we use the fact that $P_1(u^n) = u^n$, $P_1(u) = u$ since $\operatorname{div} u^n = \operatorname{div} u = 0$ in $\mathcal{D}'(\mathbb{R}^N \times (0,T))$. In the case of Dirichlet boundary conditions, the passage to the limit for $\rho^n |u^n|^2$ is shown in exactly the same way, replacing $H^{-1}(\Omega)$ by $V^{-1,2}(\Omega)$.

Once (2.58) is shown, we observe that $\sqrt{\rho^n} u^n$ converges weakly in $L^2(\Omega \times (0,T))$ to $\sqrt{\rho} u$. Indeed, in view of step 2 of the proof of part 1, $\sqrt{\rho^n}$ also solves (2.50) and converges to $\sqrt{\rho}$ in $C([0,T];L^p(\Omega))$ ($1 \le p < \infty$) because of part 1. Then, applying step 1 of the proof of part 1, we deduce our claim, namely the weak convergence of $\sqrt{\rho^n} u^n$ to $\sqrt{\rho} u$ in $L^2(\Omega \times (0,T))$.

This weak convergence, combined with (2.58), yields the strong convergence in $L^2(\Omega \times (0,T))$ of $\sqrt{\rho^n} u^n$ to $\sqrt{\rho} u$. The convergence of $\sqrt{\rho^n} u^n$ stated in part 2 of Theorem 2.4 then follows from the bounds we assumed on ρ^n, u^n and $\sqrt{\rho^n} u^n$. The final statement of part 2 concerning the convergence of u^n to u on $\{\rho > 0\}$ is shown if we show that u^n converges in measure to u on $\{\rho > 0\}$. But we deduce from the fact just shown that, extracting subsequences if necessary, $\sqrt{\rho^n} u^n$ converges a.e. on $\Omega \times (0,T)$ to $\sqrt{\rho} u$. In addition, because of part 1), we may assume that $\sqrt{\rho^n}$ converges a.e. on $\Omega \times (0,T)$ to $\sqrt{\rho}$. Hence, on the set $\{\rho > 0\}$, u^n converges a.e. to u and the proof of Theorem 2.4 is complete. □

We conclude this section by explaining how the preceding result, valid in the periodic case and in the case of Dirichlet boundary conditions, can be adapted to the case $\Omega = \mathbb{R}^N$. We begin by stating conditions and assumptions on ρ^n, u^n; of course, we no longer require ρ^n and u^n to be periodic and only request u^n to be bounded in $L^2(0,T;H^1(B_R))^N$ for all $R \in (0,\infty)$. We still assume (2.48) and (2.50) while (2.49) and (2.51) are now replaced by

$$\operatorname{div} u^n = 0 \qquad \text{a.e. on} \quad \mathbb{R}^N \times (0,T) \tag{2.59}$$

$$\rho_0^n \underset{n}{\rightarrow} \rho_0 \quad \text{in} \quad L^1(B_R), \\ u^n \underset{n}{\rightarrow} u \quad \text{weakly in } L^2(0,T; H^1(B_R)), \quad \text{for all } R \in (0,\infty) \Bigg\} \tag{2.60}$$

$$\frac{u^n}{1+|x|} \, 1_{(\rho^n \geq \delta)} = F_1^n + F_2^n, \\ F_1^n \ (\text{resp. } F_2^n) \text{ is bounded in } L^1(0,T; L^1(\mathbb{R}^N)) \\ (\text{resp. } L^1(0,T; L^\infty(\mathbb{R}^N))) \Bigg\} \tag{2.61}$$

for all $\delta > 0$. Since we shall deal mostly with situations where $\rho^n |u^n|^2$ is bounded in $L^\infty(0,T; L^1(\mathbb{R}^N))$, we only wish to observe at this stage that such a bound obviously implies that $u^n \, 1_{(\rho^n \geq \delta)}$ is bounded in $L^\infty(0,T; L^2(\mathbb{R}^N))$ and thus (2.61) holds.

Theorem 2.5. *1) Under the above conditions, ρ^n converges in $C([0,T]; L^p(B_R))$ (for all $1 \leq p < \infty$, $R \in (0,\infty)$) to the unique bounded solution ρ of*

$$\frac{\partial \rho}{\partial t} + \text{div}\,(\rho u) = 0 \quad \text{in} \quad \mathcal{D}'(\mathbb{R}^N \times (0,T)), \\ \rho|_{t=0} = \rho_0 \quad \text{a.e. in} \quad \mathbb{R}^N \Bigg\} \tag{2.62}$$

such that

$$\frac{u}{1+|x|} \, 1_{(|\rho| \geq \delta)} \in L^1(0,T; L^1(\mathbb{R}^N)) + L^1(0,T, L^\infty(\mathbb{R}^N)). \tag{2.63}$$

2) We assume in addition that $\rho^n |u^n|^2$ is bounded in $L^\infty(0,T; L^1(\mathbb{R}^N))$, ∇u^n is bounded in $L^2(\mathbb{R}^N \times (0,T))$, if $N \geq 3$ that u^n is bounded in $L^2(0,T; L^{\frac{2N}{N-2}}(\mathbb{R}^n))$ and that (2.53) holds with $\Omega = \mathbb{R}^N$ and for all $\varphi \in L^q(0,T; W^{m,q}(\mathbb{R}^N))$ such that $\text{div}\,\varphi = 0$ a.e. on $\mathbb{R}^N$. Furthermore, we assume that either u^n is bounded in $L^2(\mathbb{R}^N \times (0,T))$ or $N \geq 3$, $\rho_0^n \in L^{\frac{N}{2},\infty}(\mathbb{R}^N)$, $\rho_0 \in L^{\frac{N}{2},\infty}(\mathbb{R}^N)$ or $N = 2$ and

$$\sup_{0 \leq t \leq T} \sup_n \int_{\mathbb{R}^2} (\rho^n)^p <x>^{2(p-1)} (\log <x>)^r \, dx < \infty \\ \text{for some} \quad p \in (1,\infty] \quad \text{with} \quad r > 2p - 1. \Bigg\} \tag{2.64}$$

Then, for all $1 \leq i \leq N$, $\sqrt{\rho^n} u_i^n$ converges to $\sqrt{\rho} u_i$ in $L^p(0,T; L^p(B_R))$ for $2 < p < \infty$, $1 \leq r < \frac{2Np}{Np-4}$, $0 < R < \infty$, and u_i^n converges to u_i in $L^\theta(0,T; L^{\frac{N\theta}{N-2}}(B_R))$ on the set $\{\rho > 0\}$ for $1 \leq \theta < 2$, $0 < R < \infty$.

Remark 2.6. 1) Similar extensions to those described in Remark 2.4 (2) are possible for the preceding result.

2) Part 1 of Theorem 2.5 allows us, in fact, to extend slightly some of the uniqueness results obtained by R.J. DiPerna and P.L. Lions in [**128**]. □

Proof of part 1 of Theorem 2.5. The proof is divided in three steps.

Step 1. Truncations and consequences. We introduce $\rho_\delta^n = (\rho^n - \delta)^+$ for $\delta \in (0,1]$. Obviously, (2.61) yields for all $\delta > 0$

$$\left.\begin{array}{c} \dfrac{|u^n|}{<x>}\, 1_{(\rho_\delta^n > 0)} \quad \text{is bounded in} \\[2mm] L^1(0,T; L^\infty(\mathbb{R}^N)) + L^1(0,T; L^1(\mathbb{R}^N)) \end{array}\right\} \qquad (2.65)$$

where we define $<x> = (1 + |x|^2)^{1/2}$.

We are going to show below (in step 3) that ρ_δ^n converges in $L^p(B_R)$ uniformly in $t \in [0,T]$ to some $\overline{\rho}_\delta \geq 0$ ($\in L^\infty(\mathbb{R}^N \times (0,T))$) for all $1 \leq p < \infty$, $T \in (0,\infty)$. This will be done using, in particular, some general uniqueness results established in step 2 below that also show the uniqueness statement contained in part 1 of Theorem 2.5.

We wish to show now why such a convergence of ρ_δ^n yields the convergence of ρ^n in $C([0,T]; L^p(B_R))$ ($\forall\, 1 \leq p < \infty,\, \forall\, R \in (0,\infty)$) to some ρ which, obviously, is bounded on $\mathbb{R}^N \times (0,T)$ and solves (2.62). Then, we show why (2.63) holds in the limit.

First, we observe that we have for $n, m \geq 1$

$$|\rho^n - \rho^m| \leq |(\rho^n - \delta)^+ - (\rho^m - \delta)^+| + 2\delta.$$

This is enough to ensure that $(\rho^n)_n$ is a Cauchy sequence in $C([0,T]; L^p(B_R))$ ($\forall\, 1 \leq p < \infty,\, \forall\, R \in (0,\infty)$) and thus converges to some ρ. Obviously, $\rho = \lim_{\delta \downarrow 0_+} \uparrow \overline{\rho}_\delta$ (and one can in fact deduce a posteriori from the uniqueness statement and its proof the fact that $\overline{\rho}_\delta = (\rho - \delta)_+$).

Next, we show that (2.63) holds. To this end, we observe that in view of (2.61)

$$\left|\frac{u^n}{<x>}\, \chi_\delta(\rho^n)\right| \leq M^n(t) + F^n \qquad (2.66)$$

where $M^n \geq 0$, M^n is bounded in $L^1(0,T)$, $F^n \geq 0$ is bounded in $L^1(\mathbb{R}^N \times (0,T))$ and $\chi_\delta \in C([0,\infty),[0,\infty))$ satisfies: $0 \leq \chi_\delta \leq 1$ on $[0,\infty)$, $\chi_\delta(t) = 1$ if $t \geq \delta$, $\chi_\delta(t) = 0$ if $t \leq \delta/2$. Obviously, for all $R \in (0,\infty)$, $\chi_\delta(\rho^n)$ converges in $C([0,T]; L^p(B_R))$ ($\forall\, 1 \leq p < \infty$) and is uniformly bounded on $\mathbb{R}^N \times (0,T)$ while u^n converges to u, for example, weakly in $L^2(B_R \times (0,T))$. Therefore, $\frac{u^n}{<x>}\chi_\delta(\rho^n)$ converges weakly in $L^2(B_R \times (0,T))$ to $\frac{u}{<x>}\chi_\delta(\rho)$.

On the other hand, we may assume without loss of generality (extracting subsequences if necessary) that M^n converges weakly in the sense of measures to a non-negative bounded measure $\overline{M}$ on $[0,T]$ while F^n converges weakly in the sense of measures to a non-negative bounded measure $\overline{F}$ on $\mathbb{R}^N \times [0,T]$.

Then, we deduce from (2.66) and all these convergences

$$\left| \frac{u}{<x>} \chi_\delta(\rho) \right| \leq \overline{M} + \overline{F}. \qquad (2.67)$$

Let us then denote by M and F, respectively, the absolutely continuous parts (with respect to the Lebesgue measure) of $\overline{M}$ and $\overline{F}$. Since $\frac{u}{<x>} \chi_\delta(\rho) \in L^1(B_R \times (0,T))$ ($\forall R \in (0,\infty)$), we deduce finally

$$\frac{|u|}{<x>} 1_{(\rho \geq \delta)} \leq \left| \frac{u}{<x>} \chi_\delta(\rho) \right| \leq M + F$$

and (2.63) is shown. $\quad\square$

Step 2.   Uniqueness. We consider here f_1, f_2 bounded solutions in $C([0,T]; L^1(B_R))$ ($\forall R \in (0,\infty)$) of

$$\frac{\partial f}{\partial t} + \operatorname{div}(vf) = 0 \qquad \text{in} \quad \mathcal{D}'(\mathbb{R}^N \times (0,T)) \qquad (2.68)$$

such that $f_1(0) = f_2(0)$ a.e. on $\mathbb{R}^N$ and

$$\frac{|v|}{<x>} 1_{(|f_i| \geq \delta)} \in L^1(0,T; L^1(\mathbb{R}^N)) + L^1(0,T; L^\infty(\mathbb{R}^N)), \qquad (2.69)$$

for all $i = 1, 2$, $\delta > 0$. Then, if $\operatorname{div} v = 0$ a.e. on $\mathbb{R}^N \times (0,T)$—we could as well assume $\operatorname{div} v \in L^1(0,T; L^\infty(\mathbb{R}^N))$ as in [128]—and $v \in L^1(0,T; W^{1,1}_{\text{loc}}(\mathbb{R}^N))$, $f_1 \equiv f_2$ a.e. on $\mathbb{R}^N \times (0,T)$.

Let us first remark that $f = f_1 - f_2$ satisfies the same properties as f_1 and f_2 with, of course, $f(0) \equiv 0$ a.e. on $\mathbb{R}^N$. Indeed, we just need to observe

$$1_{(|f| \geq \delta)} \leq 1_{(|f_1| \geq \delta/2)} + 1_{(|f_2| \geq \delta/2)}.$$

Next, we use Lemma 2.3 and step 2 of the proof of Theorem 2.4 to deduce that, for all $\delta > 0$, $(|f| - \delta)^+$ satisfies exactly the same properties as f. In other words, we can assume without loss of generality that $f \geq 0$ a.e. on $\mathbb{R}^N \times (0,T)$ and

$$\frac{|v|}{<x>} 1_{(f>0)} \in L^1(0,T; L^1(\mathbb{R}^N)) + L^1(0,T; L^\infty(\mathbb{R}^N)). \qquad (2.70)$$

Indeed, observe that $f \equiv 0$ follows from $(|f| - \delta)^+ \equiv 0$ for all $\delta > 0$.

We deduce from (2.70) the following fact

$$\left.\begin{array}{l} \dfrac{|v|}{<x>}\, 1_{(f>0)} \;\leq\; M(t) + G(x,t) \quad\text{a.e. in } \mathbb{R}^N \times (0,T), \\[2mm] G \in L^1(\mathbb{R}^N\times(0,T)),\; G \geq 0,\; M \in L^1(0,T),\; M \geq 0. \end{array}\right\} \qquad (2.71)$$

Next, we consider $\varphi \in C_0^\infty(\mathbb{R})$, even, $0 \leq \varphi \leq 1$, $\varphi(x) = 1$ if $|x| \leq 1$, $\varphi(x) = 0$ if $|x| \geq 2$, φ nonincreasing on $[0,\infty)$, and we multiply (2.68) by $\varphi_n(e^{C(t)} < x >)$ where $\varphi_n(x) = \varphi(\frac{x}{n})$, $n \geq 1$, $C(t) = \int_0^t M(s)\,ds$. Integrating by parts over $\mathbb{R}^N \times (0,t)$, we find for all $t \in (0,T]$

$$\int_{\mathbb{R}^N} f(t)\varphi\Big(e^{C(t)}\frac{<x>}{n}\Big)\,dx + \int_0^t \int_{\mathbb{R}^N} ds\,dx \left(-\varphi'\Big(e^{C(s)}\frac{<x>}{n}\Big)\right)$$

$$\cdot \Big\{e^{C(s)}\frac{1}{n}\Big\} \cdot \Big\{M(s)<x> -v\cdot\frac{x}{<x>}\Big\} f \;=\; 0.$$

In view of (2.71) and the properties of φ, we deduce

$$\int_{\mathbb{R}^N} f(t)\varphi\Big(e^{C(t)}\frac{<x>}{n}\Big)\,dx \;\leq\; C \int_0^t \int_{\mathbb{R}^N} ds\,dx\, G\, 1_{(<x>\,\geq\, ne^{-C(s)})}.$$

This is enough to show that $f \equiv 0$ upon letting n go to $+\infty$ since the right-hand side goes to 0 as n goes to $+\infty$.

Step 3. Convergence of ρ_δ^n. Without loss of generality, we may assume that ρ_δ^n and $(\rho_\delta^n)^2$ converge respectively to some $\overline{\rho_\delta}$, $\overline{\rho_\delta^2}$ weakly in $L^\infty(\mathbb{R}^N \times (0,T)) - *$. Furthermore, exactly as in the proof of Theorem 2.4, we know that $\overline{\rho_\delta}$, $\overline{\rho_\delta^2}$ satisfy the equation (2.62), belong to $C([0,T]; L^p(\mathbb{R}^N)-w)$ and satisfy $\overline{\rho_\delta}|_{t=0} = (\rho_0 - \delta)^+$, $\overline{\rho_\delta^2}|_{t=0} = (\rho_0 - \delta)^{+2} = (\overline{\rho_\delta})^2|_{t=0}$ a.e. in $\mathbb{R}^N$. In addition, by Lemma 2.2 and step 2 of the proof of Theorem 2.4, we see that $(\overline{\rho_\delta})^2$ also satisfies (2.62).

Therefore, if we show that $\frac{|u|}{<x>}1_{(\overline{\rho_\delta}\geq\alpha)}$, $\frac{|u|}{<x>}1_{(\overline{\rho_\delta^2}\geq\alpha)} \in L^1(0,T; L^\infty(\mathbb{R}^N))$ $+ L^1(0,T; L^1(\mathbb{R}^N))$ for all $\alpha > 0$, we deduce from step 2 above that $\overline{\rho_\delta^2} = (\overline{\rho_\delta})^2$. Hence, ρ_δ^n converges to $\overline{\rho_\delta}$ in $L^2(B_R \times (0,T))$ and thus in $L^p(B_R \times (0,T))$ for all $R \in (0,\infty)$, $1 \leq p < \infty$. Since the bounds on $\frac{|u|}{<x>}1_{(\overline{\rho_\delta}\geq\alpha)}$ and on $\frac{|u|}{<x>}1_{(\overline{\rho_\delta^2}\geq\alpha)}$ are proven in exactly the same way, we just show them for $\frac{|u|}{<x>}1_{(\overline{\rho_\delta}\geq\alpha)}$. We first remark that (2.61) implies that we have

$$\frac{|u^n|}{<x>}\, 1_{(\rho_n\geq\delta)} \;\leq\; M^n(t) + F^n(x,t) \qquad \text{a.e. in}\quad \mathbb{R}^N \times (0,T)$$

where $M^n, F^n \geq 0$, M^n is bounded in $L^1(0,T)$, F^n is bounded in $L^1(\mathbb{R}^N \times (0,T))$. In particular, we deduce for some C independent of n (and δ)

$$\frac{1}{<x>}\,|u^n\rho_\delta^n| \;\leq\; C\,\{M^n(t) + F^n(x,t)\} \qquad \text{a.e. in}\quad \mathbb{R}^N \times (0,T).$$

Then, step 1 of the proof of Theorem 2.4 shows that $u^n \rho_\delta^n$ converges to $u\overline{\rho_\delta}$ weakly, say, in $L^2(B_R \times (0,T))$ for all $R \in (0,\infty)$, and we deduce as in step 1 above

$$\frac{|u|}{<x>} \, \overline{\rho_\delta} \leq C\{M+F\} \qquad \text{a.e. in} \quad \mathbb{R}^N \times (0,T) \qquad (2.72)$$

where $M \in L^1(0,T)$, $F \in L^1(\mathbb{R}^N \times (0,T))$ are respectively the absolutely continuous parts (with respect to Lebesgue measure) of the weak limits (in the sense of measures) of M^n, F^n. In particular, (2.72) yields the desired fact: $\frac{|u|}{<x>} 1_{(\overline{\rho_\delta} \geq \alpha)} \in L^1(0,T; L^\infty(\mathbb{R}^N)) + L^1(0,T; L^1(\mathbb{R}^N))$ for all $\alpha > 0$.

At this stage, we have shown the convergence of ρ_δ^n to $\overline{\rho_\delta}$ in $L^p(B_R \times (0,T))$ for all $1 \leq p < \infty$, $R \in (0,\infty)$. In order to show the convergence in $C([0,T]; L^p(B_R))$ for all $R \in (0,\infty)$, $1 \leq p < \infty$, it is enough to consider the case when $p = 2$ for instance. Then, we fix R in $(0,\infty)$ and we choose $\varphi \in C_0^\infty(\mathbb{R}^N)$, $\varphi \geq 0$, $\varphi \equiv 1$ on B_R. We claim that we have, for all $n \geq 1$ and for all $t \in [0,T]$,

$$\left.\begin{array}{l} \displaystyle\int_{\mathbb{R}^N} |\varphi \, \rho_\delta^n(t)|^2 \, dx = \int_{\mathbb{R}^N} |\varphi(\rho_0^n - \delta)^+|^2 \, dx \\[3mm] \displaystyle\qquad\qquad + \int_0^t \int_{\mathbb{R}^N} dx \, ds \, \{(\rho_\delta^n)^2 u^n \cdot \nabla \varphi^2\}, \end{array}\right\} \qquad (2.73)$$

$$\left.\begin{array}{l} \displaystyle\int_{\mathbb{R}^N} |\varphi \, \overline{\rho_\delta}(t)|^2 \, dx = \int_{\mathbb{R}^N} |\varphi(\rho_0 - \delta)^+|^2 \, dx \\[3mm] \displaystyle\qquad\qquad + \int_0^t \int_{\mathbb{R}^N} dx \, ds \, \{(\overline{\rho_\delta})^2 u \cdot \nabla \varphi^2\}. \end{array}\right\} \qquad (2.74)$$

Indeed, using step 2 of the proof of Theorem 2.4 once more, $(\rho_\delta^n)^2$ and $(\overline{\rho_\delta})^2$ satisfy respectively the same equations as ρ_δ^n and ρ_δ and belong to $C([0,T]; L^p(B_{R'}) - w)$ for all $1 < p < \infty$, $R' \in (0,\infty)$. Then, (2.73), (2.74) follow upon multiplication by φ^2, and show, by the way, that $\overline{\rho_\delta} \in C([0,T]; L^p(B_{R'}) - w)$ for all $R' \in (0,\infty)$, $1 < p < \infty$. In addition, we check as in the proof of Theorem 2.4 that ρ_δ^n converges to $\overline{\rho_\delta}$ in $C([0,T]; L^2(B_{R'}) - w)$ for all $R'(0,\infty)$ and thus, in particular, $\varphi \rho_\delta^n$ converges to $\varphi \rho_\delta$ in $C([0,T]; L^2(\mathbb{R}^N) - w)$. This fact, combined with (2.73) and (2.74), shows that $\varphi \rho_\delta^n$ converges to $\varphi \rho_\delta$ in $C([0,T]; L^2(\mathbb{R}^N))$ provided we show

$$\left.\begin{array}{l} \displaystyle\int_0^t \int_{\mathbb{R}^N} dx \, ds \, \{(\rho_\delta^n)^2 u^n \cdot \nabla \varphi^2\} \\[4mm] \displaystyle\underset{n}{\to} \int_0^t \int_{\mathbb{R}^N} dx \, ds \, \{(\overline{\rho_\delta})^2 u \cdot \nabla \varphi^2\}, \quad \text{uniformly on } [0,T]. \end{array}\right\} \qquad (2.75)$$

This is easy since $(\rho_\delta^n)^2$ converges to $(\overline{\rho}_\delta)^2$ in $L^2(B_{R'} \times (0,T))$ ($\forall R' \in (0,\infty)$) and u^n converges to u weakly in $L^2(B_{R'} \times (0,T))$ ($\forall R' \in (0,\infty)$).

Therefore, $\varphi \rho_\delta^n$ converges to $\varphi \overline{\rho}_\delta$ in $C([0,T]; L^2(\mathbb{R}^N))$ and thus ρ_δ^n converges to $\overline{\rho}_\delta$ in $C([0,T]; L^2(B_R))$ for all $R \in (0,\infty)$, and the proof of part 1 is complete. $\square$

Proof of part 2 of Theorem 2.5. Let us first observe that step 1 implies easily that $\sqrt{\rho^n} u^n$ converges weakly to $\sqrt{\rho} u$ in $L^2(\mathbb{R}^N \times (0,T))$ (for instance). Thus, exactly as in the proof of Theorem 2.4, we have only to show that we have for all $R \in (0,\infty)$

$$\int_0^T dt \int_{B_R} dx \, \rho^n |u^n|^2 \underset{n}{\rightarrow} \int_0^T dt \int_{B_R} dx \, \rho |u|^2. \qquad (2.76)$$

The proof of (2.76) is divided into four steps. First, we show that $P(\rho^n u^n)$ converges to $P(\rho u)$ in $L^2(B_R \times (0,T))$ ($\forall R \in (0,\infty)$) where we denote by P the orthogonal projection in $L^2(\mathbb{R}^N)^N$ onto divergence-free vector fields ($P = P_1$ with the notation of section 2.1). Then (step 2), we show that (2.76) holds in the case when u^n is bounded in $L^2(\mathbb{R}^N \times (0,T))$. Step 3 is devoted to the proof of (2.76) in the case when $N \geq 3$, $\rho \in L^{\frac{N}{2},\infty}(\mathbb{R}^N)$, and we treat the case $N = 2$ with the condition (2.64) in step 4.

Step 1. Compactness of $P(\rho^n u^n)$ in $L^2(B_R \times (0,T))$ ($\forall R \in (0,\infty)$). Since $P = \mathrm{Id} + \nabla(-\Delta)^{-1}\mathrm{div}$ is bounded on each Sobolev space $W^{m,q}(\mathbb{R}^N)$ ($m \geq 0$, $1 < q < \infty$), we deduce from (2.53) that $\frac{\partial}{\partial t} P(\rho^n u^n)$ is bounded in $L^q(0,T; W^{-m,q'}(\mathbb{R}^N))$. On the other hand, by assumption, $P(\rho^n u^n)$ is bounded in $L^\infty(0,T; L^2(\mathbb{R}^N))$. Hence, Appendix C implies that $P(\rho^n u^n)$ converges to $P(\rho u)$ in $C([0,T]; L^2(\mathbb{R}^N - w))$. We then wish to show that, for each fixed $R \in (0,\infty)$,

$$\int_0^T dt \int_{B_R} dx \, |P(\rho^n u^n)|^2 \underset{n}{\rightarrow} \int_0^T dt \int_{B_R} dx \, |P(\rho u)|^2 \, dx. \qquad (2.77)$$

We then write $\varphi = 1_{B_R}$.

Then, we take as in Lemma 2.3 a regularizing kernel ω_ε ($\varepsilon \in (0,1]$) and we observe that $\{\varphi P(\rho^n u^n)\} * \omega_\varepsilon$ converges to $\{\varphi P(\rho u)\} * \omega_\varepsilon$ in $C([0,T]; L^2(\mathbb{R}^N))$. Therefore, we have for all $\varepsilon \in (0,1]$

$$\int_0^T dt \int_{\mathbb{R}^N} dx \, \{\varphi P(\rho^n u^n)\} \cdot (\{P(\rho^n u^n)\} * \omega_\varepsilon)$$

$$\underset{n}{\rightarrow} \int_0^T dt \int_{\mathbb{R}^N} dx \, \{\varphi P(\rho u)\} \cdot (\{P(\rho u)\} * \omega_\varepsilon),$$

since $P(\rho^n u^n)$ converges weakly in $L^2(\mathbb{R}^N \times (0,T))$ to $P(\rho u)$.

In view of this convergence, it only remains to show that the following integral can be made arbitrarily small for ε small enough uniformly in n

$$\int_0^T dt \int_{\mathbb{R}^N} dx \; \varphi P(\rho^n u^n)\}\{P(\rho^n u^n)\} - P(\rho^n u^n) * \omega_\varepsilon)$$

$$= \int_0^T dt \int_{\mathbb{R}^N} dx \; P(\varphi P(\rho u))\{(\rho^n u^n) - (\rho^n u^n) * \omega_\varepsilon\}.$$

First of all, we remark that $P(\rho^n u^n)$ is bounded in $L^\infty(0,T;L^2(\mathbb{R}^N))$ and thus $P(\varphi P(\rho^n u^n))$ is also bounded in $L^\infty(0,T;L^2(\mathbb{R}^N))$. Furthermore, from the definition of P, we deduce easily the following bound

$$|P(\varphi P(\rho^n u^n))| \leq \frac{C}{|x|^N} \qquad \text{a.e.} \quad |x| \geq R+1, \quad t \in (0,T).$$

These two facts imply that it is enough to show

$$\left.\begin{array}{c} \lim_{\varepsilon \to 0} \sup_n \|\rho^n u^n - (\rho^n u^n) * \omega_\varepsilon\|_{L^1(0,T;L^2(B_M))} = 0, \\[2mm] \text{for all} \quad M \in (0,\infty). \end{array}\right\} \qquad (2.78)$$

To this end, we drop the superscript n and write

$$(\rho u) * \omega_\varepsilon - \rho u = (\rho * \omega_\varepsilon) u - \rho u$$
$$+ \int_0^1 \int_{\mathbb{R}^N} \nabla u(x + \lambda(y-x)) \cdot (y-x) \, \rho(y) \, \omega_\varepsilon(x-y) \, dy \, d\lambda.$$

Next, we remark that we have

$$\left| \int_0^1 \int_{\mathbb{R}^N} \nabla u(x + \lambda(y-x)) \cdot (y-x) \, \rho(y) \, \omega_\varepsilon(x-y) \, dy \, d\lambda \right|$$

$$\leq \|\rho\|_{L^\infty(\mathbb{R}^N)} \varepsilon \int_0^1 \int_{\mathbb{R}^N} |\nabla u(x + \lambda(y-x))| \, \omega_\varepsilon(x-y) \, dy \, d\lambda$$

and if we take the L^2 norm on $\mathbb{R}^N$ of the last integral, we can estimate its square, using the Cauchy–Schwarz inequality, by

$$\int_0^1 d\lambda \int_{\mathbb{R}^N} dy \left\{ \int_{\mathbb{R}^N} |\nabla u(x + \lambda(y-x)|^2 \, \omega_\varepsilon(y-x) \, dx \right\}$$

$$= \int_0^1 d\lambda \int_{\mathbb{R}^N} \omega_\varepsilon(z) \, dz \int_{\mathbb{R}^N} dx |\nabla u(x + \lambda z)|^2 = \int_{\mathbb{R}^N} |\nabla u(x)|^2 \, dx.$$

On the other hand, we have by Sobolev embeddings for some $p = p(N) \in (1, \infty)$

$$\| |(\rho^n * \omega_\varepsilon) - \rho^n| |u^n| \|_{L^2(B_M)} \leq C \|\rho^n * \omega_\varepsilon - \rho^n\|_{L^2(0,T;L^p(B_M))}.$$

This is enough to prove (2.78) and to conclude since the compactness shown in part 1 yields

$$\lim_{\varepsilon \to 0} \sup_{n \geq 1} \|\rho^n * \omega_\varepsilon - \rho^n\|_{L^2(0,T;L^p(B_M))} = 0.$$

Step 2. The case when u^n is bounded in L^2. We complete here the proof of part 2 in the case when we assume that u^n is bounded in $L^2(\mathbb{R}^N \times (0,T))$. By definition of P, we know there exists $\pi^n \in L^2(0,T;\mathcal{D}^{1,2}(\mathbb{R}^N))$ $(\tilde{\mathcal{D}}^{1,2}(\mathbb{R}^2)$ if $N = 2$—see Appendices A and B for more details on these spaces) such that

$$\rho^n u^n = P(\rho^n u^n) + \nabla \pi^n. \tag{2.79}$$

Since u^n is divergence free, we deduce for all $\delta > 0$

$$u^n = \frac{1}{\rho^n + \delta} \left\{ P(\rho^n u^n) + \delta u^n + \nabla \pi^n \right\} \tag{2.80}$$

$$\left. \begin{array}{l} \pi^n = \pi_\delta^n + \hat{\pi}_\delta^n, \quad \pi_\delta^n, \hat{\pi}_\delta^n \in L^2(0,T;\mathcal{D}^{1,2}(\mathbb{R}^N)) \\ (L^2(0,T;\tilde{\mathcal{D}}^{1,2}(\mathbb{R}^2))) \quad \text{if} \quad N = 2 \end{array} \right\} \tag{2.81}$$

$$\text{div}\left(\frac{1}{\rho^n + \delta} \left\{ \nabla \pi_\delta^n + P(\rho^n u^n) \right\} \right) = 0 \quad \text{in} \quad \mathcal{D}' \tag{2.82}$$

$$\text{div}\left(\frac{1}{\rho^n + \delta} \left\{ \nabla \hat{\pi}_\delta^n + \delta u^n \right\} \right) = 0 \quad \text{in} \quad \mathcal{D}', \tag{2.83}$$

and we write with obvious notation similar decompositions for u, ρ that involve π, π_δ and $\hat{\pi}_\delta$.

We next remark that we have

$$\int_{\mathbb{R}^N} \rho^n \left| \frac{\delta u^n + \nabla \hat{\pi}_\delta^n}{\rho^n + \delta} \right|^2 \leq \int_{\mathbb{R}^N} \frac{1}{\rho^n + \delta} |\delta u^n + \nabla \hat{\pi}_\delta^n|^2 \, dx$$

$$\leq \int_{\mathbb{R}^N} \frac{\delta^2}{\rho^n + \delta} |u^n|^2 \, dx \leq \delta \|u^n\|_{L^2(\mathbb{R}^N)}^2$$

since we deduce easily from (2.83) (using the density of $C_0^\infty(\mathbb{R}^N)$ into $\mathcal{D}^{1,2}(\mathbb{R}^N)$ or $\tilde{\mathcal{D}}^{1,2}(\mathbb{R}^2)$ if $N = 2$; see Appendix A)

$$\int_{\mathbb{R}^N} \frac{1}{\rho^n + \delta} \left\{ |\nabla \hat{\pi}_\delta^n|^2 + \delta u^n \cdot \nabla \hat{\pi}_\delta^n \right\} dx = 0.$$

In particular, we obtain

$$\lim_{\delta \to 0} \sup_{n \geq 1} \left\| \sqrt{\rho^n} u^n - \sqrt{\rho^n} \left\{ \frac{1}{\rho^n + \delta} (P(\rho^n u^n) + \nabla \pi_\delta^n) \right\} \right\|_{L^2(\mathbb{R}^N \times (0,T))} = 0$$

(2.84)

and similarly

$$\lim_{\delta \to 0} \left\| \sqrt{\rho} u - \sqrt{\rho} \left\{ \frac{1}{\rho + \delta} (P(\rho u) + \nabla \pi_\delta) \right\} \right\|_{L^2(\mathbb{R}^N \times (0,T))} = 0.$$

(2.85)

Therefore, in order to complete the proof, we need only to show that for each $\delta > 0$

$$\sqrt{\rho^n} \left(\frac{1}{\rho^n + \delta} \right) \{ P(\rho^n u^n) + \nabla \pi_\delta^n \} \xrightarrow{n} \sqrt{\rho} \left(\frac{1}{\rho + \delta} \right) \{ P(\rho u) + \nabla \pi_\delta \}$$

$$\text{in} \quad L^2(B_R \times (0,T)), \quad \forall\, R \in (0, \infty).$$

But we know from part 1 that ρ^n and thus $\sqrt{\rho^n}$, $\frac{1}{\rho^n + \delta}$ converge respectively to ρ, $\sqrt{\rho}$, $\frac{1}{\rho + \delta}$ in $C([0,T]; L^p(B_R))$ ($\forall\, 1 \leq p < \infty$, $\forall\, R \in (0, \infty)$). In addition, we also know from step 1 above that $P(\rho^n u^n)$ converges to $P(\rho u)$ in $L^2(B_R \times (0,T))$, $\forall\, R \in (0, \infty)$. Therefore, it only remains to show that $\nabla \pi_\delta^n$ converges to $\nabla \pi_\delta$ in $L^2(B_R \times (0,T))$.

As seen from the following result, the above convergence is, in fact, a consequence of (2.82) and the convergences we just recalled.

Lemma 2.4. *Let h^n be bounded in $L^2(\mathbb{R}^N)^N$, let $(a_{ij}^n)_{1 \leq i,j \leq N}$ be bounded in $L^\infty(\mathbb{R}^N)$. We assume*

$$\exists\, \nu > 0, \; \forall\, n \geq 1, \text{a.e. in } x \in \mathbb{R}^N, \; \forall\, \xi \in \mathbb{R}^N, \quad \sum_{i,j=1}^{N} a_{ij}^n \xi_i \xi_j \geq \nu |\xi|^2,$$

(2.86)

$$a_{ij}^n \xrightarrow{n} a_{ij} \quad \text{in} \quad L^1(B_R) \quad \text{for all} \quad R \in (0, \infty),$$

(2.87)

$$h^n \xrightarrow{n} h \quad \text{in} \quad L^1(B_R) \quad \text{for all} \quad R \in (0, \infty).$$

(2.88)

We consider the unique solution $f^n \in \mathcal{D}^{1,2}(\mathbb{R}^N)$ (if $N = 2$, $f^n \in \tilde{\mathcal{D}}^{1,2}(\mathbb{R}^2)$ with $\int_{B_1} f^n\, dx = 0$) of

$$-\sum_{i,j=1}^{N} \frac{\partial}{\partial x_i} \left(a_{ij}^n \left(\frac{\partial f^n}{\partial x_j} + h_j^n \right) \right) = 0 \quad \text{in} \quad \mathcal{D}'(\mathbb{R}^N)$$

(2.89)

and we denote by f the solution of (2.89) with a_{ij}^n, h_j^n replaced by a_{ij}, h. Then, f^n converges in $H^1(B_R)$ to f for all $R \in (0, \infty)$.

Proof of Lemma 2.4. First, we observe that (by the density of $C_0^\infty(\mathbb{R}^N)$ in $\mathcal{D}^{1,2}(\mathbb{R}^N)$, or $C_0^\infty(\mathbb{R}^2)$ in $\tilde{\mathcal{D}}^{1,2}(\mathbb{R}^2)$—see the argument of Appendix A) f^n is bounded in $\mathcal{D}^{1,2}(\mathbb{R}^N)$ (resp. $\tilde{\mathcal{D}}^{1,2}(\mathbb{R}^2)$) and converges weakly in $\mathcal{D}^{1,2}(\mathbb{R}^N)$ ($\tilde{\mathcal{D}}^{1,2}(\mathbb{R}^2)$ if $N = 2$) to f since h^n converges to h weakly in $L^2(\mathbb{R}^N)^N$. In particular we deduce, from the Rellich–Kondrakov theorem, that f^n converges to f in $L^2(B_R)$ for all $R \in (0, \infty)$. Then, for $R \in (0, \infty)$ fixed, we consider $\varphi \in C_0^\infty(\mathbb{R}^N)$ satisfying: $\varphi \geq 0$ on $\mathbb{R}^N$, $\varphi \equiv 1$ on B_R; and we multiply (2.89) by φf_n (or φf). We then obtain

$$\left.\begin{aligned}
&\int_{\mathbb{R}^N} \left(\sum_{i,j=1}^N a_{ij}^n \frac{\partial f^n}{\partial x_i} \frac{\partial f^n}{\partial x_j} \right) \varphi \, dx \\
&= \int_{\mathbb{R}^N} \sum_{i,j=1}^N a_{ij}^n \frac{\partial f^n}{\partial x_j} \frac{\partial \varphi}{\partial x_i} f^n \, dx - \int_{\mathbb{R}^N} \sum_{i,j=1}^N a_{ij}^n h_j^n \frac{\partial}{\partial x_i} (\varphi f^n) \, dx
\end{aligned}\right\} \quad (2.90)$$

$$\left.\begin{aligned}
&\int_{\mathbb{R}^N} \left(\sum_{i,j=1}^N a_{ij} \frac{\partial f}{\partial x_i} \frac{\partial f}{\partial x_j} \right) \varphi \, dx \\
&= -\int_{\mathbb{R}^N} \sum_{i,j=1}^N a_{ij} \frac{\partial f}{\partial x_j} \frac{\partial \varphi}{\partial x_i} f \, dx - \int_{\mathbb{R}^N} \sum_{i,j=1}^N a_{ij} h \frac{\partial}{\partial x_i} (\varphi f) \, dx.
\end{aligned}\right\} \quad (2.91)$$

We then claim that the right-hand side converges, as n goes to $+\infty$, to the right-hand side of (2.91). Indeed, a_{ij}^n is, uniformly in n, bounded and converges in $L^p(\operatorname{Supp}\varphi)$ for $1 \leq p < \infty$ while f^n and h^n converge in $L^2(\operatorname{Supp}\varphi)$ therefore $a_{ij}^n \frac{\partial \varphi}{\partial x_i} f^n$, $a_{ij}^n h_j^n$ converge in $L^2(\operatorname{Supp}\varphi)$ to, respectively, $a_{ij} \frac{\partial \varphi}{\partial x_i} f$, $a_{ij} h_j$. In addition, $\frac{\partial f^n}{\partial x_j}$ and $\frac{\partial}{\partial x_i} (\varphi f^n)$ converge weakly in $L^2(\mathbb{R}^N)$ to, respectively, $\frac{\partial f}{\partial x_j}$ and $\frac{\partial}{\partial x_i} (\varphi f)$, and this shows our claim.

Hence, the right-hand side of (2.90) converges to the right-hand side of (2.91). But we also have

$$\nu \int_{\mathbb{R}^N} |\nabla(f^n - f)|^2 \, dx$$

$$\leq \int_{\mathbb{R}^N} \sum_{i,j=1}^N a_{ij}^n \left(\frac{\partial f^n}{\partial x_i} - \frac{\partial f}{\partial x_i} \right) \left(\frac{\partial f^n}{\partial x_j} - \frac{\partial f}{\partial x_j} \right) dx$$

$$= \int_{\mathbb{R}^N} \sum_{i,j=1}^N a_{ij}^n \frac{\partial f^n}{\partial x_i} \frac{\partial f^n}{\partial x_j} \, dx + \int_{\mathbb{R}^N} \sum_{i,j=1}^N a_{ij}^n \frac{\partial f}{\partial x_i} \frac{\partial f}{\partial x_j} \, dx$$

$$- \int_{\mathbb{R}^N} \sum_{i,j=1}^N a_{ij}^n \left(\frac{\partial f^n}{\partial x_i} \frac{\partial f}{\partial x_j} + \frac{\partial f^n}{\partial x_j} \frac{\partial f}{\partial x_i} \right) dx,$$

and the lemma is shown if we prove that this upper bound goes to 0 as n goes to $+\infty$. This is the case since we just proved that the first term goes to $\int_{\mathbb{R}^N} \sum_{i,j=1}^{N} a_{ij} \frac{\partial f}{\partial x_i} \frac{\partial f}{\partial x_j}\, dx$, while the second term converges obviously to the same quantity in view of the properties of a_{ij}^n. Finally since $a_{ij}^n \frac{\partial f}{\partial x_i}$, or $a_{ij}^n \frac{\partial f}{\partial x_j}$ converges in $L^2(\mathbb{R}^N)$ to $a_{ij} \frac{\partial f}{\partial x_i}$, $a_{ij} \frac{\partial f}{\partial x_j}$ respectively (by Lebesgue's lemma), the last integral converges to $2\int_{\mathbb{R}^N} \sum_{i,j=1}^{N} a_{ij} \frac{\partial f}{\partial x_i} \frac{\partial f}{\partial x_j}\, dx$, and we conclude. □

Step 3. The case when $N \geq 3$, $\rho^n, \rho \in L^{\frac{N}{2},\infty}(\mathbb{R}^N)$. Let us first observe that the proof of part 1 of Theorem 2.5 and the proof of step 2 of Theorem 2.4 immediately yield the fact that meas $\{\rho^n \geq \lambda\}$, meas $\{\rho \geq \lambda\}$ are independent of $t \in [0,T]$ for all $\lambda > 0$ and thus $\rho^n, \rho \in L^\infty(0,T; L^{\frac{N}{2},\infty}(\mathbb{R}^N))$. We are going to use the results of Appendix A and more precisely Theorem 2. To this end, we introduce the solutions $u_{R,\varepsilon}^n, u_{R,\varepsilon}$ for a.e. $t \in (0,T)$, $R \in (0,\infty)$, $\varepsilon \in (0,1]$ of, respectively

$$\left. \begin{aligned} -\Delta u_{R,\varepsilon}^n + \frac{1}{\varepsilon}\rho^n u_{R,\varepsilon}^n + \nabla p_{R,\varepsilon}^n &= -\Delta u^n + \frac{1}{\varepsilon}\rho^n u^n \quad \text{in } \mathcal{D}'(B_R), \\ u_{R,\varepsilon}^n \in H_0^1(B_R), \quad \operatorname{div} u_{R,\varepsilon}^n &= 0 \quad \text{a.e. in } B_R; \end{aligned} \right\} \quad (2.92)$$

$$\left. \begin{aligned} -\Delta u_{R,\varepsilon} + \frac{1}{\varepsilon}\rho u_{R,\varepsilon} + \nabla p_{R,\varepsilon} &= -\Delta u + \frac{1}{\varepsilon}\rho u \quad \text{in } \mathcal{D}'(B_R), \\ u_{R,\varepsilon} \in H_0^1(B_R), \quad \operatorname{div} u_{R,\varepsilon} &= 0 \quad \text{a.e. in } B_R. \end{aligned} \right\} \quad (2.93)$$

We may then apply Theorem 2 in Appendix A with $f^n = \frac{1}{\varepsilon}\rho^n$, the assumptions required in Theorem 2 being satisfied in our case in particular because of part 1 of Theorem 2.5 proven above, and we obtain, for all $\varepsilon \in (0,1]$,

$$\left. \begin{aligned} \int_{B_R} \rho^n |u_{R,\varepsilon}^n|^2\, dx &+ \varepsilon \int_{B_R} |\nabla u_{R,\varepsilon}^n|^2\, dx \\ \leq \int_{B_R} \rho^n u_{R,\varepsilon}^n u^n\, dx &+ \varepsilon \int_{B_R} \nabla u^n \cdot \nabla u_{R,\varepsilon}^n\, dx, \\ \text{a.e. } t \in (0,T), \end{aligned} \right\} \quad (2.94)$$

$$u_{R,\varepsilon}^n \to u^n, \quad u_{R,\varepsilon} \to u \quad \text{in } L^2(0,T;\mathcal{D}^{1,2}(\mathbb{R}^N)) \quad \text{as } R \to +\infty, \quad (2.95)$$

$$\left. \begin{aligned} \sup_n \|u^n - u_{R,\varepsilon}^n\|_{L^2(B_M \times (0,T))} \to 0 \quad \text{as } R \to +\infty, \\ \text{for all } M \in (0,\infty). \end{aligned} \right\} \quad (2.96)$$

Finally, we also have

$$\left. \begin{aligned} u_{R,\varepsilon}^n \rightharpoonup u_{R,\varepsilon} \quad \text{weakly in } L^2(0,T;H^1(B_R)), \\ \text{for all } R \in (0,\infty), \ \varepsilon \in (0,1]. \end{aligned} \right\} \quad (2.97)$$

Indeed, we deduce $H^1(B_R)$ bounds on $u^n_{R,\varepsilon}$ from (2.94)—recall that $\rho^n |u^n|^2$ is bounded in $L^\infty(0,T; L^1(\mathbb{R}^N))$ by assumption. Then, (2.97) follows from the uniqueness of the equation (2.93) passing to the limit in (2.92) (recall that, by part 1, ρ^n converges to ρ in $C([0,T]; L^p(B_R))$ for all $1 < p < \infty$).

We then write, for all $M \in (0,\infty)$ fixed

$$\|\sqrt{\rho^n} u^n - \sqrt{\rho} u\|_{L^2(B_M \times (0,T))} \leq C \sup_{n \geq 1} \|u^n - u^n_{R,\varepsilon}\|_{L^2(B_M \times (0,T))}$$
$$+ \|\sqrt{\rho^n} u^n_{R,\varepsilon} - \sqrt{\rho} u\|_{L^2(B_M \times (0,T))},$$

where C denotes various positive constants independent of n, R, ε. Using (2.95) and (2.96), we deduce

$$\left.\begin{array}{l} \|\sqrt{\rho^n} u^n - \sqrt{\rho} u\|_{L^2(B_M \times (0,T))} \\[2mm] \leq \ \|\sqrt{\rho^n} u^n_{R,\varepsilon} - \sqrt{\rho} u\|_{L^2(B_M \times (0,T))} + \omega_\varepsilon(R) \end{array}\right\} \qquad (2.98)$$

where we denote by $\omega_\varepsilon(R)$ various positive constants that depend only on ε and R, and such that $\omega_\varepsilon(R) \to 0$ as $R \to +\infty$, for each $\varepsilon \in (0,T]$.

Next, we remark that (2.94) yields

$$\int_0^T \int_{B_R} \rho^n |u^n_{R,\varepsilon}|^2 \, dx \, dt \ \leq \ \int_0^T \int_{B_R} \rho^n u^n \cdot u^n_{R,\varepsilon} \, dx \, dt + C\varepsilon \qquad (2.99)$$

while we have

$$\left.\begin{array}{c} \displaystyle\int_0^T \int_{B_R} \rho u \cdot u_{R,\varepsilon} \, dx \, dt \ , \ \int_0^T \int_{B_R} \rho |u_{R,\varepsilon}|^2 \, dx \, dt \\[4mm] \displaystyle \to \int_0^T \int_{\mathbb{R}^N} \rho |u|^2 \, dx \, dt \\[3mm] \text{as } R \text{ goes to } +\infty, \text{ for all } \varepsilon \in (0,1]. \end{array}\right\} \qquad (2.100)$$

Indeed, $\rho \in L^\infty(0,T; L^{\frac{N}{2},\infty}(\mathbb{R}^N))$ and by the (sharp in Lorentz spaces) Sobolev embeddings (2.95) implies that $u_{R,\varepsilon}$ converges to u in $L^2(0,T; L^{\frac{2N}{N-2}}(\mathbb{R}^N))$.

We then claim that we have for all $R \in (0,\infty)$, $\varepsilon \in (0,1]$

$$\int_0^T \int_{B_R} \rho^n u^n \cdot u^n_{R,\varepsilon} \, dx \, dt \ \underset{n}{\to} \ \int_0^T \int_{B_R} \rho u \cdot u_{R,\varepsilon} \, dx \, dt. \qquad (2.101)$$

Indeed, since $u^n_{R,\varepsilon}$ and $u_{R,\varepsilon}$ are divergence-free vector fields and vanish outside B_R, each integral can be rewritten as an integral over $\mathbb{R}^N$ (or B_R) of $P(\rho^n u^n) \cdot u^n_{R,\varepsilon}$, $P(\rho u) \cdot u_{R,\varepsilon}$ respectively. Then, in view of step 1 above,

$P(\rho^n u^n)$ converges to $P(\rho u)$ in $L^2(B_R \times (0,T))$ while $u_{R,\varepsilon}^n$ converges weakly to $u_{R,\varepsilon}$ in $L^2(B_R \times (0,T))$ because of (2.97) as n goes to $+\infty$. This proves the claim (2.101).

Combining (2.99), (2.100) and (2.101), we obtain

$$\|\sqrt{\rho^n} u_{R,\varepsilon}^n - \sqrt{\rho} u_{R,\varepsilon}\|_{L^2(B_R \times (0,T))} \leq C\varepsilon + \omega_\varepsilon(R)$$

$$+ \int_0^T dt \int_{B_R} dx \, \rho^n u^n \cdot u_{R,\varepsilon}^n + \rho u \cdot u_{R,\varepsilon} - 2\sqrt{\rho^n} \sqrt{\rho} u_{R,\varepsilon}^n \cdot u$$

and

$$\overline{\lim_n} \, \|\sqrt{\rho^n} u_{R,\varepsilon}^n - \sqrt{\rho} u\|_{L^2(B_R \times (0,T))} \leq C\varepsilon + \omega_\varepsilon(R) \qquad (2.102)$$

since $u_{R,\varepsilon}^n$ converges weakly to $u_{R,\varepsilon}$ while $\sqrt{\rho^n}$ converges strongly to $\sqrt{\rho}$.

Adding up (2.98) and (2.102), we finally deduce

$$\overline{\lim_n} \, \|\sqrt{\rho^n} u^n - \sqrt{\rho} u\|_{L^2(B_M \times (0,T))} \leq C\varepsilon + \omega_\varepsilon(R),$$

and we conclude upon letting first R go to $+\infty$ and then ε go to 0.

Step 4. The case when $N = 2$, (2.64) holds. We first remark that, because of part 1 of Theorem 2.5, (2.64) yields

$$\sup_{0 \leq t \leq T} \int_{\mathbb{R}^2} \rho^p <x>^{2(p-1)} (\log <x>)^r \, dx < \infty. \qquad (2.103)$$

We are going to show in that case that $\sqrt{\rho^n} u^n$ converges to $\sqrt{\rho} u$ in $L^2(\mathbb{R}^2 \times (0,T))$. To this end, we use the notation of Appendix A. We observe first that $\tilde{T}_R(u^n)$ is bounded in $L^2(0,T; \tilde{\mathcal{D}}^{1,2}(\mathbb{R}^2))$ uniformly in n, R, $\tilde{T}_R(u)$ is bounded in $L^2(0,T; \tilde{\mathcal{D}}^{1,2}(\mathbb{R}^2))$ and that $\tilde{T}_R(u^n)$ converges weakly in $L^2(0,T; \tilde{\mathcal{D}}^{1,2}(\mathbb{R}^2))$ (or $L^2(0,T; H_0^1(B_R))$) to $\tilde{T}_R(u)$ as n goes to $+\infty$ ($\forall R \in (0,\infty)$). In particular, Theorem 1 of Appendix A yields for all $R_0 \in (0,\infty)$

$$\left. \begin{array}{c} \displaystyle\sup_n \left| \int_{B_{R_0}} \int_0^T \rho^n |u^n|^2 - \rho^n u^n \cdot \tilde{T}_R(u^n) \, dx \, dt \right| \\[3mm] + \displaystyle\left| \int_{B_{R_0}} \int_0^T \rho |u|^2 - \rho u \cdot \tilde{T}_R(u) \, dx \, dt \right| \to 0 \\[3mm] \text{as} \quad R \to +\infty. \end{array} \right\} \qquad (2.104)$$

Next, we claim that we have

$$\left. \begin{array}{c} \displaystyle\sup_n \left(\int_{|x| \geq R_0} dx \int_0^T dt \, \rho^n |u^n|^2 - \rho^n |u^n| \, |\tilde{T}_R(u^n)| \right) \\[3mm] + \displaystyle\int_{|x| \geq R_0} dx \int_0^T dt \, \rho |u|^2 - \rho |u| \, |\tilde{T}_R(u)| \to 0 \\[3mm] \text{as} \quad R \to +\infty. \end{array} \right\} \qquad (2.105)$$

Indeed, we choose $s \in (2p-1, r)$ and we use Hölder's inequality to obtain

$$\int_{|x| \geq R_0} dx \int_0^T dt \, f |v| \, |w|$$

$$\leq \left(\int_{|x| \geq R_0} dx \, f^p < x >^{2(p-1)} (\log < x >)^s \right)^{1/p}$$

$$\cdot \int_0^T dt \left(\int_{\mathbb{R}^2} dx \, \frac{|v|^{2p'}}{<x>^2} (\log < x >)^{-\frac{s}{p-1}} \right)^{\frac{1}{p'}}$$

$$\cdot \int_0^T dt \left(\int_{\mathbb{R}^2} dx \, \frac{|w|^{2p'}}{<x>^2} (\log < x >)^{-\frac{s}{p-1}} \right)^{\frac{1}{p'}}$$

$$\leq \left(\log \sqrt{1+R_0^2} \right)^{-\frac{r-s}{p}} \left(\int_{\mathbb{R}^2} dx \, f^p < x >^{2(p-1)} (\log < x >)^r \right)^{\frac{1}{p}}$$

$$\cdot \|v\|_{L^2(0,T;\tilde{\mathcal{D}}^{1,2}(\mathbb{R}^2))} \|w\|_{L^2(0,T;\tilde{\mathcal{D}}^{1,2}(\mathbb{R}^2))}$$

where $f = \rho^n, \rho$, $v = u^n, u$, $w = u^n, \tilde{T}_R(u^n), u, \tilde{T}_R(u)$.

The limit (2.105) is shown using the bounds (2.64) and (2.103) on ρ^n, ρ and the bounds recalled above on $\tilde{T}_R(u^n)$, $\tilde{T}_R(u)$.

Finally, we notice that we have, since $\tilde{T}_R(u^n)$, $\tilde{T}_R(u)$ are divergence-free and vanish outside B_R

$$\int_0^T \int_{\mathbb{R}^2} \rho^n u^n \cdot \tilde{T}_R(u^n) \, dt \, dx = \int_0^T \int_{\mathbb{R}^2} P(\rho^n u^n) \cdot \tilde{T}_R(u^n) \, dt \, dx$$

$$= \int_0^T \int_{B_R} P(\rho^n u^n) \cdot \tilde{T}_R(u^n) \, dt \, dx \underset{n}{\rightarrow} \int_0^T \int_{B_R} P(\rho u) \cdot \tilde{T}_R(u) \, dt \, dx$$

$$= \int_0^T \int_{\mathbb{R}^2} P(\rho u) \cdot \tilde{T}_R(u) \, dt \, dx = \int_0^T \int_{\mathbb{R}^2} \rho u \cdot \tilde{T}_R(u) \, dt \, dx,$$

$$\text{for all} \quad R \in (0, \infty).$$

The convergence is a consequence of the strong convergence in $L^2(B_R \times (0,T))$ of $P(\rho^n u^n)$ towards $P(\rho u)$ as shown in step 1 above, and of the weak convergence in $L^2(B_R \times (0,T))$ of $\tilde{T}_R(u^n)$ to $\tilde{T}_R(u)$ recalled above.

Combining the preceding convergence with (2.104) and (2.105), we deduce easily

$$\int_0^T \int_{\mathbb{R}^2} \rho^n |u^n|^2 \, dx \, dt \underset{n}{\rightarrow} \int_0^T \int_{\mathbb{R}^2} \rho |u|^2 \, dx \, dt,$$

and we conclude the proof of Theorem 2.5. □

Remarks 2.7. 1) It is plausible that the additional bounds assumed upon u^n on ρ^n in part 2 of Theorem 2.5 are superfluous. They are used to allow us to pass to the limit in $\rho^n|u^n|^2$ and if this passage to the limit were true (without these extra conditions), more general existence results than Theorem 2.1 in the case when $\Omega = \mathbb{R}^N$ would be possible. We indicate in the next two remarks different arguments that can be used to pass to the limit but which, unfortunately, do not give better results than the methods introduced above.

2) We begin with a method taken (and adapted) from [16] (see also the references therein). This argument requires u^n to be bounded in $L^2(0, T; H^1(\mathbb{R}^N))^N$, $N = 2$ or 3 and some conditions on ρ explained below. We just sketch the argument, letting (ρ^n, u^n) be a sequence of (weak) solutions of (2.1)–(2.2) satisfying, uniformly in n, the a priori estimates described at the beginning of this section. The method of proof consists in integrating (2.2) in time between t and $t + h$ and multiplying by $u^n(t+h) - u^n(t)$: this yields, using the a priori estimates,

$$
\int_{\mathbb{R}^N} \left(\rho^n(t+h) u^n(t+h) - \rho^n(t) u^n(t) \right) \cdot \left(u^n(t+h) - u^n(t) \right) dx
$$
$$
\leq Ch^{1/2} \left(\|u^n(t+h)\|_{H^1(\mathbb{R}^N)} + \|u^n(t)\|_{H^1(\mathbb{R}^N)} \right)
$$
$$
+ C \left(\int_t^{t+h} ds \int_{\mathbb{R}^2} dx \ \sqrt{\rho^n} |u^n|^4 \right)^{1/2}
$$
$$
\cdot \left(\|\nabla u^n(t+h)\|_{L^2(\mathbb{R}^N)} + \|\nabla u^n(t)\|_{L^2(\mathbb{R}^N)} \right).
$$

Next, if $N = 2$ or $N = 3$

$$
\int_t^{t+h} ds \int_{\mathbb{R}^N} dx \ \sqrt{\rho^n} |u^n| \, |u^n|^3
$$
$$
\leq \int_t^{t+h} ds \left(\int_{\mathbb{R}^N} \rho^n |u^n|^2 \, dx \right)^{1/2} \left(\int_{\mathbb{R}^N} |u^n|^6 \, dx \right)^{1/2}
$$
$$
\leq C \int_t^{t+h} ds \ \|u^n\|_{H^1(\mathbb{R}^N)}^{3/2} \leq Ch^{1/4}.
$$

Hence, we find for all $h \in (0, 1)$

$$
\int_0^T dt \int_{\mathbb{R}^N} \left(\rho^n(t+h) u^n(t+h) - \rho^n(t) u^n(t) \right) \cdot \left(u^n(t+h) - u^n(t) \right) dx \leq Ch^{1/4}.
$$

It is in fact possible to extend this argument to $N \geq 4$ multiplying by $u^n(t+h) * \omega_\varepsilon - u^n(t) * \omega_\varepsilon$ (instead of $u^n(t+h) - u^n(t)$). This leads to a bound like $\left(Ch^{1/2} + \frac{C}{\varepsilon^N} h + C\varepsilon \right) = Ch^{1/(N+1)}$ if $\varepsilon = h^{1/(N+1)}$ using the fact that $\|u^n - u^n * \omega_\varepsilon\|_{L^2(\mathbb{R}^N \times (0,T))} \leq C\varepsilon$.

Next, we deduce from the preceding inequality

$$\int_0^T dt \int_{\mathbb{R}^N} \left| \sqrt{\rho^n(t+h)} u^n(t+h) - \sqrt{\rho^n(t)} u^n(t) \right|^2 dx$$

$$\leq Ch^{1/4} + \int_0^T dt \int_{\mathbb{R}^N} dx \, u^n(t) \cdot u^n(t+h) \left| \sqrt{\rho^n(t)} - \sqrt{\rho^n(t+h)} \right|^2$$

$$\leq Ch^{1/4} + \int_0^T dt \int_{\mathbb{R}^N} |u^n(t)| \, |u^n(t+h)| \, |\rho^n(t) - \rho^n(t+h)|$$

$$\leq \omega(h)$$

where $\omega(h) \to 0$ as $h \to 0_+$ and $\omega(h)$ does not depend on n. The last inequality requires some assumptions on ρ^n like for instance $\rho_0^n = \overline{\rho} + f_0^n$ with $\overline{\rho} > 0$, $f_0^n \in L^q(\mathbb{R}^N)$ for some $q \in (\frac{N}{2}, \infty)$, $f_0^n \xrightarrow{n} f_0$ in $L^q(\mathbb{R}^N)$. This condition as explained above yields $L^2(0, T; H^1)$ bounds and one deduces easily from part 1 of Theorem 2.5 (and its proof) that $\rho^n = \overline{\rho} + f^n$ where f^n converges in $C([0, T]; L^q(\mathbb{R}^N))$. This is enough to yield the above bound.

The above "time-continuity in L^2" of $\sqrt{\rho^n} u^n$ allows us to obtain compactness in $L^2(B_R \times (0, T))$ of $\sqrt{\rho^n} u^n$, using the compactness of ρ^n (see above and part 1 of Theorem 2.5) and the fact that u^n is bounded in $L^2(0, T; H^1(\mathbb{R}^N))$.

3) Another method of proof consists of using some particular projections: we introduce P_R, the projection from $L^2(\mathbb{R}^N)^N$ onto $\{v \in L^2(\mathbb{R}^N)^N, \ v = 0$ a.e. on B_R^c, $\operatorname{div} v = 0$ in $\mathcal{D}'(\mathbb{R}^N)\}$; notice that necessarily $v \cdot n = 0$ on ∂B_R by trace theorems. Then, if we consider ρ^n, u^n as in the preceding remark, it is not difficult to check that for all $R_0 \in (1, \infty)$, there exists $R_n \in (R_0, R_0+1)$, such that $\frac{\partial}{\partial t} \{P_{R_n}(\rho^n u^n)\}$ is bounded in $L^q(0, T; W^{-1,q}(B_{R_n}))$ for some $q > 1$. This is enough to ensure that $P_{R_n}(\rho^n u^n)$ is relatively compact in $L^2(B_{R_n} \times (0, T))$ (for instance) and converges to $P_R(\rho u)$ if R_n (or a subsequence) converges to some $R \in [R_0, R_0+1]$. On the one hand, we have then

$$\rho^n u^n = P_{R_n}(\rho^n u^n) + \nabla \pi^n \quad \text{in} \quad B_{R_n},$$

$$\rho^n |u^n|^2 = P_{R_n}(\rho^n u^n) u^n + \operatorname{div}(u^n \pi^n) \quad \text{in} \quad B_{R_n}$$

for some $\pi^n \in L^\infty(0, T; H^1(B_{R_n}))$ (which we can normalize by $\int_{B_{R_n}} \pi^n \, dx = 0$), and, if we let n go to $+\infty$, we obtain

$$\rho \overline{|u|^2} = P_R(\rho u) \cdot u + \operatorname{div}(\mu) \quad \text{in} \quad B_R$$

where $\overline{|u|^2}$ is the weak limit of $|u^n|^2$ and μ is the weak limit of $u^n \pi^n$. Notice that $\mu \in L^2(0, T; L^{\frac{N}{N-1}, 1}(B_R))$ (in fact, it is bounded in that space uniformly in R_0) if u^n is bounded in $L^2(\mathbb{R}^N \times (0,T))$ or if ρ^n is bounded in $L^{\frac{N}{2},\infty}(\mathbb{R}^N)$ (assume for instance $N \geq 3$). On the other hand, we also have

$$\rho|u|^2 = P_R(\rho u) \cdot u + \text{div}\,(\tilde{\mu}) \qquad \text{in} \quad B_R$$

with the same bounds on $\tilde{\mu}$. In particular, we deduce, upon letting R go to $+\infty$,

$$\text{div}\,(\overline{\mu}) = \rho(\overline{|u|^2} - |u|^2) \in L^\infty(0,T;L^1(\mathbb{R}^N)) \geq 0$$

where $\mu - \tilde{\mu} \in L^2\big(0,T;L^{\frac{N}{N-1},1}(\mathbb{R}^N)\big)$. It is then easy to conclude that $\overline{\mu} \equiv 0$ and thus $\sqrt{\rho^n}u^n$ converges to ρu in $L^2(B_R \times (0,T))$ ($\forall\, R \in (0,\infty)$).

Let us finally observe that this method of proof—which requires many rather technical justifications that we leave to the reader—requires either u^n to be bounded in $L^2(\mathbb{R}^N \times (0,T))$ or ρ^n to be bounded in $L^\infty\big(0,T; L^{\frac{N}{2},\infty}(\mathbb{R}^N)\big)$, i.e. ρ_0^n to be bounded in $L^{\frac{N}{2},\infty}(\mathbb{R}^N)$. $\square$

2.4 Existence proofs

In this section, we give complete proofs of the existence part of Theorem 2.1. We split the argument into three steps. In the first one, we solve an approximated problem and thus construct approximated solutions. Next, in step 2, we use the a priori estimates and the compactness results obtained in the preceding section to pass to the limit and build solutions of (2.1)–(2.2), and this will prove Theorem 2.1 in the cases when Ω is bounded, namely the periodic case or the case of Dirichlet boundary conditions. Finally, in a third step, we treat (and deduce) the case when $\Omega = \mathbb{R}^N$.

Step 1. Construction of approximated solutions. Our goal here is to construct solutions of the following approximated system

$$\frac{\partial \rho}{\partial t} + \text{div}\,(u_\varepsilon \rho) = 0 \qquad \text{in} \quad \mathcal{D}' \tag{2.106}$$

$$\left.\begin{array}{c} \dfrac{\partial \rho u}{\partial t} + \text{div}\,(\rho u_\varepsilon \otimes u) - \text{div}\,(2\mu_\varepsilon d) + \nabla p = \rho f_\varepsilon \\[2mm] \text{in } \mathcal{D}'\,, \quad \text{div}\, u = 0 \text{ in } \mathcal{D}'. \end{array}\right\} \tag{2.107}$$

If we consider the periodic case, then (2.106)–(2.107) hold in $\mathcal{D}'(\mathbb{R}^N \times (0,\infty))$ and all unknowns are assumed to be periodic of period $T_i > 0$ in x_i, for each $i \in \{1,\dots,N\}$. Let us recall that we define in this case $\Omega = \prod_{i=1}^N (0,T_i)$, while, if we treat the case of Dirichlet conditions, (2.106)–(2.107) hold in $\mathcal{D}'(\Omega \times (0,\infty))$ and we require u to vanish on $\partial\Omega \times (0,\infty)$.

We now have to explain the real meaning of (2.106), (2.107) or, in other words, the precise definition of u_ε and μ_ε which are regularizations of u and $\mu(\rho)$ respectively, depending upon a parameter $\varepsilon \in (0,1]$. In the periodic case, we simply take $u_\varepsilon = u * \omega_\varepsilon$, $\mu_\varepsilon = \mu^\varepsilon(\rho) * \omega_\varepsilon$, $f_\varepsilon = f * \omega_\varepsilon$, where ω_ε is a regularizing kernel as in the preceding section and μ^ε is defined below. In fact, for technical reasons, we take $f_\varepsilon = (f * \omega_\varepsilon)\zeta_\varepsilon(t)$ where $\zeta_\varepsilon \in C^\infty([0,T])$, $\zeta_\varepsilon(t) = 1$ if $t \geq 2\varepsilon$, $0 \leq \zeta_\varepsilon(t) \leq 1$ if $t \in [0,T]$, $\zeta_\varepsilon = 0$ if $t \leq \varepsilon$. Observe that we still have div $u_\varepsilon = 0$.

In the case of Dirichlet boundary conditions, we set $\bar{\mu}(\rho) = \mu^\varepsilon(\rho)$ in Ω, $= 1$ in Ω^c and define $\mu_\varepsilon = \widetilde{\mu^\varepsilon(\rho)} * \omega_\varepsilon|_\Omega$. The definition of u_ε in that case is a bit more delicate since we wish to smoothe u, while keeping the Dirichlet conditions and the divergence-free property. One possible (explicit) way is the following: if $u \in L^2(0,T; H_0^1(\Omega))$ (for example) where $T \in (0,\infty)$ is fixed, we set $\bar{u}_\varepsilon$ to be the truncation in Ω_ε of u (extended by 0 to Ω) as defined in Appendix A and we define u_ε by $\bar{u}_\varepsilon * \omega_{\varepsilon/2}$. Clearly, u_ε, which vanishes near $\partial\Omega$, is smooth in x (recall from Appendix A that $\bar{u}_\varepsilon \in L^2(0,T; H_0^1(\Omega))$) and satisfies div $u_\varepsilon = 0$ in $\mathbb{R}^N$. Finally, we set $f_\varepsilon = \zeta_\varepsilon(f \, 1_{(d>2\varepsilon)}) * \omega_\varepsilon$ where $d = \text{dist}(x, \partial\Omega)$.

We would like also to make a simplification on $\mu(\cdot)$: since, anyway, all values of solutions ρ remain uniformly bounded (typically in an interval $[0, \|\rho_0\|_{L^\infty}]$), we can assume without loss of generality that $\mu(t)$ is constant for $t \geq 0$ large and in particular that $(t \mapsto \mu(t))$ is bounded on $[0, +\infty)$. Then, μ^ε is a $C^\infty([0,\infty[)$ function, bounded away from 0, which is constant for $t \geq 0$ large and such that $\sup_{[0,\infty)} |\mu^\varepsilon - \mu| \leq \varepsilon$.

We now discuss the initial conditions associated to (2.106)–(2.107), namely

$$\rho|_{t=0} = \rho_0^\varepsilon, \quad \rho u|_{t=0} = m_0^\varepsilon \quad \text{on} \quad \Omega \qquad (2.108)$$

where $\rho_0^\varepsilon = (\rho_0)_\varepsilon + \varepsilon$, m_0^ε is defined below using $\overline{m}_0^\varepsilon = (m_0 \rho_0^{-1/2})_\varepsilon (\rho_0^{1/2})_\varepsilon$. In the periodic case, for $f = \rho_0, \rho_0^{1/2}, m_0 \rho_0^{-1/2}$, we define $f_\varepsilon = f * \omega_\varepsilon$. In the case of Dirichlet conditions, $(\rho_0)_\varepsilon = \tilde{\rho}_0 * \omega_\varepsilon|_\Omega$, $(\rho_0^{1/2})_\varepsilon = \tilde{\rho}_0^{1/2} * \omega_\varepsilon|_\Omega$ and $(m_0 \rho_0^{-1/2})_\varepsilon = (m_0 \rho_0^{-1/2} \, 1_{(d>2\varepsilon)}) * \omega_\varepsilon$ where $\tilde{\rho}_0 = \rho_0$ on Ω, $= 1$ on Ω^c and $d = \text{dist}(x, \partial\Omega)$. Obviously, $\rho_0^\varepsilon \in C^\infty(\overline{\Omega})$, $\overline{m}_0^\varepsilon \in C_0^\infty(\Omega)$. Let us immediately remark that we have for some $C_0 \geq 0$ independent of ε

$$\varepsilon \leq \rho_0^\varepsilon \leq C_0 \qquad (2.109)$$

$$\left. \begin{array}{l} \rho_0^\varepsilon \xrightarrow{\varepsilon} \rho_0 \quad \text{in} \quad L^p(\Omega) \quad (1 \leq p < \infty), \\[2mm] \overline{m}_0^\varepsilon \xrightarrow{\varepsilon} m_0 \quad \text{in} \quad L^2(\Omega), \quad \overline{m}_0^\varepsilon (\rho_0^\varepsilon)^{-1/2} \xrightarrow{\varepsilon} m_0 \rho_0^{-1/2} \quad \text{in} \quad L^2(\Omega). \end{array} \right\} \qquad (2.110)$$

The last convergence in (2.110) is easily deduced from the following facts: $(\rho_0^{1/2})_\varepsilon \leq (\rho_0)_\varepsilon^{1/2}$ in $\overline{\Omega}$, $\left(\frac{m_0}{\rho_0^{1/2}}\right)_\varepsilon \xrightarrow{\varepsilon} \frac{m_0}{\rho_0^{1/2}}$ in $L^2(\Omega)$.

We finally build m_0^ε. First of all, we decompose, as in section 2.2, $\overline{m}_0^\varepsilon$ in the following way

$$\left.\begin{array}{l} \overline{m}_0^\varepsilon \;=\; \rho_0^\varepsilon \overline{u}_0^\varepsilon + \nabla q_0^\varepsilon, \quad \overline{u}_0^\varepsilon, q_0^\varepsilon \in C^\infty(\overline{\Omega}), \\[4pt] \operatorname{div} \overline{u}_0^\varepsilon = 0 \quad \text{in } \Omega, \quad \overline{u}_0^\varepsilon \cdot n = 0 \quad \text{on } \partial\Omega \end{array}\right\} \tag{2.111}$$

(denoting by n the unit outward normal to $\partial\Omega$). Let us observe that q_0^ε is determined, up to an additive constant, by the equation

$$\operatorname{div}\left\{ \frac{1}{\rho_0^\varepsilon}\left(\nabla q_0^\varepsilon - \overline{m}_0^\varepsilon\right)\right\} \;=\; 0 \quad \text{in } \Omega, \qquad \frac{\partial q_0^\varepsilon}{\partial n} = 0 \quad \text{on } \partial\Omega, \tag{2.112}$$

and we finally set

$$\left.\begin{array}{l} m_0^\varepsilon \;=\; \rho_0^\varepsilon u_0^\varepsilon + \nabla q_0^\varepsilon, \qquad \text{where} \quad u_0^\varepsilon \in C_0^\infty(\Omega), \\[4pt] \|u_0^\varepsilon - \overline{u}_0^\varepsilon\|_{L^2(\Omega)} \;\leq\; \varepsilon, \quad \operatorname{div} u_0^\varepsilon \;=\; 0 \quad \text{in } \Omega. \end{array}\right\} \tag{2.113}$$

We then deduce from (2.109) and (2.110)

$$m_0^\varepsilon \overset{\varepsilon}{\rightharpoonup} m_0 \quad \text{in } L^2(\Omega), \quad m_0^\varepsilon (\rho_0^\varepsilon)^{-1/2} \overset{\varepsilon}{\rightharpoonup} m_0 \rho_0^{-1/2} \quad \text{in } L^2(\Omega). \tag{2.114}$$

Observe that we have $m_0^\varepsilon = \overline{m}_0^\varepsilon + \rho_0^\varepsilon(u_0^\varepsilon - \overline{u}_0^\varepsilon)$, $m_0^\varepsilon(\rho_0^\varepsilon)^{-1/2} = \overline{m}_0^\varepsilon(\rho_0^\varepsilon)^{-1/2} + (\rho_0^\varepsilon)^{1/2}(u_0^\varepsilon - \overline{u}_0^\varepsilon)$.

In fact, as we explained in section 2.1, (2.108) is not really meaningful since (2.107) shows that ρu is determined "up to a gradient" and thus the initial condition, contained in (2.108), on $\rho u|_{t=0}$ really means an equality modulo a gradient. Since ρ_0^ε satisfies (2.109), and $\operatorname{div} u_0^\varepsilon = 0$ in Ω, we may—see also section 2.1—impose

$$\rho|_{t=0} \;=\; \rho_0^\varepsilon \quad \text{in } \Omega, \quad u|_{t=0} \;=\; u_0^\varepsilon \quad \text{in } \Omega. \tag{2.115}$$

We then state and prove the following existence result.

Theorem 2.6. *With the above notation and assumptions, there exists a solution (ρ, u) of (2.106)–(2.107) and (2.115) such that $\rho, u \in C^\infty(\mathbb{R}^N \times [0,\infty))$, ρ, u periodic in the periodic case; $\rho, u \in C^\infty(\overline{\Omega} \times [0,\infty))$, $u = 0$ on $\partial\Omega \times [0,\infty)$.*

Remarks 2.8. 1) The regularization procedure we are using is directly inspired by J. Leray's original work on (homogeneous) incompressible Navier–Stokes equations ([**284**]).

2) Obviously, we have $\varepsilon \leq \rho \leq C_0$ on $\overline{\Omega} \times [0,\infty)$ (since $\operatorname{div} u_\varepsilon = 0$ in $\Omega \times (0,\infty)$, and u_ε is periodic or vanishes on $\partial\Omega$).

3) It is in fact possible to prove the uniqueness of (ρ, u), using for instance the type of arguments developed in section 2.5. □

Proof of Theorem 2.6. We are going to show the existence of a solution by a fixed point argument. In fact, this fixed point argument will yield a solution (ρ, u) with the following regularity: $\rho \in C(\overline{\Omega} \times [0, \infty))$, $u \in L^2(0, T; H^2(\Omega)) \cap C([0, T]; H_0^1(\Omega))$, $\frac{\partial u}{\partial t} \in L^2(\Omega \times (0, T))$ for all $T \in (0, \infty)$ in the case of Dirichlet boundary conditions and a similar regularity in the periodic case. To limit the length of the proof, we only treat the case of Dirichlet boundary conditions: the proof in the periodic case follows the same line of arguments and is in fact much simpler. Finally, we fix $T \in (0, \infty)$ and work on $[0, T]$.

We now define the mapping whose fixed point will yield a solution. Let C be the convex set in $C(\overline{\Omega} \times [0, T]) \times L^2(0, T; H_0^1(\Omega))$ defined by

$$C = \{(\overline{\rho}, \overline{u}) \in C(\overline{\Omega} \times [0, T]) \times L^2(0, T; H_0^1(\Omega) \ /$$
$$\varepsilon \leq \overline{\rho} \leq C_0 \text{ in } \overline{\Omega} \times [0, T] , \text{ div } \overline{u} = 0 \text{ a.e. on } \Omega \times (0, T),$$
$$\|\overline{u}\|_{L^2(0,T;H_0^1(\Omega))} \leq R_0\} \quad \text{where } R_0 > 0 \text{ is to be determined.}$$

We define a map F from C into itself as follows: $F(\overline{\rho}, \overline{u}) = (\rho, u)$ as defined below. First of all, we solve

$$\frac{\partial \rho}{\partial t} + \text{div } (\overline{u}_\varepsilon \rho) = 0 \quad \text{in } \Omega \times (0, T), \ \rho|_{t=0} = \rho_0^\varepsilon \text{ in } \Omega, \qquad (2.116)$$

where $\overline{u}_\varepsilon$ is constructed from $\overline{u}$ as u_ε was from u above. Observe in particular that $\overline{u}_\varepsilon \in L^2(0, T; C^k(\overline{\Omega}))$ for all $k \geq 0$, div $\overline{u}_\varepsilon = 0$ in $\Omega \times (0, T)$, $\overline{u}_\varepsilon$ vanishes near $\partial\Omega$ (a.e. $t \in (0, T)$). The solution of (2.116) by classical (and elementary) considerations on (divergence-free) transport equations is given by a simple integration along "particle paths", i.e. solutions of the following ordinary differential equation

$$\frac{dX}{ds} = \overline{u}_\varepsilon(X, s), \quad X(s; x, t) = x, \ x \in \overline{\Omega}, \ t \in [0, T]. \qquad (2.117)$$

In view of the properties of $\overline{u}_\varepsilon$, there exists a unique solution X of (2.117), continuous in $(s, t) \in [0, T]^2$, smooth in $x \in \overline{\Omega}$ such that $\partial_x^\alpha X \in C([0, T] \times \overline{\Omega} \times [0, T])$ for all α and $X(s; x, t) \in \overline{\Omega}$ for all $(s, t) \in [0, T]^2$, $x \in \overline{\Omega}$. Then, we have

$$\rho(x, t) = \rho_0^\varepsilon(X(0; x, t)), \quad \forall \, x \in \overline{\Omega} , \ \forall \, t \in [0, T]. \qquad (2.118)$$

Obviously, $\varepsilon \leq \rho \leq C_0$ in $\overline{\Omega} \times [0, T]$, $\rho \in C([0, T]; C^k(\overline{\Omega}))$ for all $k \geq 0$ and in view of (2.116) $\frac{\partial \rho}{\partial t} \in L^2(0, T; C^k(\overline{\Omega}))$ for all $k \geq 0$. Furthermore, ρ and

$\frac{\partial \rho}{\partial t}$ are bounded in these spaces uniformly in $(\overline{\rho}, \overline{u}) \in C$. In particular, the set of ρ built in this way is clearly compact in $C(\overline{\Omega} \times [0, T])$.

We now build u: first of all, we set $\mu_\varepsilon = \mu(\rho)_\varepsilon$ with the same construction as above and we wish to solve the following problem

$$\left. \begin{array}{l} \rho \dfrac{\partial u}{\partial t} + \rho \overline{u}_\varepsilon \cdot \nabla u - \operatorname{div}(2\mu_\varepsilon d) + \nabla p = \rho f_\varepsilon \quad \text{in} \quad \Omega \times (0, T), \\[2mm] \operatorname{div} u = 0 \quad \text{in} \quad \Omega \times (0, T), \quad u|_{t=0} = u_0^\varepsilon \quad \text{in} \quad \Omega, \end{array} \right\} \quad (2.119)$$

$$\left. \begin{array}{l} u \in L^2(0, T; H^2(\Omega)) \cap C([0, T]; H_0^1(\Omega)); \\[2mm] \nabla p, \dfrac{\partial u}{\partial t} \in L^2(\Omega \times (0, T)). \end{array} \right\} \quad (2.120)$$

This is nothing but an inhomogeneous (linear) Stokes equation with rather smooth coefficients: the regularity of $\rho, \overline{u}_\varepsilon$ has been discussed above, $\mu_\varepsilon \in C^\infty(\overline{\Omega} \times [0, T])$, $f_\varepsilon \in C_0^\infty(\Omega \times (0, T))$ and $u_0^\varepsilon \in C_0^\infty(\Omega)$, $\operatorname{div} u_0^\varepsilon = 0$ in Ω.

We postpone the discussion of this problem and admit temporarily that there exists a unique solution u of (2.119)–(2.120) (depending continuously on data). This fact is established in Proposition 2.1 below. Then, when $(\overline{\rho}, \overline{u}) \in C$, u is bounded in $L^2(0, T; H^2(\Omega))$ while $\frac{\partial u}{\partial t}$ is bounded in $L^2(\Omega \times (0, T))$. Therefore, u is compact in $L^2(0, T; H_0^1(\Omega))$. This shows that the mapping F is compact on C.

Hence, if we wish to use the Schauder theorem in order to conclude the existence of a fixed point, we have only to choose R_0 in such a way that $\|u\|_{L^2(0,T;H_0^1(\Omega))} \leq R_0$. To this end, we multiply (2.119) by u, integrate by parts using (2.17) and obtain easily (all manipulations are justified by the regularity of ρ and u) for all $t \in [0, T]$

$$\int_\Omega \rho \frac{|u|^2}{2}(x, t)\, dx + \mu \int_0^t \int_\Omega |\nabla u|^2 (x, s)\, dx\, ds$$

$$\leq C_0 \|f\|_{L^2(\Omega \times (0,T))} \left(\int_0^t \int_\Omega |u|^2 (x, s)\, dx\, ds \right)^{1/2},$$

hence,

$$\sup_{t \in [0,T]} \|u(t)\|_{L^2(\Omega)} + \|u\|_{L^2(0,T;H_0^1(\Omega))} \leq C_1 \qquad (2.121)$$

where C_1 depends only on $\overline{\mu}, f, C$ but not on $R_0, \overline{\rho}, \overline{u}$. We then choose $R_0 \geq C_1$.

In order to conclude, we still have to show that a fixed point (ρ, u) is in fact smooth. This is easily done by a bootstrap argument that we only sketch. First of all, we observe that $u_\varepsilon \in C([0, T]; C^k(\overline{\Omega}))$, $\rho, \mu_\varepsilon \in C^{0,1/2}([0, T]; C^k(\overline{\Omega}))$ for all $k \geq 0$, and using L^p-theory (see V.A. Solonnikov [444],[445] for instance), or direct proofs similar to the proof of

Proposition 2.1, we deduce from (2.119)–(2.120) that $u \in L^p(0, T; W^{2,p}(\Omega))$, $\frac{\partial u}{\partial t} \in L^p(\Omega \times (0, T))$ for all $1 < p < \infty$. With this regularity on u, we can bootstrap and gain more time regularity on u_ε then ρ and thus more regularity (in (x, t)) on u.

Before stating Proposition 2.1 which fills the only gap left in the above proof, we first observe that (2.119) may be written as

$$
\left.
\begin{aligned}
c \frac{\partial u_i}{\partial t} + b_i \cdot \nabla u - a\Delta u_i + \frac{\partial p}{\partial x_i} = g_i \quad \text{in } \Omega \times (0, T), \ 1 \le i \le N, \\
\text{div } u = 0 \quad \text{in } \Omega \times (0, T), \quad u|_{t=0} = u^0 \quad \text{in } \Omega,
\end{aligned}
\right\}
\tag{2.122}
$$

where $g \in L^2(\Omega \times (0, T))$, $c \in L^\infty(\Omega \times (0, T))$, $a \in L^\infty(0, T; W^{1,\infty}(\Omega))$, $\frac{\partial a}{\partial t} \in L^1(0, T; L^\infty(\Omega))$, $b \in L^2(0, T; L^\infty(\Omega))$, $c \ge \delta$, $a \ge \delta$ a.e. on $\Omega \times (0, T)$ for some $\delta > 0$, $u^0 \in H_0^1(\Omega)$.

Proposition 2.1. *There exists a unique solution u of (2.120)–(2.122).*

Proof of Proposition 2.1. We only prove that the a priori estimates contained in (2.120) hold. The proof will show the uniqueness of solutions, and the existence follows in a straightforward way from a priori estimates by standard arguments that we leave to the reader.

Next, in order to prove a priori estimates, we multiply (2.122) by $\frac{\partial u_i}{\partial t}$, sum over i and integrate (by parts) over Ω to find for almost all $t \in (0, T)$

$$
\delta \int_\Omega \left| \frac{\partial u}{\partial t} \right|^2 dx + \frac{1}{2} \int_\Omega a \frac{d}{dt} |\nabla u|^2 \, dx
$$
$$
\le \int_\Omega |g| \left| \frac{\partial u}{\partial t} \right| + |b| \, |\nabla u| \left| \frac{\partial u}{\partial t} \right| + |\nabla a| \, |\nabla u| \left| \frac{\partial u}{\partial t} \right| dx.
$$

Hence, using the Cauchy–Schwarz inequality, we find

$$
\frac{\delta}{2} \int_\Omega \left| \frac{\partial u}{\partial t} \right|^2 dx + \frac{1}{2} \frac{d}{dt} \left(\int_\Omega a |\nabla u|^2 \, dx \right)
$$
$$
\le C \left(1 + \|b\|_{L^\infty(\Omega)}^2 + \|\nabla a\|_{L^\infty(\Omega)}^2 + \left\| \frac{\partial a}{\partial t} \right\|_{L^\infty(\Omega)} \right) \int_\Omega a |\nabla u|^2 \, dx.
$$

Since $b \in L^2(0, T; L^\infty(\Omega))$, $\nabla a \in L^2(0, T; L^\infty(\Omega))$, $\frac{\partial a}{\partial t} \in L^1(0, T; L^\infty(\Omega))$, we deduce from Grönwall's inequality an a priori estimate on u in $L^\infty(0, T; H_0^1(\Omega))$ and on $\frac{\partial u}{\partial t}$ in $L^2(\Omega \times (0, T))$ depending only on the data c, a, b, g, u_0.

In particular, we may then write (2.122) as

$$
-a\Delta u + \nabla p = h \quad \text{in } \Omega, \quad \text{div } u = 0 \quad \text{in } \Omega, \quad u \in H_0^1(\Omega)
\tag{2.123}
$$

for almost all $t \in (0, T)$. In addition, $h \in L^2(\Omega \times (0, T))$ and its norm is bounded in terms of the data. From here on, whenever we say bounded, it means that the bound depends only on the data. From the previous bound on u, we deduce in particular that

$$\nabla p = h + \operatorname{div}(a\nabla u) - \nabla a \cdot \nabla u$$

and thus is bounded in $L^2(0, T; H^{-1}(\Omega))$. Therefore, if we normalize p by imposing

$$\int_{\Omega} p \, dx = 0, \quad \text{a.e.} \quad t \in (0, T),$$

we deduce that p is bounded in $L^2(\Omega \times (0, T))$ (see [472] for instance).

Then, we write (2.123) as a usual Stokes problem, namely

$$-\Delta u + \nabla \tilde{p} = \tilde{h} \quad \text{in } \Omega, \quad \operatorname{div} u = 0 \quad \text{in } \Omega, \quad u \in H_0^1(\Omega), \tag{2.124}$$

where $\tilde{h} = \frac{h}{a} - \frac{\nabla a p}{a^2}$ is bounded in $L^2(\Omega \times (0, T))$, $\tilde{p} = \frac{p}{a}$, and we conclude that u is bounded in $L^2(0, T; H^2(\Omega))$ by classical regularity results on Stokes equation (see [472] for example).

This completes the proof of Proposition 2.1 and of Theorem 2.6.   □

Step 2. Passage to the limit. First of all, we collect a priori estimates and follow the arguments of section 2.3. Since $\operatorname{div} u_\varepsilon = 0$, we immediately obtain for all $\beta \in C(\mathbb{R}, \mathbb{R})$

$$\int_{\Omega} \beta(\rho^\varepsilon) \, dx = \int_{\Omega} \beta(\rho_0^\varepsilon) \, dx \qquad \text{for all } t \in (0, \infty). \tag{2.125}$$

Here and below, we denote by $(u^\varepsilon, \rho^\varepsilon)$ the solution built in step 1 (observe that u_ε is the regularization of u^ε, namely $(\overline{u^\varepsilon})_\varepsilon * \omega_{\varepsilon/2}$ in the case of Dirichlet boundary conditions).

Next, exactly as in section 2.3, we obtain the analogue of (2.34)–(2.36), namely

$$\|u^\varepsilon\|_{L^2(0,T;H^1(\Omega))} \le C, \tag{2.126}$$

$$\sup_{0 \le t \le T} \||\rho^\varepsilon|u^\varepsilon|^2\|_{L^1(\Omega)} \le C \tag{2.127}$$

where C denotes various positive constants independent of ε.

Because of (2.110), we may then apply part 1 of Theorem 2.4 to deduce that ρ^ε converges, up to the extraction of subsequences, to some ρ in $C([0, T]; L^p(\Omega))$ ($\forall\, 1 \le p < \infty$, $\forall\, T \in (0, \infty)$) which is bounded, satisfies (2.125) with ρ^ε and ρ_0^ε replaced respectively by ρ and ρ_0, and thus satisfies (2.17). Furthermore, ρ satisfies (2.1) (and is periodic in the periodic case) where u is a weak limit in $L^2(0, T; H^1(\Omega))$ ($\forall\, T \in (0, \infty)$) of u^ε. Of course, u is periodic in the periodic case and $u \in L^2(0, T; H_0^1(\Omega))$ in the case of Dirichlet boundary conditions.

In particular, this convergence implies, in view of the construction of μ_ε, that we have

$$\left.\begin{array}{c} \mu_\varepsilon \to \mu(\rho) \quad \text{in} \quad C([0,T];L^p(\Omega)) \\[4pt] (\forall\, 1 \le p < \infty, \ \forall\, T \in (0,\infty)), \text{ as } \varepsilon \to 0, \end{array}\right\} \tag{2.128}$$

$$\rho^\varepsilon f_\varepsilon \to \rho f \quad \text{in} \quad L^2(\Omega \times (0,T)) \quad (\forall\, T \in (0,\infty)), \text{ as } \varepsilon \to 0. \tag{2.129}$$

In addition, from the results shown in Appendix A, u_ε is also bounded in $L^2(0,T;H^1(\Omega))$ and u_ε converges weakly in $L^2(0,T;H^1(\Omega))$ to u—this is obvious in the periodic case.

These bounds imply that (2.53) holds with $q = 2$, $m = \max\left(\frac{N}{2} - 1, 1\right)$: indeed, $\mu_\varepsilon d^\varepsilon$ is bounded in $L^2(\Omega \times (0,T))$ while $\rho^\varepsilon u^\varepsilon \otimes u_\varepsilon$ is bounded in $L^2(0,T;L^p(\Omega))$ with $p \in [1,2)$ if $N = 2$, $p = \frac{N}{N-1}$ if $N \ge 3$, and thus is bounded in $L^2(0,T;H^{-s}(\Omega))$ with $s > 0$ if $N = 2$, $s = \frac{N}{2} - 1$ if $N \ge 3$.

We then deduce from part 2 of Theorem 2.4 that $\sqrt{\rho^\varepsilon}u^\varepsilon$ converges to $\sqrt{\rho}u_i$ in $L^p(0,T;L^r(\Omega))$ for $2 < p < \infty$, $1 \le r < \frac{2Np}{Np-4}$ and thus $\rho^\varepsilon u^\varepsilon$ converges to ρu_i in $L^p(0,T;L^r(\Omega))$ for the same (p,r).

These convergences allow us to recover (2.2) from (2.107) upon letting ε go to 0. In fact, we recover (2.12) (the weak formulation of (2.2)) provided we show in the case of Dirichlet boundary conditions that

$$\int_\Omega \rho_0^\varepsilon u_0^\varepsilon \cdot \phi\, dx \ \to\ \int_\Omega m_0 \cdot \phi\, dx, \quad \text{as} \quad \varepsilon \to 0,$$

for all $\phi \in C_0^\infty(\Omega)^n$ such that div $\phi = 0$. This is clear in view of (2.113)–(2.114) since we have

$$\int_\Omega \rho_0^\varepsilon u_0^\varepsilon \cdot \phi\, dx = \int_\Omega m_0^\varepsilon \cdot \phi\, dx \ \to\ \int_\Omega m_0 \cdot \phi\, dx, \quad \text{as} \quad \varepsilon \to 0,$$

for all $\phi \in L^2(\Omega)^N$ with div $\phi = 0$ in $\mathcal{D}'(\Omega)$, $\phi \cdot u = 0$ on $\partial\Omega$.

The only fact left in order to complete the proof of Theorem 2.1 is the energy inequalities (2.13)–(2.14). This is in fact relatively easy since $(\rho^\varepsilon, u^\varepsilon)$ also satisfies some energy identities obtained as in section 2.3 by multiplying (2.107) by u^ε and integrating over Ω, using (2.106) and the boundary conditions. We find then for all $t \ge 0$

$$\frac{d}{dt}\int_\Omega \rho^\varepsilon |u^\varepsilon|^2\, dx + \int_\Omega \mu_\varepsilon (\partial_i u_j^\varepsilon + \partial_j u_i^\varepsilon)^2\, dx \ = \ 2\int_\Omega \rho^\varepsilon f_\varepsilon \cdot u^\varepsilon\, dx. \tag{2.130}$$

We have seen above that $\sqrt{\rho^\varepsilon}u^\varepsilon$ converges in $L^2(\Omega \times (0,T))$ (in particular) to $\sqrt{\rho}u$, μ_ε converges in $C([0,T];L^p(\Omega))$ ($\forall\, 1 \le p < \infty$) and is uniformly bounded on $\Omega \times (0,\infty)$, while f^ε converges to f in $L^2(\Omega \times (0,T))$, for all $T \in (0, \infty)$. This is enough to imply (2.13) provided we show for all $\varphi \in C_0^\infty(0, \infty)$, $\varphi \geq 0$

$$\left.\begin{aligned}
\varliminf_\varepsilon \int_0^\infty dt \int_\Omega dx\, \varphi(t)\, \mu_\varepsilon(\rho^\varepsilon)(\partial_i u_j^\varepsilon + \partial_j u_i^\varepsilon)^2 \\
\geq \int_0^\infty dt \int_\Omega dx\, \varphi(t)\, \mu(\rho)(\partial_i u_j + \partial_j u_i)^2.
\end{aligned}\right\} \tag{2.131}$$

In order to show (2.131), we observe that we have

$$\begin{aligned}
0 &\leq \int_0^\infty dt \int_\Omega dx\, \varphi(t)\, \mu_\varepsilon(\partial_i(u_j^\varepsilon - u_j) + \partial_j(u_i^\varepsilon - u_i))^2 \\
&= \int_0^\infty dt \int_\Omega dx\, \varphi(t)\, \mu_\varepsilon(\partial_i u_j^\varepsilon + \partial_j u_i^\varepsilon)^2 \\
&\quad + \int_0^\infty dt \int_\Omega dx\, \varphi(t)\, \mu_\varepsilon(\partial_i u_j + \partial_j u_i)^2\, dx \\
&\quad - \int_0^\infty dt \int_\Omega dx\, \varphi(t)\, \mu_\varepsilon(\partial_i u_j + \partial_j u_i)(\partial_i u_j^\varepsilon + \partial_j u_i^\varepsilon).
\end{aligned}$$

Since $\varphi(t)(\partial_i u_j + \partial_j u_i)^2 \in L^1(\Omega \times (0, \infty))$ and μ_ε is uniformly bounded and converges in measure on $\Omega \times \mathrm{Supp}\,(\varphi)$ to $\mu(\rho)$, we deduce easily that $\varphi \mu_\varepsilon (\partial_i u_j + \partial_j u_i)^2$ and $\varphi^{1/2} \mu_\varepsilon (\partial_i u_j + \partial_j u_i)$ converge, respectively, to $\varphi \mu(\rho) (\partial_i u_j + \partial_j u_i)^2$ in $L^1(\Omega \times (0, \infty))$ and to $\varphi^{1/2} \mu(\rho)(\partial_i u_j + \partial_j u_i)$ in $L^2(\Omega \times (0, \infty))$. In addition, $\varphi^{1/2}(\partial_i u_j^\varepsilon + \partial_j u_i^\varepsilon)$ converges weakly in $L^2(\Omega \times (0, \infty))$ to $\varphi^{1/2}(\partial_i u_j + \partial_j u_i)$. Therefore, the two last integrals converge, as ε goes to 0, to $\int_0^\infty dt \int_\Omega dx\, \varphi(t)\, \mu(\rho)\, \mu(\rho)(\partial_i u_j + \partial_j u_i)^2$. This implies (2.131).

Next, in order to prove (2.14), we first integrate (2.130) between 0 and t to find

$$\left.\begin{aligned}
\int_\Omega \rho^\varepsilon |u^\varepsilon|^2\, dx(t) + \int_0^t ds \int_\Omega \mu_\varepsilon(\partial_i u_j^\varepsilon + \partial_j u_i^\varepsilon)^2\, dx \\
= 2 \int_0^t ds \int_\Omega \rho^\varepsilon f_\varepsilon \cdot u^\varepsilon\, dx + \int_\Omega \rho_0^\varepsilon |u_0^\varepsilon|^2\, dx
\end{aligned}\right\} \tag{2.132}$$

for all $t \geq 0$. Then, we observe that we have

$$\begin{aligned}
\int_\Omega \rho_0^\varepsilon |u_0^\varepsilon|^2\, dx &= \int_\Omega \frac{1}{\rho_0^\varepsilon} |m_0^\varepsilon - \nabla q_0^\varepsilon|^2\, dx \\
&= \int_\Omega \frac{|m_0^\varepsilon|^2}{\rho_0^\varepsilon} + \frac{|\nabla q_0^\varepsilon|^2}{\rho_0^\varepsilon} - \frac{2}{\rho_0^\varepsilon}(\rho_0^\varepsilon u_0^\varepsilon + \nabla q_0^\varepsilon) \cdot \nabla q_0^\varepsilon\, dx \\
&= \int_\Omega \frac{|m_0^\varepsilon|^2}{\rho_0^\varepsilon} - 2 u_0^\varepsilon \cdot \nabla q_0^\varepsilon - \frac{|\nabla q_0^\varepsilon|^2}{\rho_0^\varepsilon}\, dx.
\end{aligned}$$

Since $u_0^\varepsilon = 0$ on $\partial\Omega$ (in the case of Dirichlet boundary conditions) and $\operatorname{div} u_0^\varepsilon = 0$, we finally obtain

$$\int_\Omega \rho_0^\varepsilon |u_0^\varepsilon|^2 \, dx + \int_\Omega \frac{|\nabla q_0^\varepsilon|^2}{\rho_0^\varepsilon} \, dx = \int_\Omega \frac{|m_0^\varepsilon|^2}{\rho_0^\varepsilon} \, dx. \qquad (2.133)$$

Since $\frac{m_0^\varepsilon}{(\rho_0^\varepsilon)^{1/2}}$ converges in $L^2(\Omega)$ to $\frac{m_0}{\rho_0^{1/2}}$, we deduce (2.14) from (2.132) exactly as before. $\square$

Remark 2.9. In fact, it is often possible to sharpen a little the energy inequality (2.14), replacing $\frac{|m_0|^2}{\rho_0}$ by $\rho_0|u_0|^2$ for some u_0 to be determined satisfying $\operatorname{div} u_0 = 0$. However, we cannot do it in full generality and we have to make some assumptions on ρ_0.

The *first case* we can treat is when $\operatorname{infess}_\Omega \rho_0 > 0$. Then, exactly as in section 2.1, we can check that u_0^ε converges in $L^2(\Omega)$ to $u_0 = P_{\rho_0}(m_0)$; in the case of Dirichlet boundary conditions, $\nabla q_0 = m_0 - u_0$ is determined by the elliptic equation

$$\left.\begin{array}{l} \operatorname{div}\left(\dfrac{\nabla q_0 - m_0}{\rho_0}\right) = 0 \quad \text{in } \Omega, \quad \nabla q_0 \in L^2(\Omega), \\[2mm] (\nabla q_0 - m_0) \cdot n = 0 \quad \text{on } \partial\Omega, \end{array}\right\} \qquad (2.134)$$

and it is clear that (2.14) holds (in fact for all $t \geq 0$ since $u \in C([0,T]; L^2_w)$ for all $T \in (0,\infty)$, see section 2.1) with $\frac{|m_0|^2}{\rho_0}$ replaced by $\rho_0|u_0|^2$.

The *second case* allows ρ_0 to vanish. For instance, we assume that Ω is connected, $\rho_0 = 0$ a.e. on $\Omega - \omega$, $\rho_0 \geq \delta > 0$ a.e. in ω where $\overline\omega \in \Omega$, ω is smooth, and we only consider the case of Dirichlet boundary conditions. First of all, we observe that $|\nabla q_0^\varepsilon| = \frac{|\nabla q_0^\varepsilon|}{(\rho_0^\varepsilon)^{1/2}} (\rho_0^\varepsilon)^{1/2}$ is bounded in $L^2(\Omega)$. Next, we can normalize q_0^ε in such a way that $\int_{\partial\Omega} q_0^\varepsilon \, dS = 0$. Therefore, if we extract subsequences if necessary, q_0^ε converges weakly in $H^1(\Omega)$ to q_0 satisfying $\int_{\partial\Omega} q_0 dS = 0$, $\nabla q_0 = 0$ on $\Omega - \omega$ and thus $q_0 = 0$ on $\Omega - \omega$. Hence, $q_0 \in H_0^1(\omega)$. In addition, u_0^ε is bounded in $L^2_{\text{loc}}(\omega)$ and we may assume that u_0^ε converges weakly in $L^2(K)$ ($\forall K$ compact $\subset \omega$) to some $u_0 \in L^2_{\text{loc}}(\omega)$ such that $\rho_0|u_0|^2 \in L^1(\omega)$, and we have

$$\int_\omega \rho_0|u_0|^2 \, dx + \int_\omega \frac{|\nabla q_0|^2}{\rho_0} \, dx \leq \int_\Omega \frac{|m_0|^2}{\rho_0} \, dx = \int_\omega \frac{|m_0|^2}{\rho_0} \, dx \qquad (2.135)$$

$$m_0 = \rho_0 u_0 + \nabla q_0 \quad \text{in } \omega, \quad \operatorname{div}\left(\frac{\nabla q_0 - m_0}{\rho_0}\right) = 0 \quad \text{in } \omega. \qquad (2.136)$$

We next claim that there exists a unique $q_0 \in H_0^1(\omega)$ which satisfies (2.136) (assuming that $m_0 \in L^2(\omega)$, $\frac{|m_0|}{\rho_0} \in L^2(\omega)$), and we have $\int_\omega \frac{|\nabla q_0|^2}{\rho_0} \, dx =$

$\int_\omega \frac{\nabla q_0 \cdot m_0}{\rho_0}\, dx$. If this claim were established, we would deduce that (2.135) is in fact an equality and thus $\rho_0^\varepsilon |u_0^\varepsilon|^2$, $\frac{|\nabla q_0^\varepsilon|^2}{\rho_0^\varepsilon}$ converge in $L^1(\Omega)$ to $\rho_0 |u_0|^2$, $\frac{|\nabla q_0|^2}{\rho_0}$, respectively, where we extend these functions to Ω by 0 outside ω. This is enough to conclude that (2.14) holds with $\rho_0 |u_0|^2$ replacing $\frac{|m_0|^2}{\rho_0}$.

In order to show the above claim, we have only to show that for any solution $q_0 \in H_0^1(\omega)$ of

$$\operatorname{div}\left(\frac{\nabla q_0 - m_0}{\rho_0}\right) = 0 \quad \text{in}\quad \omega$$

we have $\int_\omega \frac{|\nabla q_0|^2}{\rho_0}\, dx = \int_\omega \frac{\nabla q_0 \cdot m_0}{\rho_0}\, dx$. Then, we multiply the equation by $q_0 \zeta(\frac{d}{\varepsilon})$ where $d = \operatorname{dist}(x, \partial\omega)$, $\varepsilon > 0$, $\zeta \in C^\infty([0, \infty])$, $\zeta(t) = 0$ if $t \le 1/2$, $\zeta(t) = 1$ if $t \ge 1$, $0 \le \zeta(t) \le 1$ on $[0, \infty)$, and we obtain

$$\int_\omega \frac{(\nabla q_0 - m_0) \cdot \nabla q_0}{\rho_0} \zeta\left(\frac{d}{\varepsilon}\right) dx + \int_\omega \frac{\nabla q_0 - m_0}{\rho_0} \cdot \frac{\nabla d}{\varepsilon} \zeta'\left(\frac{d}{\varepsilon}\right) q_0\, dx = 0.$$

It only remains to show that the second integral goes to 0 as ε goes to 0. Indeed, we have

$$\left|\int_\omega \frac{\nabla q_0 - m_0}{\rho_0} \cdot \frac{\nabla d}{\varepsilon} \zeta'\left(\frac{d}{\varepsilon}\right) q_0\, dx\right|$$

$$\le \frac{C}{\delta^{1/2}} \left\|\frac{\nabla q_0 - m_0}{\rho_0^{1/2}}\right\|_{L^2(\omega)} \left(\int_{(0 < d < \varepsilon)} \frac{q_0^2}{d^2}\, dx\right)^{1/2} \underset{\varepsilon}{\to} 0.$$

In fact, the above analysis can be extended to the case when $\rho \ge \alpha\, d(x)^\gamma$, $\rho \le \beta\, d(x)^\gamma$ a.e. in ω for some $\alpha, \beta > 0$, $\gamma \ge 0$. This is possible in view of the following variant of Hardy's inequality

$$\int_\omega \frac{|q|^2}{d^{2+\gamma}}\, dx \le C \int_\omega \frac{|\nabla q|^2}{d^\gamma}\, dx.$$

This inequality follows easily from the following computation: we have for all $f \in C_0^\infty(0, \infty)$

$$0 = \int_0^\infty \frac{d}{dx}\left\{\frac{1}{x^{\gamma+1}} f^2(x)\right\} dx = 2 \int_0^\infty \frac{f f'}{x^{\gamma+1}}\, dx - (\gamma+1) \int_0^\infty \frac{f^2}{x^{\gamma+2}}\, dx$$

hence $\int_0^\infty \frac{f^2}{x^{\gamma+2}}\, dx \le \frac{4}{(\gamma+1)^2} \int_0^\infty \frac{(f')^2}{x^\gamma}\, dx$. $\quad\Box$

Remark 2.10. We observe here without proof that the existence and compactness results can be extended to the case when $\rho_0\ (\ge 0)$ is assumed to be in $L^p(\Omega)$ where $p > 1$ if $N = 2$ and $p = \frac{N}{2}$ if $N \geq 3$. We still assume that $\frac{|m_0|^2}{\rho_0} \in L^1(\Omega)$ ($m_0 = 0$ a.e. on $\{\rho_0 = 0\}$). Then weak solutions are defined exactly as in section 2.1 except that $\rho \in C([0,\infty); L^p(\Omega))$, and one can adapt the preceding proofs to show that Theorem 2.1 holds in that case □

Step 3. Existence in the case when $\Omega = \mathbb{R}^N$. We use the existence results we just proved with $\Omega = B_R$ and Dirichlet boundary conditions on ∂B_R. We then obtain approximated solutions (ρ_R, u_R) and we let R go to $+\infty$. More precisely, we denote by (ρ_R, u_R) a global weak solution of (2.1)–(2.2) in B_R where $R \in (0,\infty)$ with the boundary condition (2.6), restricting of course ρ_0, m_0 to B_R. Recall that, as explained in Remark 2.1 (7), we may assume without loss of generality that $u_\infty = 0$.

Next, we observe that all the estimates shown in section 2.3 in the case when $\Omega = \mathbb{R}^N$ hold uniformly in R large. We may then apply Theorem 2.5 to deduce the relative compactness of ρ_R, $\sqrt{\rho_R} u_R$ and $\rho_R u_R$ in $C([0,T]; L^p(B_M))$ ($1 < p < \infty$), $L^p(0,T; L^r(B_M))$ ($2 < p < \infty$, $1 \leq r < \frac{2Np}{Np-4}$) respectively (for all $T \in (0,\infty)$, $M \in (0,\infty)$). Finally, this compactness allows us to prove the existence of solutions in $\mathbb{R}^N$ letting R go to $+\infty$ and using the same arguments as in step 2 above. □

2.5 Uniqueness: weak = strong

In this section, we show that any global weak solution coincides with a more regular solution as long as such a "strong" solution exists. More precisely, we prove that a weak solution is equal to a strong solution whenever the latter exists. It is not difficult to check that smooth solutions exist for a certain time interval—at least if ρ_0 does not vanish—and the result that we are going to present then implies that any weak solution is equal to the smooth one on this time interval.

In order to simplify the presentation, we only treat the periodic case and the case of Dirichlet boundary conditions even if similar results can be obtained in the case when $\Omega = \mathbb{R}^N$ by convenient adaptations of the arguments below. We then consider a global weak solution u of (2.1)–(2.2) and (2.6) as built in Theorem 2.1. We assume (for instance) that $f \in L^2(0,T; L^\infty(\Omega))$ and fix $T \in (0,\infty)$. We next assume that there exists a solution $\overline{p}, \overline{u} \in C(\overline{\Omega} \times [0,T])$ (resp., in the periodic case, $C(\mathbb{R}^N \times [0,T])$ periodic) of (2.1)–(2.2) in Ω (resp. in $\mathbb{R}^N$) with $\nabla \overline{u} \in L^2(0,T; L^\infty(\Omega))$, $\nabla \overline{p} \in L^2(0,T; L^\infty(\Omega))$, $\frac{\partial \overline{u}}{\partial t} \in L^2(0,T; L^\infty(\Omega))$) and with $\overline{u} = 0$ on $\partial\Omega \times (0,T)$. Furthermore, we assume that μ is locally Lipschitz on $[0,\infty)$ and that $\overline{p}, \overline{u}$ satisfy

$$\overline{p}|_{t=0} = \rho_0 \quad \text{in } \Omega, \quad \overline{p}\,\overline{u}|_{t=0} = m_0 \quad \text{in } \Omega. \tag{2.137}$$

Notice that this equality implies in fact that $m_0 = \rho_0 \overline{u}(0)$ with div $\overline{u}(0) = 0$ in Ω. Let us notice, that, of course, (2.2) holds with some pressure field $\overline{p}$ that belongs to $L^1(0, T; L^\infty(\Omega)) + L^2(0, T; W^{-1,\infty}(\Omega))$.

Theorem 2.7. *Assume in addition that $\rho \not\equiv 0$. Then we have $u \equiv \overline{u}$ a.e. in $\Omega \times (0, T)$.*

Proof of Theorem 2.7. We first recall that we have for (almost) all $t \in (0, T)$

$$\left. \begin{aligned} &\frac{1}{2} \int_\Omega \rho |u|^2 \, dx + \frac{1}{2} \int_0^t \int_\Omega \mu(\rho)(\partial_i u_j + \partial_j u_i)^2 \, dx \, ds \\ &\leq \int_0^t \int_\Omega \rho f \cdot u \, dx \, ds + \frac{1}{2} \int_\Omega \frac{|m_0|^2}{\rho_0} \, dx. \end{aligned} \right\} \tag{2.14}$$

Next, we remark that, in view of the regularity of $\overline{u}$, we deduce from the weak formulation (2.12) of (2.2) the following equality

$$\left. \begin{aligned} &\int_\Omega \rho u \cdot \overline{u} \, dx + \frac{1}{2} \int_0^t \int_\Omega \mu(\rho)(\partial_i u_j + \partial_j u_i)(\partial_i \overline{u}_j + \partial_j \overline{u}_i) \, dx \, ds \\ &= \int_\Omega m_0 \cdot \overline{u}(0) \, dx + \int_0^t \int_\Omega \rho f \cdot \overline{u} \, dx \, ds + \int_0^t \int_\Omega \rho u \cdot \left\{ \frac{\partial \overline{u}}{\partial t} + u \cdot \nabla \overline{u} \right\} dx \, ds \end{aligned} \right\} \tag{2.138}$$

a.e. $t \in (0, T)$. Then we write

$$\left. \begin{aligned} &\rho \frac{\partial \overline{u}}{\partial t} + \rho u \cdot \nabla \overline{u} - \operatorname{div}(2\mu(\rho)\overline{d}) + \nabla p \\ &= \overline{p} f + (\rho - \overline{p})\left(\frac{\partial \overline{u}}{\partial t} + \overline{u} \cdot \nabla \overline{u} \right) + \rho(u - \overline{u}) \cdot \nabla \overline{u} - \operatorname{div}(2(\mu(\rho) - \mu(\overline{p}))\overline{d}). \end{aligned} \right\} \tag{2.139}$$

If we first multiply (2.139) by u and integrate over $\Omega \times (0, t)$, we find

$$\left. \begin{aligned} &\int_0^t \int_\Omega \left\{ \rho \frac{\partial \overline{u}}{\partial t} + \rho u \cdot \nabla \overline{u} \right\} u \, dx \, ds \\ &\quad + \frac{1}{2} \int_0^t \int_\Omega \mu(\rho)(\partial_i \overline{u}_j + \partial_j \overline{u}_i)(\partial_i u_j + \partial_j u_i) \, dx \, ds \\ &= \int_0^t \int_\Omega \overline{p} f \cdot u + (\rho - \overline{p})\left(\frac{\partial \overline{u}}{\partial t} + \overline{u} \cdot \nabla \overline{u} \right) \cdot u + \rho(u - \overline{u}) \cdot \nabla \overline{u} \cdot u \\ &\quad + \frac{1}{2} \int_0^t \int_\Omega (\mu(\rho) - \mu(\overline{p}))(\partial_i \overline{u}_j + \partial_j \overline{u}_i)(\partial_i u_j + \partial_j u_i) \, dx \, ds. \end{aligned} \right\} \tag{2.140}$$

Combining (2.138) and (2.140), and using (2.137), we obtain for almost all $t \in (0, T)$

$$
\left.
\begin{aligned}
&\int_\Omega \rho u \cdot \overline{u} \, dx + \int_0^t \int_\Omega \mu(\rho)(\partial_i u_j + \partial_j u_i)(\partial_i \overline{u}_j + \partial_j \overline{u}_i) \, dx \, ds \\
&= \int_\Omega \frac{|m_0|^2}{\rho_0} + \int_0^t \int_\Omega \rho f \cdot \overline{u} + \overline{\rho} f \cdot u \\
&\quad + (\rho - \overline{\rho})\Big(\frac{\partial \overline{u}}{\partial t} + \overline{u} \cdot \nabla \overline{u}\Big) \cdot u + \rho(u - \overline{u}) \cdot \nabla \overline{u} \cdot u \, dx \, ds \\
&\quad + \frac{1}{2} \int_0^t \int_\Omega (\mu(\rho) - \mu(\overline{\rho}))(\partial_i \overline{u}_j + \partial_j \overline{u}_i)(\partial_i u_j + \partial_j u_i) \, dx \, ds.
\end{aligned}
\right\} \quad (2.141)
$$

Finally, we multiply (2.139) by $\overline{u}$ and integrate over $\Omega \times (0, t)$ to find

$$
\left.
\begin{aligned}
&\frac{1}{2} \int_\Omega \rho |\overline{u}|^2 \, dx + \frac{1}{2} \int_0^t \int_\Omega \mu(\rho)(\partial_i \overline{u}_j + \partial_j \overline{u}_i)^2 \, dx \, ds \\
&= \frac{1}{2} \int_\Omega \frac{|m_0|^2}{\rho_0} + \int_0^t \int_\Omega \overline{\rho} f \cdot u \\
&\quad + (\rho - \overline{\rho})\Big(\frac{\partial \overline{u}}{\partial t} + \overline{u} \cdot \nabla \overline{u}\Big) \cdot \overline{u} + \rho(u - \overline{u}) \cdot \nabla \overline{u} \cdot \overline{u} \\
&\quad + \frac{1}{2}(\mu(\rho) - \mu(\overline{\rho}))(\partial_i \overline{u}_j + \partial_j \overline{u}_i)(\partial_i \overline{u}_j + \partial_j \overline{u}_i) \, dx \, ds.
\end{aligned}
\right\} \quad (2.142)
$$

Then, if we add up (2.14) and (2.142) and substract (2.104), we obtain

$$
\begin{aligned}
&\frac{1}{2} \int_\Omega \rho |u - \overline{u}|^2 \, dx + \frac{1}{2} \int_0^t \int_\Omega \mu(\rho)(\partial_i(u_j - \overline{u}_j) + \partial_j(u_i - \overline{u}_i))^2 \, dx \, ds \\
&\leq \int_0^t \int_\Omega f \cdot (u - \overline{u})(\rho - \overline{\rho}) \, dx \, ds \\
&\quad + \frac{1}{2} \int_0^t \int_\Omega (\mu(\rho) - \mu(\overline{\rho}))(\partial_i \overline{u}_j + \partial_j \overline{u}_i)(\partial_i(\overline{u}_j - u_j) + \partial_j(\overline{u}_i - u_i)) \, dx \, ds \\
&\quad + \int_0^t \int_\Omega (\rho - \overline{\rho})\Big(\frac{\partial \overline{u}}{\partial t} + \overline{u} \cdot \nabla \overline{u}\Big) \cdot (\overline{u} - u) - \rho(u - \overline{u}) \cdot \nabla(u - \overline{u}) \, dx \, ds.
\end{aligned}
$$

Hence, we deduce from the assumptions made upon $\overline{u}$ that we have for almost all $t \in (0, T)$ and for all $\varepsilon > 0$

$$
\left.
\begin{aligned}
&\int_\Omega \rho |u - \overline{u}|^2 \, dx + \int_0^t \int_\Omega |\nabla(u - \overline{u})|^2 \, dx \, ds \\
&\leq \int_0^t \int_\Omega C(s)\rho |u - \overline{u}|^2 + \varepsilon |u - \overline{u}|^2 + C_\varepsilon(s)|\rho - \overline{\rho}|^2 \, dx \, ds,
\end{aligned}
\right\} \quad (2.143)
$$

where C, C_ε denote various non-negative measurable functions in $L^1(0, T)$. Next, we wish to estimate $\|\rho - \overline{\rho}\|_{L^2(\Omega)}$. We write

$$\frac{\partial}{\partial t}(\rho - \overline{\rho}) + \operatorname{div}\{u(\rho - \overline{\rho})\} = (\overline{u} - u) \cdot \nabla \overline{\rho}$$

and deduce easily (see section 2.3 for related arguments) for all $t \in [0, T]$

$$\left.\begin{aligned}
&\int_\Omega \rho |u - \overline{u}|^2 + |\rho - \overline{\rho}|^2 \, dx + \int_0^t \int_\Omega |\nabla(u - \overline{u})|^2 \, dx \, ds \\
&\leq \int_0^t ds \, C(s) \int_\Omega \rho |u - \overline{u}|^2 + |\rho - \overline{\rho}|^2 + \varepsilon \int_0^t ds \int_\Omega |u - \overline{u}|^2 \, dx.
\end{aligned}\right\}$$
(2.144)

Next, we observe that there exists $\varepsilon > 0$ such that we have for all $v \in H^1(\Omega)$ and for all $\rho \in L^\infty(\Omega)$ such that $\int_\Omega \rho \, dx = \int_\Omega \rho_0 \, dx > 0$, $\|\rho\|_{L^\infty(\Omega)} \leq \|\rho_0\|_{L^\infty(\Omega)}$

$$\varepsilon \int_\Omega |v|^2 \, dx \leq \frac{1}{2} \int_\Omega |\nabla v|^2 \, dx + \frac{1}{2} \int_\Omega \rho |v|^2 \, dx.$$

Indeed, if this were not the case, we would find v_n, ρ_n satisfying

$$\left.\begin{aligned}
&\int_\Omega |\nabla v_n|^2 \, dx + \int_\Omega \rho_n |v_n|^2 \, dx \xrightarrow[n]{} 0, \\
&\fint_\Omega |v_n|^2 \, dx = 1, \quad \rho_0 \geq 0, \\
&\rho_n \xrightarrow[n]{} \rho \, w - L^\infty(\Omega) - *, \quad \int_\Omega \rho \, dx = \int_\Omega \rho_0 \, dx.
\end{aligned}\right\}$$

Hence v_n converges to 1 in $H^1(\Omega)$, and $\rho_n |v_n|^2 \xrightarrow[n]{} \rho \, w - L^1(\Omega)$. The contradiction proves our claim.

Inserting the above inequality in (2.144), we then conclude that $u = \overline{u}$, $\rho = \overline{\rho}$ a.e. in $\Omega \times (0, T)$, by applying Grönwall's inequality. □

Remark 2.11. Modifying a little the above proof (using Sobolev's inequality), one can extend the preceding result to the case when $\frac{\partial \overline{u}}{\partial t}$, f, $\nabla \overline{\rho} \in L^2(0, T; L^p(\Omega))$, $\nabla \overline{u} \in L^2(0, T; L^\infty(\Omega))$ where $p = N$ if $N \geq 3$, $p > 2$ if $N = 2$. □

3

NAVIER–STOKES EQUATIONS

This chapter is devoted to the classical Navier–Stokes equations in the homogeneous, incompressible case. The system, described in section 1.2, can be deduced from (2.1)–(2.2) by setting $\rho \equiv \overline{\rho}$ where $\overline{\rho}$ is a positive constant and by introducing the kinematic viscosity $\nu = \mu(\overline{\rho})/\overline{\rho}$ and a reduced pressure field $p/\overline{\rho}$. We then obtain

$$\frac{\partial u}{\partial t} + u \cdot \nabla u - \nu \Delta u + \nabla p = f, \quad \operatorname{div} u = 0 \quad \text{in} \quad \Omega \times (0, T) \qquad (3.1)$$

where $T > 0$ is fixed and f is given on $\Omega \times (0, T)$. Of course, (3.1) is complemented with boundary conditions (the same as in chapter 2) and an initial condition

$$u|_{t=0} = u_0 \quad \text{in} \quad \Omega. \qquad (3.2)$$

Without loss of generality—otherwise we simply subtract a gradient term from u—we may always assume that we have

$$\operatorname{div} u_0 = 0 \quad \text{in} \quad \Omega. \qquad (3.3)$$

3.1 A brief review of known results

We begin with the celebrated results due to J. Leray [283] (see also [472], [293] and the bibliography for more references on the subject) concerning the global existence of weak solutions. In order to simplify the presentation and notation, we denote by H^1 (H^s, $W^{m,p}$) the usual Sobolev space $H^1(\mathbb{R}^N)$ in the case when $\Omega = \mathbb{R}^N$, or $H^1_{\text{per}} = \{u \in H^1_{\text{loc}}(\mathbb{R}^N), u \text{ periodic}\}$ in the periodic case and by H^{-1} the dual space (H^{-s}, $W^{-m,p'}$). In the results which follow, we assume

$$u_0 \in L^2(\Omega), \qquad f \in L^2(0, T; H^{-1}). \qquad (3.4)$$

In the case of Dirichlet boundary conditions, we assume in addition

$$u_0 \cdot n = 0 \quad \text{on} \quad \partial\Omega. \tag{3.5}$$

Recall that (3.5) is meaningful since (3.3) holds and $u_0 \in L^2(\Omega)$ (hence $u_0 \cdot n \in H^{-1/2}(\partial\Omega)$). Again this is not a restrictive assumption since we can always decompose (uniquely and continuously) any $\tilde{u}_0 \in L^p(\Omega)$ into a gradient term (in $L^p(\Omega)$) and a divergence-free vector field in $L^p(\Omega)$ satisfying (3.5) (for all $1 < p < \infty$).

In the case of Dirichlet boundary conditions, we need to introduce some functional spaces for rather delicate reasons to which we shall come back in detail later on. We set for $1 < p < \infty$,

$$V^{0,p}(\Omega) = \{u \in L^p(\Omega),\ \text{div}\, u = 0 \ \text{in}\ \Omega,\ u \cdot n = 0 \ \text{on}\ \partial\Omega\}$$
$$V^{1,p}(\Omega) = \{u \in W_0^{1,p}(\Omega),\ \text{div}\, u = 0 \ \text{in}\ \Omega\},$$

and we recall that $\tilde{\mathcal{D}}(\Omega) = \{\varphi \in C_0^\infty(\Omega),\ \text{div}\,\varphi = 0 \ \text{in}\ \Omega\}$ is dense in $V^{0,p}(\Omega)$, $V^{1,p}(\Omega)$ respectively for the L^p, $W^{1,p}$ norms. Finally, we denote by $V^{-1,p}$ the dual space of $V^{1,p'}$ where $\frac{1}{p} + \frac{1}{p'} = 1$ ($1 < p < \infty$).

Next, we recall the weak formulation of (3.1) as given in chapter 2 for a more general system without checking that all terms written below make sense, since this point will be a straightforward consequence of the regularity we assume for weak solutions: we request that we have for all $\varphi \in C^\infty(\Omega \times [0,T])$ such that $\text{div}\,\varphi = 0$ and with compact support in $\Omega \times [0,T)$

$$\int_0^T \int_\Omega dt\, dx \left\{ \nu \nabla u \cdot \nabla\varphi - u_i u_j \partial_i \varphi_j - u \cdot \frac{\partial\varphi}{\partial t} \right\}$$
$$= \int_0^T <f, \varphi>_{H^{-1} \times H_0^1} dt + \int_\Omega u_0 \cdot \varphi,\quad \text{div}\, u = 0 \ \text{in}\ \mathcal{D}'(\Omega \times (0,T)). \tag{3.6}$$

In the periodic case (or in the case $\Omega = \mathbb{R}^N$) we replace H_0^1 by H^1 and φ is then assumed to satisfy: $\varphi \in C^\infty(\mathbb{R}^N \times [0,T])$, $\text{div}\,\varphi = 0$, φ is periodic in x for all $t \in [0,T]$.

This formulation implies that (3.1) holds in the sense of distributions for some pressure field which is a distribution. Observe also that the term $u_i u_j \partial_j \varphi_j$ can be written as $-u_j \partial_j u_i \varphi_i$ (as soon as $\nabla u \in L^2(\Omega \times (0,T))$, $u \in L^2(\Omega \times (0,T))$ for example).

In the case of Dirichlet boundary conditions, (3.6) is also equivalent to a more abstract formulation involving the spaces $V^{1,p}(\Omega)$ (and $V^{-1,p}$). As we shall see, the weak solutions satisfy: $u \in L^2(0,T;V^{1,2}(\Omega))$, $u \in L^\infty(0,T;L^2(\Omega))$, $|u|^2 \in L^2(0,T;L^{\frac{N}{N-1}}(\Omega))$, and thus (3.6) implies that

$\frac{\partial u}{\partial t} \in L^2(0,T;V^{-1,\frac{N}{N-1}})$ (in fact $\frac{\partial u}{\partial t}$ should be written u' since it is considered as a (time) derivative of a function with values in some Banach space), and (3.6) is then equivalent to

$$u' + \nu Au + B(u,u) = f$$

where $f, Au \in L^2(0,T;V^{-1,2})$, $B(u,u) \in L^2(0,T;V^{-1,\frac{N}{N-1}})$ are defined by

$$<f,v> = <f,v>_{H^{-1} \times H_0^1}, \quad \forall v \in V^{1,2}$$

$$<Au,v> = \int_\Omega \nabla u \cdot \nabla v \, dx, \quad \forall v \in V^{1,2}$$

$$<B(u,u),v> = \int_\Omega (u \cdot \nabla)u \cdot v \, dx = -\int_\Omega u_i u_j \partial_i v_j \, dx, \quad \forall v \in V^{1,\frac{N}{N-1}}.$$

Observe finally that, since Ω is bounded, $V^{1,N} \subset V^{1,2}$ and we deduce that $V^{-1,2} \subset V^{-1,\frac{N}{N-1}}$ (identifying L^2 with its dual as usual).

We now state some global existence results of weak solutions: the first two results concern the case when $\Omega = \mathbb{R}^N$ (or the periodic case) in two dimensions ($N = 2$) and in dimensions $N \geq 3$ respectively, while the next two results are devoted to Dirichlet boundary conditions with $N = 2$ or $N \geq 3$ respectively.

Theorem 3.1. *($N = 2$, $\Omega = \mathbb{R}^2$ or the periodic case). There exists a unique weak solution u of (3.1)–(3.2) with the following properties: $u \in L^2(0,T;H^1) \cap C([0,T];L^2)$, $\frac{\partial u}{\partial t} \in L^2(0,T;H^{-1})$.*

Furthermore, there exists a unique $p \in L^2(B_R \times (0,T))$ (for all $R \in (0,\infty)$) such that $\nabla p \in L^2(0,T;H^{-1})$, $\int_Q p \, dx = 0$ a.e. $t \in (0,T)$ where $Q = B_1$ for example if $\Omega = \mathbb{R}^2$ or $Q = \Omega$ in the periodic case, and such that (3.1) holds in the sense of distributions.

We have for all $t \in [0,T]$

$$\left. \begin{array}{l} \dfrac{1}{2} \displaystyle\int_\Omega |u(x,t)|^2 \, dx + \nu \int_0^t \int_\Omega |\nabla u|^2 \, dx \, ds \\[3mm] \qquad = \dfrac{1}{2} \displaystyle\int_\Omega |u_0|^2 \, dx + \int_0^t <f(s),u(s)>_{H^{-1} \times H^1} ds. \end{array} \right\} \tag{3.7}$$

Theorem 3.2. *($N \geq 3$, $\Omega = \mathbb{R}^N$ or the periodic case). There exists a weak solution u of (3.1)–(3.2) and a pressure field p such that (3.1) holds in the sense of distributions, and the following properties hold: $u \in L^2(0,T;H^1) \cap C([0,T];L^2_w) \cap C([0,T];L^s(B_R))$ ($\forall 1 \leq s < 2$, $\forall R \in (0,\infty)$), $\frac{\partial u}{\partial t} \in L^2(0,T;H^{-1}) + (L^s(0,T;W^{-1,\frac{Ns}{Ns-2}}) \cap L^q(0,T;L^r))$ for $1 \leq s < \infty$,*

$1 \leq q < 2$ and $r = \frac{Nq}{Nq+q-2}$, $p \in L^2(B_R \times (0,T)) + L^s(0,T; L^{\frac{Ns}{Ns-2}})$ for $1 \leq s < \infty$, $R \in (0,\infty)$, $\nabla p \in L^2(0,T; H^{-1}) + L^q(0,T; L^r)$ for $1 \leq q < 2$ and $r = \frac{Nq}{Nq+q-2}$; and we have

$$\left.\begin{array}{l} \dfrac{1}{2} \displaystyle\int_\Omega |u(x,t)|^2 \, dx + \nu \int_0^t \int_\Omega |\nabla u|^2 \, dx \, ds \\[3mm] \qquad \leq \dfrac{1}{2} \displaystyle\int_\Omega |u_0|^2 \, dx + \int_0^t <f,u>_{H^{-1} \times H^1} \, ds, \quad \text{for all } t \geq 0. \end{array}\right\} \tag{3.8}$$

$$\frac{d}{dt}\left(\frac{1}{2}\int_\Omega |u(x,t)|^2 \, dx\right) + \nu \int_\Omega |\nabla u|^2 \, dx \leq <f,u>_{H^{-1} \times H^1} \quad \text{in } \mathcal{D}'(0,T). \tag{3.9}$$

Furthermore, if $N = 3$, there exists a solution satisfying in addition

$$\left.\begin{array}{l} \dfrac{\partial}{\partial t}\left(\dfrac{1}{2}|u|^2\right) + \text{div}\left(u\left\{\dfrac{1}{2}|u|^2 + p\right\}\right) \\[3mm] \qquad - \nu\Delta\dfrac{|u|^2}{2} + \nu|\nabla u|^2 \leq uf \quad \text{in } \mathcal{D}'. \end{array}\right\} \tag{3.10}$$

Theorem 3.3. *(N = 2, Dirichlet boundary conditions). There exists a unique weak solution u of (3.1)–(3.2) which satisfies: $u \in L^2(0,T; H_0^1(\Omega))$ (or equivalently $u \in L^2(0,T; V^{1,2})$), $u \in C([0,T]; L^2)$, $\frac{\partial u}{\partial t} \in L^2(0,T; V^{-1,2})$, and (3.7) holds.*

Theorem 3.4. *(N ≥ 3, Dirichlet boundary conditions). There exists a weak solution u of (3.1)–(3.2) satisfying (3.8),(3.9) and such that $u \in L^2(0,T; H_0^1(\Omega))$, $u \in C([0,T]; L_w^2) \cap C([0,T]; L^s(\Omega))$ for all $1 \leq s < 2$, $\frac{\partial u}{\partial t} \in L^2(0,T; V^{-1,2}) + (L^s(0,T; V^{-1, \frac{Ns}{Ns-2}}) \cap L^q(0,T; L^r))$ for $1 \leq s < \infty$, $1 \leq q < 2$ and $r = \frac{Nq}{Nq+q-2}$.*

Remarks 3.1. 1) We do not claim any originality in the results presented above except the slightly unusual presentation and, maybe, some regularity (or partial regularity) information on $\frac{\partial u}{\partial t}$ and p. The results presented in Theorems 3.1 and 3.2 are essentially contained in J. Leray [283],[284],[285] and further references can be found in R. Temam [472], J.L. Lions [293] and the bibliography for instance.

2) Let us remark that $u\{\frac{1}{2}|u|^2 + p\}$ makes sense in (3.10) (since we know that, for instance, $|u|^2 \in L^{3/2}(0,T; L^{9/5})$, $p \in L^{3/2}(0,T; L^{9/5}) + L^2(0,T; L^2)$ while $u \in L^3(0,T; L^{18/5}) \cap L^\infty(0,T; L^2)$ (by Sobolev embeddings) and $\frac{5}{9} + \frac{5}{18} = \frac{15}{18} < 1$. The meaning of uf in (3.10) has also to be clarified: since $f \in L^2(0,T; H^{-1})$, $u \in L^2(0,T; H^1)$, uf is the distribution defined by $<uf,\varphi> = <f, u\varphi>$ observing that $u\varphi \in H^1$ for smooth test functions φ.

3) We shall see below—see also sections 3.2 and 3.3—further regularity properties of weak solutions. Let us recall however that uniqueness of weak solutions is an outstanding open problem (even for more regular data f and u_0). It is clearly related to regularity issues: in particular, if we *postulate* more regularity on the weak solutions, the uniqueness follows. It is possible to show (this result is due to J. Serrin [428]) that if there exists a more regular solution then the weak solution coincides with this one. These results of the type "weak = strong" are very much in the same spirit as the one shown in section 2.5. However, the optimal—possibly the full—regularity of global weak solutions is not known: for instance, if $N = 3$, $\Omega = \mathbb{R}^3$, $f = 0$, u_0 is smooth, results due to J. Leray [283],[284],[285], L. Caffarelli, R.V. Kohn and L. Nirenberg [77] show that weak solutions are smooth except for "small sets" (zero one-dimensional Hausdorff measure in [77]) containing the possible singularities. The solution is also known to be smooth for t small and for t large, and for all t if u_0 and f are small in appropriate spaces. Of course, if $N = 2$, much more is known: for instance, if $u_0 \in H_0^1(\Omega)$ (or $H^1(\mathbb{R}^2)$), $f \in L^2(\Omega \times (0,T))$ then $u \in L^2(0,T;H^2(\Omega)) \cap C([0,T];H_0^1(\Omega))$, $\frac{\partial u}{\partial t} \in L^2(\Omega \times (0,T))$ and $p \in L^2(0,T;H^1(\Omega))$, and if f is smooth, u is smooth for $t > 0$.

Another topic which has been extensively studied concerns the space of initial conditions u_0 (take $f = 0$, $\Omega = \mathbb{R}^3$ for example) in which there exists a (unique) solution for t small or a global small solution. The most general result in that direction is probably the result of M. Cannone and Y. Meyer [80] which states that if $N = 3$, $\Omega = \mathbb{R}^3$, $u_0 \in L^3$, $f = 0$, u_0 is small in $B_\infty^{-1/2,6}$ then there exists a global solution $u \in C([0,\infty);L^3)$, and the solution is then automatically C^∞ for $t > 0$ as we shall see below.

Let us finally mention that some marginal improvements of the regularity of u_t and p shall be given in section 3.2 below, particularly in the case of Dirichlet boundary conditions.

4) We wish now to explain some of the difficulties encountered in the case of Dirichlet boundary conditions. First of all, let us remark that the information contained in Theorems 2.3–2.4 on $\frac{\partial u}{\partial t}$ does not say much if we insist upon looking at $\frac{\partial u}{\partial t}$ as a distribution. Indeed, if we can check that $V^{-1,p}(\Omega)$ is $W^{-1,p}(\Omega)/\{\nabla q \,/\, q \in L^p(\Omega)\}$, then $\frac{\partial u}{\partial t} \in L^2(0,T;V^{-1,2}(\Omega))$ (say if $N = 2$) does not imply $\frac{\partial u}{\partial t} \in L^2(0,T;H^{-1}(\Omega))$. In fact, even if $N = 2$, we do not know if $\frac{\partial u}{\partial t} \in L^2(0,T;H^{-1}(\Omega))$. Of course, there is a distribution π such that $\frac{\partial u}{\partial t} - \nabla \pi \in L^2(0,T;H^{-1}(\Omega))$, and it is possible to choose π harmonic (in x): indeed, we can write

$$\frac{\partial u}{\partial t} = -\Delta U + \nabla \pi, \quad \text{div}\, U = 0 \quad \text{in } \Omega, \quad U = 0 \quad \text{on } \partial\Omega. \qquad (3.11)$$

The regularity stated in Theorems 3.3 and 3.4 yields:

$$U \in L^2(0,T; H^1_0(\Omega)) + \left(L^s\left(0,T; W^{1,\frac{Ns}{Ns-2}}_0(\Omega)\right) \cap L^q(0,T; W^{2,r}(\Omega)) \right)$$
(3.12)

for $1 \le s < \infty$, $1 \le q < 2$, and $r = \frac{Nq}{Nq+q-2}$. From this we deduce that $\frac{\partial u}{\partial t} - \nabla \pi \in L^2(0,T; H^{-1}(\Omega)) + \left(L^s\left(0,T; W^{-1,\frac{Ns}{Ns-2}}(\Omega)\right) \cap L^q(0,T; L^q(\Omega)) \right)$ and we finally observe that (3.11) implies $\Delta \pi = 0$. In other words, up to a gradient harmonic distribution, $\frac{\partial u}{\partial t}$ has the regularity we expect. From (3.1) (which holds for a certain distribution p), we deduce that ∇p has also the same properties as $\frac{\partial u}{\partial t}$ ("natural regularity" up to the gradient of an harmonic function).

This question does not arise when $\Omega = \mathbb{R}^N$ or in the periodic case since if π is harmonic and $\nabla \pi \in H^{-1}$ (and periodic in the periodic case) then $\nabla p \equiv 0$. A related issue is the possibility of projecting (3.1) on divergence-free vector fields. Let us recall that we denote by P the projection (in L^2 say) onto divergence-free vector fields, i.e.

$$P = \mathrm{Id} + \nabla(-\Delta)^{-1}\mathrm{div}.$$

Both in the periodic case and in the case when $\Omega = \mathbb{R}^N$, P commutes with translations and thus with derivatives so it is bounded (and a projection) from any H^s into H^s ($s \in \mathbb{R}$). In fact, since P is bounded in L^p ($1 < p < \infty$), P is also bounded in any $H^{s,p}$ ($s \in \mathbb{R}$, $1 < p < \infty$). Then, we can write in those two cases

$$\frac{\partial u_i}{\partial t} + P((u \cdot \nabla)u)_i - \nu \Delta u_i = P(f)_i \tag{3.13}$$

or equivalently

$$\frac{\partial u_i}{\partial t} + \frac{\partial}{\partial x_j}(P(u_j u_i)) - \nu \Delta u_i = P(f)_i. \tag{3.14}$$

On the other hand, when we consider the case of Dirichlet boundary conditions, it does not seem to be possible to write (3.1) in such a concise form. The natural replacement for P is the orthogonal projection (in L^2) onto divergence-free vector fields $v \in L^2(\Omega)^N$ such that $v \cdot n = 0$ on $\partial\Omega$— recall that if $v \in L^2(\Omega)^N$, $\mathrm{div}\, v \in L^2(\Omega)$ then $v \cdot n \in H^{-1/2}(\partial\Omega)$. Then P is also bounded in $L^p(\Omega)$ for $1 < p < \infty$ but it no longer commutes with derivatives. In fact, it is easy to check that P is bounded in $W^{1,p}(\Omega)$ for $1 < p < \infty$ but P does not leave $W^{1,p}_0(\Omega)$ invariant (we only obtain $P(u) \cdot n = 0$ on $\partial\Omega$ and not $Pu = 0$ on $\partial\Omega$). In particular, we cannot deduce (by duality) any information on P in $W^{-1,p}(\Omega)$ including a definition.

The fact that $P(u)$ does not make sense if $u \in H^{-1}(\Omega)$ (say) can be seen from the fact that $P(-\Delta u)$ does not make sense if $u \in H^1_0(\Omega)$ even if div $u = 0$ in Ω. Indeed, if $u \in H^2(\Omega) \cap H_0^1(\Omega)$, then $P(-\Delta u) \in L^2(\Omega)$ and we claim that $P(-\Delta u)$ is not bounded in H^{-1} if u is bounded in $H_0^1(\Omega)$ even if div $u = 0$ in Ω. To prove this claim, we argue by contradiction and thus assume that $P(-\Delta u)$ extends by continuity to a continuous map from $H_0^1(\Omega)$ into $H^{-1}(\Omega)$. Then, let $u \in H^2(\Omega) \cap H_0^1(\Omega)$ be such that div $u = 0$, $\Delta u \cdot n \not\equiv 0$ on $\partial\Omega$ ($\Delta u \in L^2(\Omega)$, div $(\Delta u) = 0$ in Ω so that $\Delta u \cdot n$ makes sense in $H^{-1/2}(\partial\Omega)$); examples of such a u are not difficult to build. We next choose $u_n \in C_0^\infty(\Omega)$ such that u_n converges to u in $H_0^1(\Omega)$ and div $u_n = 0$ in Ω. Clearly, $P(-\Delta u_n) = -\Delta u_n$ converges in $H^{-1}(\Omega)$ to $-\Delta u$ while, by assumption, it should also converge to $P(-\Delta u)$, and we reach the desired contradiction since $\Delta u \cdot n \not\equiv 0$ on $\partial\Omega$ and thus $P(-\Delta u) \not\equiv -\Delta u$.

This argument shows that $P(-\Delta u)$ defined on $H^2(\Omega) \cap V^{1,2}(\Omega)$ cannot be extended by continuity to $V^{1,2}(\Omega)$. Of course, even if it does not seem a natural thing to try, we might attempt to define $P(-\Delta u)$ on $V^{1,2}(\Omega)$ in a different manner using the orthonormal basis of $V^{0,2}(\Omega)$ composed of the eigenfunctions w_i ($i \geq 1$) of the Stokes operator, namely

$$-\Delta w_i + \nabla \pi_i = \lambda_i w_i, \quad \lambda_i \in \mathbb{R}, \quad w_i \in V^{1,2}(\Omega)$$

where λ_i (> 0) are the eigenvalues. The set $\{w_i \, / \, i \geq 1\}$ is an orthogonal basis of $V^{1,2}(\Omega)$ and we have

$$\left.\begin{array}{l} u = \displaystyle\sum_{i=1}^{\infty} u_i w_i, \quad \displaystyle\int_\Omega |u|^2 \, dx = \sum_{i=1}^{\infty} |u_i|^2 \quad \text{for all } u \in V^{0,2}(\Omega), \\[2em] \displaystyle\int_\Omega |\nabla u|^2 \, dx = \sum_{i=1}^{\infty} \lambda_i |u_i|^2 \quad \text{for all } u \in V^{1,2}(\Omega), \end{array}\right\}$$

where $u_i = \int_\Omega u \, w_i \, dx$.

Then, a possible attempt to define $P(-\Delta u)$ is to consider the limit (if it exists) of $P\big(-\Delta\big(\sum_{i=1}^N u_i w_i\big)\big)$ as N goes to $+\infty$ or in other words the limit of $\sum_{i=1}^N \lambda_i u_i w_i$ since $P(-\Delta w_i) = \lambda_i w_i$. This is not possible: indeed, arguing by contradiction, if $\sum_{i=1}^N \lambda_i u_i w_i$ converges as N goes to $+\infty$ in some space, say $H^{-1}(\Omega)$ (we could also assume that it stays bounded in $H^{-1}(\Omega)$ and with a little more work we would reach a similar contradiction) for all $u \in V^{1,2}(\Omega)$, then $\sum_{i=1}^N \lambda_i u_i w_i$ converges to $T(u)$ where T is a linear mapping continuous from $V^{1,2}(\Omega)$ into $H^{-1}(\Omega)$. We claim that $T(u) = P(-\Delta u)$ if $u \in H^2(\Omega) \cap V^{1,2}(\Omega)$: if this claim is proven, we conclude easily in view of the fact shown above. Then, if $u \in H^2(\Omega) \cap V^{1,2}(\Omega)$, $\Delta u \in L^2(\Omega)$

and thus

$$\sum_{i=1}^{N} \lambda_i u_i w_i = \sum_{i=1}^{N} w_i \int_{\Omega} u\,\lambda_i w_i \, dx = \sum_{i=1}^{N} w_i \int_{\Omega} (-\Delta u) w_i \, dx$$

$$= \sum_{i=1}^{N} w_i \int_{\Omega} P(-\Delta u) w_i \, dx \;\to\; P(-\Delta u)$$

in $L^2(\Omega)$ as N goes to $+\infty$.

The specific difficulties encountered in the case of Dirichlet boundary conditions are intimately related to the simple observation already mentioned above that there exist non-trivial (non-constant) harmonic functions in L^2! Indeed, there exists $h \in L^2$, h is harmonic in Ω: hence, $\nabla h \in H^{-1}(\Omega)$ and $\operatorname{div}(\nabla h) = 0$. In the periodic case we immediately see that h is constant and thus $\nabla h \equiv 0$. If $\Omega = \mathbb{R}^N$, $T \in H^{-1}$, $\operatorname{curl} T = 0$, $\operatorname{div} T = 0$ in $\mathcal{D}'(\mathbb{R}^N)$ then we also obtain $T \equiv 0$.

5) The energy inequality (3.8) shows that $u(t)$ converges to u_0 in $L^2(\Omega)$ as t goes to 0_+. □

We now sketch the

Proof of Theorems 3.1–3.4. First of all, the *existence* of weak solutions is a particular case of Theorem 2.1 taking $\rho_0 \equiv 1$ and thus, by (2.17), $\rho \equiv 1$. Notice that Theorem 2.1 also yields the *energy inequalities* (3.9) and (3.8) for almost all $t \geq 0$. The fact that (3.8) in fact holds for all $t \geq 0$ is then a simple consequence of the continuity in time (with values in L^2_w) of u. Let us remark that the fact that $u \in C([0,T]; L^2_w)$ is a consequence of Theorem 2.3 since $P(u) = u$. Also the continuity in t with values in L^p_{loc} for $p < 2$ (or L^p in the periodic case or in the case of Dirichlet boundary conditions) follows upon decomposing u into $u_1 + u_2$ where u_1 solves: $\frac{\partial u_1}{\partial t} - \nu \Delta u_1 + \nabla p_1 = f$, $\operatorname{div} u_1 = 0$, $u_1 \in C([0,T]; L^2)$, $u_1|_{t=0} = u_0$; and where u_2 solves: $\frac{\partial u_2}{\partial t} - \nu \Delta u_2 + \nabla p_2 = -(u \cdot \nabla)u$, $\operatorname{div} u_2 = 0$, $u_2|_{t=0} = 0$, $u_2 \in W^{2,1,\frac{N+2}{2}}(\Omega \times (0,T))$ (see V.A. Solonnikov [444],[445] and section 3.3 for such estimates), therefore $u_2 \in C([0,T]; W^{\frac{2}{N+2}, \frac{N+2}{N+1}}(\Omega))$. It only remains to explain the additional regularity information on $\frac{\partial u}{\partial t}$, ∇p (step 1), the uniqueness statements if $N = 2$ (step 2), and the local energy inequality (3.10) (step 3).

There is however one more point to clarify: what we claimed above about the applications of Theorem 2.1 is not entirely correct since we need to assume that $f \in L^2(\Omega \times (0,T))$ in order to apply Theorem 2.1 while the results above only require that $f \in L^2(0,T; H^{-1}(\Omega))$. The reason why we neglect this technical point is the following: when ρ is constant, say $\rho \equiv 1$, then all the a priori estimates and passages to the limit are valid if we only assume that $f \in L^2(0,T;H^{-1}(\Omega))$ and thus the proofs already given easily adapt to that case. Another way to argue is to approximate f in $L^2(0,T;H^{-1}(\Omega))$ by $f_n \in L^2(\Omega \times (0,T))$. We then apply Theorems 2.1 and 2.3 and obtain weak solutions u_n, which as we shall prove below satisfy the properties listed in Theorems 3.1–3.4. Finally, we recover the desired results passing to the limit as n goes to $+\infty$.

Step 1. Regularity information on $\frac{\partial u}{\partial t}$, ∇p. In the periodic case or if $\Omega = \mathbb{R}^N$, we simply use (3.13) or (3.14) (which are easily deduced from the definition of weak solutions). If $N = 2$, we recall that $u \in L^4(\Omega \times (0,T))$ and thus $P(u_j u_i)$ $(\forall\, i,j) \in L^2(\Omega \times (0,T))$, hence $\frac{\partial u}{\partial t} \in L^2(0,T;H^{-1})$. Of course, this yields the regularity statements made upon p and the continuity of u in time with values in L^2. This also allows us to justify (3.7). If $N \geq 3$, we remark that $u \in L^\infty(0,T;L^2) \cap L^2(0,T;L^{\frac{2N}{N-2}})$ (by Sobolev embeddings) while $\nabla u \in L^2(\Omega \times (0,T))$. Therefore, $u \cdot \nabla u \in L^q(0,T;L^r)$ for $1 \leq q \leq 2$, $r = \frac{Nq}{Nq+q-2}$ while $u \cdot \nabla u_i$ $(\forall\, i) \in L^s(0,T;W^{-1,\frac{Ns}{Ns-2}})$ for $1 \leq s \leq \infty$ since $u \cdot \nabla u_i = \operatorname{div}(u u_i)$. The regularity for $\frac{\partial u}{\partial t}$ then follows from (3.13) and (3.14).

The regularity of p stated in Theorem 3.2 is deduced from (3.1) in the following way: we take the divergence of (3.1) and we find

$$\left.\begin{aligned} -\Delta p &= \partial_i(u_j \partial_j u_i) - \operatorname{div} f = \partial_i u_j \, \partial_j u_i - \operatorname{div} f \\ &= \partial_{ij}(u_i u_j) - \operatorname{div} f \qquad \text{in} \quad \mathbb{R}^N \end{aligned}\right\} \tag{3.15}$$

and in the periodic case p is periodic—p can be normalized by requesting that $\int_Q p\,dx = 0$ where Q is the periodic cube or $Q = B_1$ if $\Omega = \mathbb{R}^N$. The regularity of p then follows from elliptic regularity and from the bounds on $u_i u_j$, $(u \cdot \nabla)u_i$ $(1 \leq i,j \leq N)$ obtained above. Of course, we could also obtain the regularity of ∇p from equation (3.1) in view of the bounds shown above on $(u \cdot \nabla)u$ and $\frac{\partial u}{\partial t}$.

In the case of Dirichlet boundary conditions, the argument for $\frac{\partial u}{\partial t}$ is the same except that, for reasons detailed above, we have to replace $W^{-1,p}$ by $V^{-1,p}$.

Step 2. Uniqueness if $N = 2$. We only need to observe that if $u, v \in L^\infty(0,T;L^2) \cap L^2(0,T;H^1)$ and thus $u, v \in L^4(\Omega \times (0,T))$ and $\operatorname{div} v = 0$ a.e. in $\Omega \times (0,T)$, we have for all $t \in [0,T]$

$$\int_\Omega dx \int_0^t ds \, [(u \cdot \nabla)u - (v \cdot \nabla)v] \cdot (u-v)$$

$$= \int_\Omega dx \int_0^t ds \, [(u-v) \cdot \nabla u + (v \cdot \nabla)(u-v)] \cdot (u-v)$$

$$= \int_\Omega dx \int_0^t ds \, [(u-v) \cdot \nabla u] \cdot (u-v) \geq - \int_0^t \|\nabla u\|_{L^2} \|u-v\|_{L^4}^2 \, ds$$

$$\geq -C_0 \int_0^t \|\nabla u\|_{L^2} \|u-v\|_{L^2} \|\nabla(u-v)\|_{L^2} \, ds$$

for some $C_0 > 0$ independent of u, v. Then, if u, v are solutions of (3.1)–(3.2) as in Theorem 3.1 or 3.3, we deduce easily from the above inequality (and the regularity of $u, v, \frac{\partial u}{\partial t}, \frac{\partial v}{\partial t}$) for all $t \geq 0$

$$\frac{1}{2} \|u-v\|_{L^2}^2(t) + \nu \int_0^t \|\nabla(u-v)\|_{L^2}^2 \, ds$$

$$\leq C_0 \int_0^t \|\nabla u\|_{L^2} \|u-v\|_{L^2} \|\nabla(u-v)\|_{L^2} \, ds$$

$$\leq \frac{\nu}{2} \int_0^t \|(\nabla(u-v)\|_{L^2}^2 \, ds + \frac{C_0^2}{2\nu} \int_0^t \|\nabla u\|_{L^2}^2 \|u-v\|_{L^2}^2 \, ds,$$

and the uniqueness follows from Grönwall's inequality.

Step 3. The local energy inequality (3.10). In order to show (3.10), we go back to the construction of weak solutions performed in section 2.4 in our special case, namely $\rho \equiv 1$. In other words, we consider $u_\varepsilon \in C^\infty(\mathbb{R}^N \times [0,T])$ (vanishing at infinity if $\Omega = \mathbb{R}^N$, periodic in the periodic case) as a solution of

$$\left. \begin{array}{l} \dfrac{\partial u_\varepsilon}{\partial t} + (u_\varepsilon * \omega_\varepsilon) \cdot \nabla u_\varepsilon - \nu \Delta u_\varepsilon + \nabla p_\varepsilon = f_\varepsilon \quad \text{in} \quad \mathbb{R}^N \times [0,T], \\[2mm] \text{div} \, u_\varepsilon = 0 \quad \text{in} \quad \mathbb{R}^N \times [0,T] \end{array} \right\} \quad (3.16)$$

$$u_\varepsilon|_{t=0} = u_0 * \omega_\varepsilon \quad \text{in} \quad \mathbb{R}^N \qquad (3.17)$$

where $f_\varepsilon \in C^\infty(\mathbb{R}^N \times (0,T))$ converges to f in $L^2(0,T;H^{-1})$, f_ε vanishes near $t = 0$ ($\forall \, x$), f_ε is periodic in the periodic case and $f_\varepsilon \in C_0^\infty(\mathbb{R}^N \times (0,T))$ if $\Omega = \mathbb{R}^N$. Let us emphasize that this is essentially the original approximation of (3.1) introduced in J. Leray [**283**]. We also know that u_ε converges weakly—extracting subsequences if necessary in $L^2(0,T;H^1)$—to a weak solution u satisfying the conditions listed in Theorem 3.2. In addition, u_ε converges to u in $L^s(0,T;L^q(B_R))$ for $2 < s < \infty$, $q < \frac{2Ns}{Ns-4}$ and for all $R \in (0,\infty)$. In particular, $|u_\varepsilon|^2$ converges to $|u|^2$ in $L^r(0,T;L^q(B_R))$ for $1 < r < \infty$, $1 \leq q < \frac{Nr}{Nr-2}$, $R \in (0,\infty)$.

If $N = 3$, we deduce that $|u_\varepsilon|^2$ converges to $|u|^2$ in $L^{\frac{3}{2}}(0, T; L^q(B_R))$ for $q < \frac{9}{5}$ while u_ε converges to u in $L^3(0, T; L^q(B_R))$ for $q < \frac{18}{5}$ ($\forall\, R \in (0, \infty)$). Hence, $(u_\varepsilon * \omega_\varepsilon)|u_\varepsilon|^2$ converges to $u|u|^2$ in $L^1(B_R \times (0, T))$ for all $R \in (0, \infty)$.

In addition, the bounds obtained in step 1 on $\frac{\partial u}{\partial t}$ and p are easily shown to hold for (3.16) and are uniform in $\varepsilon \in (0, 1)$. In particular, p_ε converges weakly to p in $L^2(0, T; L^q(B_R))$ for $q < \frac{3}{2}$ while u_ε converges to u in $L^2(0, T; L^q(B_R))$ for $q < 6$.

Next, we multiply (3.16) by u_ε and we obtain on $\mathbb{R}^N \times [0, T]$

$$\frac{\partial}{\partial t}\left(\frac{1}{2}|u_\varepsilon|^2\right) + \mathrm{div}\left((u_\varepsilon * \omega_\varepsilon)\left\{\frac{1}{2}|u_\varepsilon|^2 + p_\varepsilon\right\}\right) - \nu\Delta\frac{|u_\varepsilon|^2}{2} + \nu|\nabla u_\varepsilon|^2 = u_\varepsilon f_\varepsilon.$$

Without loss of generality, we may assume that $|\nabla u_\varepsilon|^2$ converges weakly (in the sense of measures) to a bounded non-negative measure D on $\mathbb{R}^N \times [0, T]$. By standard functional analysis considerations, we deduce that $D \geq \nu|\nabla u|^2$. This fact, together with the convergences established above, allows us to pass to the limit as ε goes to 0 in the above equality to recover the inequality (3.10), thus concluding our proof. $\quad\square$

We conclude this section with an observation on the regularity of solutions of Navier–Stokes equations if $N \geq 3$: we *postulate* the existence of weak solutions u of (3.1)–(3.2) with the properties listed in the above existence results and such that $u \in C([0, T]; L^N(\Omega))$. The result that follows shows that if f is smooth then u is smooth for $t > 0$. More precisely, we have the following classical result

Theorem 3.5. *Let $N \geq 3$, let $f \in L^2(\Omega \times (0, T)) \cap L^r(\Omega \times (0, T))$ for some $r \in [2, N)$. Let u be a weak solution of (3.1)–(3.2) as given by Theorem 3.2 or 3.4. We assume that $u \in C([0, T]; L^N(\Omega))$. Then, for each $\varepsilon > 0$, $u \in L^q(\varepsilon, T; W^{2,q}(\Omega))$, $p \in L^q(\varepsilon, T; W^{1,q}(\Omega))$ and $\frac{\partial u}{\partial t} \in L^q(\varepsilon, T; L^q(\Omega))$ for $2 \leq q \leq r$.*

Remarks 3.2. 1) If u_0 is smooth enough (in the appropriate Besov space), the argument below shows that we can take $\varepsilon = 0$ in the preceding result.

2) If f is smooth, then it is straightforward to deduce from the above result the fact that u is smooth for $t > 0$.

3) The proof presented below can be adapted to the case when u is assumed to satisfy $u \in L^\alpha(0, T; L^\beta(\Omega))$ where $2 \leq \alpha < \infty$, $\beta = \frac{N\alpha}{\alpha - 2}$ (or even more generally $L^{\alpha_1}(0, T; L^{\beta_1}(\Omega)) + L^{\alpha_2}(0, T; L^{\beta_2}(\Omega)) + C([0, T]; L^N(\Omega))$ where $2 \leq \alpha_1, \alpha_2 < \infty$, $\beta_i = \frac{N\alpha_i}{\alpha_i - 2}$, $i = 1, 2$). The above result remains valid.

4) It is possible to extend slightly the above result by requiring u to satisfy the following property: for all $\varepsilon > 0$, $u = u_1 + u_2$ where $\|u_1\|_{L^\infty(0, T; L^{N,\infty}(\Omega))} \leq \varepsilon$, $u_2 \in L^\infty(\Omega \times (0, T))$. $\quad\square$

Proof of Theorem 3.5. We first observe that for all $\varepsilon > 0$ there exist (u_1, u_2) such that

$$u = u_1 + u_2, \quad \|u_1\|_{L^\infty(0,T;L^N(\Omega))} \le \varepsilon, \quad u_2 \in L^\infty(\Omega \times (0,T)), \quad (3.18)$$

and we may even assume u_2 to be smooth on $\overline{\Omega} \times [0,T]$ (periodic in the periodic case or in $C_0^\infty(\Omega \times (0,T))$ in the other cases).

We next wish to make a few remarks on the following linear equations

$$\left.\begin{array}{l} \dfrac{\partial v}{\partial t} + u \cdot \nabla v - \nu \Delta v + \nabla p = g, \\[2mm] \operatorname{div} v = 0 \quad \text{in } \Omega \times (0,T), \quad v|_{t=0} = 0 \quad \text{in } \Omega \end{array}\right\} \quad (3.19)$$

with the same boundary conditions for v as for u. We first claim that if $g \in L^r(\Omega \times (0,T)) \cap L^2(\Omega \times (0,T))$ for some $r \in [2, N)$ then there exists a unique solution $v \in L^q(0,T;W^{2,q}(\Omega))$, $\frac{\partial v}{\partial t} \in L^q(0,T;L^q(\Omega))$, $\nabla p \in L^q(0,T;L^q(\Omega))$ of (3.19) for all $q \in [2,r]$. The existence (and uniqueness) follows from the a priori estimates we explain now. First of all, we have (multiplying by v) a priori estimates in $L^2(0,T;H^1) \cap L^\infty(0,T;L^2)$. Next, we remark that we have by Sobolev embeddings and because of (3.18)

$$\|u_1 \cdot \nabla v\|_{L^q(\Omega \times (0,T))} \le \varepsilon \|\nabla v\|_{L^q\left(0,T;L^{\frac{Nq}{N-q}}(\Omega)\right)} \le C\varepsilon \|D^2 v\|_{L^q(\Omega \times (0,T))},$$

$$\begin{aligned} \|u_2 \cdot \nabla v\|_{L^q(\Omega \times (0,T))} &\le C_\varepsilon \|\nabla v\|_{L^q(\Omega)} \\ &\le C_\varepsilon \|v\|_{L^q(\Omega \times (0,T))}^{1/2} \|D^2 v\|_{L^q(\Omega \times (0,T))}^{1/2} \\ &\le C_\varepsilon \|v\|_{L^2(\Omega \times (0,T))}^{\theta/2} \|v\|_{L^s(\Omega \times (0,T))}^{\frac{1-\theta}{2}} \|D^2 v\|_{L^q(\Omega \times (0,T))}^{1/2} \\ &\le C_\varepsilon \|v\|_{L^2(\Omega \times (0,T))}^{\gamma} \left\{ \left\|\frac{\partial v}{\partial t}\right\|_{L^q(\Omega \times (0,T))} + \|D^2 v\|_{L^q(\Omega \times (0,T))} \right\}^{1-\gamma} \end{aligned}$$

where, above and below, C denotes various positive constants independent of v, ε, q in $[2, r]$, C_ε denotes various positive constants independent of v, q in $[2, r]$, $\gamma = \frac{\theta}{2} \in (0, \frac{1}{2}]$, $\frac{\theta}{2} + \frac{1-\theta}{s} = \frac{1}{q}$, $s = \frac{(N+2)q}{N+2-2q}$ (if $q < \frac{N+2}{2}$, $s = +\infty$ if $q > \frac{N+2}{2}$, s arbitrary in $(2, \infty)$ if $q = \frac{N+2}{2}$).

Then, we use $L^q(\Omega \times (0,T))$ estimates for linear (Stokes) equations due to V.A. Solonnikov [444],[445] (which are in fact valid in all dimensions) and we deduce

$$\begin{aligned} &\|v\|_{L^q(0,T;W^{2,q}(\Omega))} + \left\|\frac{\partial v}{\partial t}\right\|_{L^q(\Omega \times (0,T))} + \|\nabla p\|_{L^q(\Omega \times (0,T))} \\ &\le C\varepsilon \|D^2 v\|_{L^q(\Omega \times (0,T))} + C_\varepsilon \|v\|_{L^2(\Omega \times (0,T))}^{\gamma} \\ &\quad \cdot \left\{ \left\|\frac{\partial v}{\partial t}\right\|_{L^q(\Omega \times (0,T))} + \|D^2 v\|_{L^q(\Omega \times (0,T))} \right\}^{1-\gamma} + C\|g\|_{L^q(\Omega \times (0,T))}. \end{aligned}$$

Since we already have a priori estimates on v in $L^2(\Omega \times (0,T))$, the desired a priori estimates are shown.

The next step consists in showing that there exists a unique weak solution $(\in L^2(0,T;H^1) \cap L^\infty(0,T;L^2))$ of (3.19) or in other words that if $q \equiv 0$ then $v \equiv 0$. To this end, we first observe that $u \cdot \nabla v \in L^2(0,T;L^{\frac{2N}{N+2}})$ and thus $\frac{\partial u}{\partial t} \in L^2(0,T;H^{-1})$ (or $L^2(0,T;V^{-1,2})$ in the case of Dirichlet boundary conditions). Therefore, $v \in C([0,T];L^2)$ and we multiply (3.19) by v to obtain for all $t \in [0,T]$

$$\int_\Omega \frac{1}{2}|v(t)|^2\,dx + \nu \int_0^t \int_\Omega |\nabla v|^2\,dx\,ds$$
$$= -\int_0^t \int_\Omega (u \cdot \nabla v) \cdot v\,dx\,ds = -\int_0^t \int_\Omega u \cdot \nabla \frac{|v|^2}{2}\,dx\,ds = 0,$$

the last computations being easy to justify since $u \in C([0,T];L^N)$, $u \cdot \nabla v \in L^2(0,T;L^{\frac{2N}{N-2}})$, $v \in L^2(0,T;L^{\frac{2N}{N-2}})$ (argue by density on v for instance).

We may now complete the proof of Theorem 3.5 by observing that $v_1 = tu$ ($p_1 = tp$) solves (3.19) with $g = tf + u \in L^{r_1}(\Omega \times (0,T)) \cap L^2(\Omega \times (0,T))$ with $r_1 = \min(p,4)$ hence $v_1 \in L^q(0,T;W^{2,q}(\Omega))$, $\nabla p_1 \in L^q(\Omega \times (0,T))$ for $2 \le q \le r_1$. If $p \le 4$, we conclude, while if $p > 4$ (hence $N \ge 5$), we observe that $v_2 = t^2u$ solves (3.19) with $g = t^2f + 2v_1 \in L^{r_2}(\Omega \times (0,T)) \cap L^2(\Omega \times (0,T))$ while $r_2 = \min\left(p, \frac{4(N+2)}{N-6}\right)$ if $N \ge 7$, $r_2 = p$ if $N = 5$ or 6 (since $u_1 \in L^{4\frac{N+2}{N-6}}(\Omega \times (0,T))$ by the regularity just established). If $p \le 4\frac{N+2}{N-6}$, we conclude. If $p > 4\frac{N+2}{N-6}$ (hence $N \ge 11$), we consider $v_3 = t^3u$ and reiterating the preceding argument, we prove Theorem 3.5. $\square$

The same type of technique can be used to prove the regularity of weak solutions of stationary Navier–Stokes equations if $N = 4$ (such results are classical if $N = 2$ or 3). More precisely, we consider stationary weak solutions of Navier–Stokes equations, in the case of Dirichlet boundary conditions to fix ideas, namely solutions of

$$\left.\begin{array}{ll} -\nu\Delta u + (u \cdot \nabla)u + \nabla p = f & \text{in } \Omega, \\ u \in H_0^1(\Omega), \quad \operatorname{div} u = 0 & \text{in } \Omega. \end{array}\right\} \tag{3.20}$$

Then, if $f \in H^{-1}(\Omega)$, there exists at least one solution u of (3.20) (see for instance R. Temam [472]). We claim that if $N = 4$, any such solution belongs to $W^{2,q}(\Omega)$ if $f \in L^q(\Omega)$ and $q \in \left[\frac{2N}{N+2}, N\right)$, and in addition $\nabla p \in L^q(\Omega)$. In particular, if this claim is shown, then $f \in L^\infty(\Omega)$ implies $u \in W^{2,q}(\Omega)$ for $q < N$ and thus $u \in C^\alpha(\overline{\Omega})$ for all $\alpha \in (0,1)$. By standard regularity results, we then deduce $u \in W^{2,q}(\Omega)$ for all $q < \infty$, and if $f \in C^\alpha(\overline{\Omega})$ ($\alpha \in (0,1)$) then $u \in C^{2,\alpha}(\overline{\Omega})$.

In order to prove the above claim, we argue as in the proof of Theorem 3.5 and we remark that, by Sobolev embeddings, $u \in L^4(\Omega)$ (recall that $N = 4$) hence for all $\varepsilon > 0$

$$u = u_1 + u_2, \quad u_2 \in L^\infty(\Omega), \quad \|u_1\|_{L^4(\Omega)} \leq \varepsilon.$$

Next, for any $g \in L^r(\Omega) \cap L^{\frac{2N}{N+2}}(\Omega)$, there exists a unique solution v of

$$\left. \begin{array}{l} -\nu \Delta v + (u \cdot \nabla)v + \nabla p = g \quad \text{in} \quad \Omega, \\ v \in H_0^1(\Omega), \quad \text{div}\, v = 0 \quad \text{in}\ \Omega \end{array} \right\} \tag{3.21}$$

and $v \in W^{2,q}(\Omega)$ for $q \in \left[\frac{2N}{N+2}, r\right]$, where $r \in \left[\frac{2N}{N+2}, \infty\right)$.

The proof of this claim relies upon the above decomposition of u and follows the same lines as the corresponding argument in the proof of Theorem 3.5. In particular, taking $g = f$, $r = q$, u has to be this unique solution and thus has the claimed regularity.

3.2 Refined regularity of weak solutions via Hardy spaces

In this section, we review some results due to R. Coifman, P.L. Lions, Y. Meyer and S. Semmes [95] which concern some (marginal) improvements of the known regularity of weak solutions. They rely upon multi-dimensional Hardy spaces and they are valid in the periodic case or in the case when $\Omega = \mathbb{R}^N$. We shall discuss corresponding results in the case of Dirichlet boundary conditions in the next section. In order to simplify the presentation, we only consider the case $\Omega = \mathbb{R}^N$ since the adaptations to the periodic case are straightforward.

We first recall the definition and some of the main properties of Hardy spaces introduced by E. Stein and G. Weiss [455] (for more facts on these spaces see C. Fefferman and E. Stein [149], R. Coifman and G. Weiss [96]). The Hardy space, denoted by $\mathcal{H}^1(\mathbb{R}^N)$ to avoid confusion with Sobolev spaces, is a closed subspace of $L^1(\mathbb{R}^N)$ defined by

$$\mathcal{H}^1(\mathbb{R}^N) = \left\{ f \in L^1(\mathbb{R}^N) \,/\, \sup_{t \geq 0} |h_t * f| \in L^1(\mathbb{R}^N) \right\} \tag{3.22}$$

where $h_t = \frac{1}{t^N} h\left(\frac{\cdot}{t}\right)$, $h \in C_0^\infty(\mathbb{R}^N)$, $h \geq 0$, Supp $h \subset B(0,1)$; in fact, it can be shown that this space is independent of the choice of h. Also, $\mathcal{H}^1$ can be characterized in terms of Riesz transforms R_j as

$$\mathcal{H}^1(\mathbb{R}^N) = \{ f \in L^1(\mathbb{R}^N) \,/\, \forall\, 1 \leq j \leq N,\ R_j f \in L^1(\mathbb{R}^N) \}. \tag{3.23}$$

In addition, we have

$$R_j \quad \text{is bounded from} \quad \mathcal{H}^1 \quad \text{into} \quad \mathcal{H}^1, \qquad (3.24)$$

provided we equip $\mathcal{H}^1$ with a norm taken to be, for instance, $\| \sup_{t \geq 0} |h_t *$
$f| \|_{L^1(\mathbb{R}^N)}$. $\mathcal{H}^1$ is a separable Banach space whose dual is $BMO(\mathbb{R}^N)$
and which is the dual of $VMO(\mathbb{R}^N)$—the "completion of $C_0^\infty(\mathbb{R}^N)$ for the
BMO norm" ($\sup_Q f_Q |b - f_Q b|$ where the supremum is taken over all cubes
of $\mathbb{R}^N$).

In [95] it was shown that Hardy spaces can be used to analyse the regularity of the various nonlinear quantities identified by the compensated
compactness theory due to L. Tartar [468], [469] and F. Murat [349], [350],
[351]. In particular, it was shown that $E \cdot B \in \mathcal{H}^1(\mathbb{R}^N)$ if $E, B \in L^2(\mathbb{R}^N)$,
curl $E = $ div $B = 0$ in $\mathcal{D}'(\mathbb{R}^N)$ and we have for some $C > 0$ independent of
E, B

$$\|E \cdot B\|_{\mathcal{H}^1(\mathbb{R}^N)} \leq C \|E\|_{L^2(\mathbb{R}^N)} \|B\|_{L^2(\mathbb{R}^N)}. \qquad (3.25)$$

If $u \in H^1(\mathbb{R}^N)^N$, div $u = 0$ a.e. in $\mathbb{R}^N$, we can use this result to deduce

$$\left. \begin{array}{c} \|(u \cdot \nabla) u_i\|_{\mathcal{H}^1(\mathbb{R}^N)} \leq C \|u\|_{L^2(\mathbb{R}^N)} \|\nabla u_i\|_{L^2(\mathbb{R}^N)}, \\ \text{for all } 1 \leq i \leq N, \end{array} \right\} \qquad (3.26)$$

$$\left. \begin{array}{c} \|\partial_i u_j\, \partial_j u_k\|_{\mathcal{H}^1(\mathbb{R}^N)} \leq C \|\nabla u\|_{L^2(\mathbb{R}^N)}^2, \\ \text{for all } 1 \leq i, k \leq N. \end{array} \right\} \qquad (3.27)$$

We then consider a weak solution u of the Navier–Stokes equations (3.1)–
(3.2) (in the case $\Omega = \mathbb{R}^N$) and we recall that we always assume $u_0 \in$
$L^2(\mathbb{R}^N)$, $f \in L^2(0, T; H^{-1}(\mathbb{R}^N))$ (at least) while u satisfies the conditions
listed in Theorems 3.1–3.2. Let us also observe that we can assume without
loss of generality that the force term f satisfies

$$\text{div } f = 0 \quad \text{in} \quad \mathcal{D}'(\mathbb{R}^N \times (0, T)). \qquad (3.28)$$

Indeed, we can always decompose $f = f_1 + \nabla \pi$ where div $f_1 = 0$, $f_1, \nabla \pi \in$
$L^2(0, T; H^{-1}(\mathbb{R}^N))$, incorporate π to the pressure p and replace f by f_1.
Notice that for "most" functional spaces $X = L^p(0, T; H^{s,q}(\mathbb{R}^N))$ ($s \in \mathbb{R}$,
$q \in (1, \infty)$, $p \in [1, \infty]$) if $f \in X$ then $f_1 \in X$.

With this normalization of f, we have

Theorem 3.6. *The following properties hold*

$$\left. \begin{array}{ll} \partial_{ij} p \in L^1(0, T; \mathcal{H}^1(\mathbb{R}^N)) & (1 \leq i, j \leq N), \\ \nabla p \in L^1\left(0, T; L^{\frac{N}{N-1}, 1}(\mathbb{R}^N)\right)^N, & \\ p \in L^1\left(0, T; L^{\frac{N}{N-2}, 1}(\mathbb{R}^N)\right) & \text{if } N \geq 3, \\ p \in L^1(0, T; C_0(\mathbb{R}^N)) & \text{if } N = 2, \end{array} \right\} \qquad (3.29)$$

$$(u \cdot \nabla)u, \ \nabla p \in L^2(0,T; \mathcal{H}^1(\mathbb{R}^N))^N, \tag{3.30}$$

$$\left. \begin{aligned} &u \in L^1(0,T; C_0(\mathbb{R}^N)) \quad \text{if} \quad N = 3 \quad \text{and} \\ &f \in L^1\left(0,T; L^{\frac{3}{2},1}(\mathbb{R}^3)\right), \end{aligned} \right\} \tag{3.31}$$

$$\left. \begin{aligned} &\text{If } Du_0 \text{ is a bounded measure on } \mathbb{R}^N, \\ &\text{and if } Df \text{ is a bounded measure on } \mathbb{R}^N \times [0,T) \\ &\text{then } Du \in L^\infty(0,T; L^1(\mathbb{R}^N)). \end{aligned} \right\} \tag{3.32}$$

Remarks 3.3. 1) The case $N = 2$ of (3.29) for the regularity of p is due to L. Tartar [**470**]. The regularity shown in (3.31) was first obtained by L. Tartar (unpublished), and C. Foias, C. Guillopé and R. Temam [**152**] by different methods. Finally, (3.32) is a slight improvement of a result originally proven by P. Constantin [**99**], where it was shown that curl $u \in L^\infty(0,T; L^1(\mathbb{R}^N))$ if $N = 3$.

2) If $N \geq 4$, (3.31) may be replaced by $u \in L^1(0,T; L^{\frac{N}{N-3},1}(\mathbb{R}^N))$ with a similar proof.

3) Since p is defined up to a constant, (3.29) really means that we normalize p in such a way that p goes to 0 as $|x|$ goes to $+\infty$ (in $L^{\frac{N}{N-2},1}(\mathbb{R}^N)$ sense if $N \geq 3$).

4) If we go back to the above modification of f ($f = f_1 + \nabla \pi$; div $f_1 = 0$; $f_1, \nabla \pi \in L^2(0,T; H^{-1}(\mathbb{R}^N))$), let us observe that (3.32) holds if Df_1 is a bounded measure on $\mathbb{R}^N \times [0,T)$.

5) Theorem 3.6 is stated in [**95**] but the proofs of (3.31)–(3.32) are only sketched there. This is why we give below a complete proof of this result. □

Proof of Theorem 3.6

Step 1. Proof of (3.29)–(3.30). Recall that p satisfies (3.15) and that f satisfies (3.28). Then, using (3.26),(3.27) and the known regularity of u, we deduce immediately that $(u \cdot \nabla)u \in L^2(0,T; \mathcal{H}^1(\mathbb{R}^N))^N$, $\Delta p \in L^1(0,T; \mathcal{H}^1(\mathbb{R}^N))$ and $\Delta p = \partial_i h_i$ in $\mathbb{R}^N$ where $h_i \in L^2(0,T; \mathcal{H}^1(\mathbb{R}^N))^N$ ($1 \leq i \leq N$). Since $\partial_{ij} p = R_i R_j(-\Delta p)$, $\partial_i p = -R_i R_k h_k$ ($1 \leq i,j \leq N$), we deduce from (3.24) that $\partial_{ij} p \in L^1(0,T; \mathcal{H}^1(\mathbb{R}^N))$, $\nabla p \in L^2(0,T; \mathcal{H}^1(\mathbb{R}^N))^N$ ($1 \leq i,j \leq N$). Hence, (3.30) is proven while the rest of (3.29) follows from the regularity of $D^2 p$ and Sobolev embeddings.

Step 2. Proof of (3.31). If $N = 3$, $f \in L^1\left(0,T; L^{\frac{3}{2},1}(\mathbb{R}^3)\right)$, we claim that we have for each $i \in \{1, \ldots, N\}$

$$\frac{\partial u_i}{\partial t} - \nu \Delta u_i \in L^1\left(0,T; L^{\frac{3}{2},1}(\mathbb{R}^3)\right). \tag{3.33}$$

Indeed, in view of (3.29), we have only to show that $(u \cdot \nabla) u_i \in L^1(0, T; L^{\frac{3}{2},1}(\mathbb{R}^3))$, and this follows from Sobolev embeddings since they imply that $u \in L^2(0, T; L^{6,2}(\mathbb{R}^3))^3$ while we have $\nabla u_i \in L^2(0, T; L^2(\mathbb{R}^3))^3$. Next, we remark that the solution $\tilde{u}_i$ of

$$\frac{\partial \tilde{u}_i}{\partial t} - \nu \Delta \tilde{u}_i = 0 \quad \text{in } \mathbb{R}^3 \times (0, T), \quad \tilde{u}_i|_{t=0} = u_i^0 \quad \text{in } \mathbb{R}^3$$

is given by $\tilde{u}_i(t) = u_i^0 * ((4\pi\nu t)^{-3/2} C^{-\frac{|x|^2}{4\nu t}})$. Hence, $\|\tilde{u}_i(t)\|_{L^\infty(\mathbb{R}^3)} \leq \|u_i^0\|_{L^2(\mathbb{R}^3)} \frac{(2\pi\nu t)^{-3/4}}{8}$ and we deduce easily that $\tilde{u}_i \in L^p(0, T; C_0(\mathbb{R}^3))$ for all $1 \leq p < 4/3$. Therefore, (3.31) follows from

Lemma 3.1. *Let $N \geq 3$, let $g \in L^1(0, T; L^{\frac{N}{2},1}(\mathbb{R}^N))$ and let v be the solution of*

$$\frac{\partial v}{\partial t} - \nu \Delta v = g \quad \text{in } \mathbb{R}^N \times (0, T), \quad v|_{t=0} = 0 \quad \text{in } \mathbb{R}^N. \tag{3.34}$$

Then, for almost all $t \in (0, T)$, $v(t) \in C_0(\mathbb{R}^3)$ and $v \in L^1(0, T; C_0(\mathbb{R}^3))$.

Proof of Lemma 3.1. By density, it is enough to show that we can estimate $\|v\|_{L^1(0,T;L^\infty(\mathbb{R}^3))}$ in terms of $\|g\|_{L^1(0,T;L^{\frac{N}{2},1}(\mathbb{R}^N))}$. Using the density of functions piecewise constant in t, we see that it is enough to show such an estimate when $g = 1_{(0,t_0)} h$ where $t_0 \in (0, T)$, $h \in L^{\frac{N}{2},1}(\mathbb{R}^N)$. Then, rescaling (t, x) (i.e. considering $v(\frac{x}{\sqrt{t_0}}, \frac{t}{t_0})$), it is enough to consider the case when $t_0 = 1$ provided we obtain an estimate of v in $L^1(0, \infty; L^\infty(\mathbb{R}^N))$. In addition, replacing h by $h^\sharp$, the Schwarz spherical decreasing rearrangement, we increase, for all $t > 0$, $\|v(t)\|_{L^\infty(\mathbb{R}^N)}$ (see, for instance, C. Bandle [20], A. Alvino, P.L. Lions and G. Trombetti [8],[9]) and thus, without loss of generality, we may assume that v and h are non-negative, spherically symmetric and nonincreasing with respect to $|x|$. Then, $\int_0^\infty \|v(t)\|_{L^\infty(\mathbb{R}^N)} dt = \int_0^\infty v(0, t) dt$.

On the other hand, $V(x) = \int_0^\infty v(x, t) dt$ solves

$$-\nu \Delta V = h \quad \text{in } \mathbb{R}^N \quad \text{or equivalently} \quad V = \frac{c_N}{\nu} \frac{1}{|x|^{N-2}} * h,$$

and $V \in C_0(\mathbb{R}^N)$ since $h \in L^{\frac{N}{2},1}(\mathbb{R}^N)$, $\frac{1}{|x|^{N-2}} \in L^{\frac{N}{N-2},\infty}(\mathbb{R}^N)$. Hence,

$$V(0) = \|V\|_{L^\infty(\mathbb{R}^N)} \leq C\|h\|_{L^{\frac{N}{2},1}(\mathbb{R}^N)} \qquad \left(\text{with } C = \frac{c_N}{\nu}\right)$$

and the proof of the lemma is complete. $\quad\square$

Step 3. Proof of (3.32). We essentially follow an argument introduced by P. Constantin [98]. Formally, we differentiate (3.1) and we obtain for all $1 \le i, k \le N$ on $\mathbb{R}^N \times (0, T)$

$$\frac{\partial}{\partial t}(\partial_k u_i) + (u \cdot \nabla)(\partial_k u_i) - \nu \Delta(\partial_k u_i) = \partial_k f_i - \partial_k \partial_i p - \partial_k u_j \, \partial_j u_i.$$

In view of the assumptions made upon f, (3.29) and the fact that $u \in L^2(0, T; H^1(\mathbb{R}^N))$, the right-hand side of the above equation is a bounded measure on $\mathbb{R}^N \times [0, T)$. Still arguing formally, we deduce

$$\frac{\partial}{\partial t}|\partial_k u_i| + (u \cdot \nabla)|\partial_k u_i| - \nu \Delta|\partial_k u_i| \le m$$

where m is a bounded non-negative measure on $\mathbb{R}^N \times [0, T)$, and integrating over $\mathbb{R}^N \times [0, t]$ we obtain a uniform (in t) bound on $\int_{\mathbb{R}^N} |\partial_k u_i(x, t)| \, dx$.

It only remains to justify the above argument for any weak solution. To this end, we consider, for $h \in (0, 1]$, $i, k \in \{1, \dots, N\}$ fixed, $v_h(x, t) = \frac{1}{h}(u_i(x + he_k, t) - u_i(x, t))$. We have obviously on $\mathbb{R}^N \times (0, T)$

$$\left.\begin{aligned}
\frac{\partial v_h}{\partial t} + u \cdot \nabla v_h - \nu \Delta v_h &= \frac{1}{h}\left(f_i(\cdot + he_k) - f_i - \partial_i p(\cdot + he_k) + \partial_i p\right) \\
&\quad + \frac{1}{h}\left(u_j(\cdot + he_k) - u_j\right) \cdot \partial_j u_i(\cdot + he_k).
\end{aligned}\right\}$$
$$(3.35)$$

Exactly as above, we deduce that the right-hand side, denoted by m_h, is bounded in $L^1(\mathbb{R}^N \times (0, T))$ uniformly in $h \in (0, 1]$. Since $u \in L^2(0, T; H^1(\mathbb{R}^N))$ and $v_h \in L^\infty(0, T; L^2(\mathbb{R}^N)) \cap L^2(0, T; L^{\frac{2N}{N-2}}(\mathbb{R}^N))$, we deduce from Lemma 2.3 (section 2.3) that we have

$$\frac{\partial}{\partial t}(v_h * \omega_\varepsilon) + u \cdot \nabla(v_h * \omega_\varepsilon) - \nu \Delta(v_h * \omega_\varepsilon) = m_h * \omega_\varepsilon + r_h^\varepsilon$$

where $r_h^\varepsilon \to_\varepsilon 0$ in $L^2(0, T; L^1(\mathbb{R}^N)) \cap L^1(0, T; L^{\frac{N}{N-1}}(\mathbb{R}^N))$ for each $h > 0$ fixed. Then, we write, recalling the classical convexity inequality ($-\Delta|f| \le (-\Delta f)\,\mathrm{sign}\, f$ in $\mathcal{D}'$),

$$\frac{\partial}{\partial t}|v_h * \omega_\varepsilon| + (u \cdot \nabla)|v_h * \omega_\varepsilon| - \nu \Delta|v_h * \omega_\varepsilon| \le |m_h * \omega_\varepsilon| + |r_h^\varepsilon|,$$

and we recover, letting ε go to 0_+,

$$\left.\begin{aligned}
\frac{\partial |v_h|}{\partial t} + (u \cdot \nabla)|v_h| - \nu \Delta|v_h| &\le |m_h| \quad \text{in} \quad \mathbb{R}^N \times (0, T), \\
|v_h|\Big|_{t=0} &= w_h^0, \quad |v_h| \in L^2(0, T; H^1(\mathbb{R}^N)) \cap L^\infty(0, T; L^2(\mathbb{R}^N))
\end{aligned}\right\}$$
$$(3.36)$$

where $w_h^0 = \frac{1}{h}\,|u_i^0(\cdot + he_k) - u_i^0| \in L^1(\mathbb{R}^N) \cap L^2(\mathbb{R}^N)$ and is bounded in $L^1(\mathbb{R}^N)$ uniformly in $h \in (0,1]$.

Next, we multiply (3.36) by $\varphi(\frac{\cdot}{n})$ where $\varphi \in C_0^\infty(\mathbb{R}^N)$, $\varphi \equiv 1$ on B_1, $\varphi(x) = 0$ if $|x| \geq 2$, $0 \leq \varphi \leq 1$ on $\mathbb{R}^N$ and we find, integrating over $\mathbb{R}^N \times [0,t]$ for all $t \in [0,T]$, denoting by C various non-negative constants independent of t, n and h,

$$
\int_{\mathbb{R}^N} |v_h(x,t)|\,\varphi\Big(\frac{x}{n}\Big)\,dx \;\leq\; C + \frac{\nu}{n^2} \int_0^t \int_{\mathbb{R}^N} |v_h|\Big(-\Delta\varphi\Big(\frac{x}{n}\Big)\Big)\,dx\,ds
$$
$$
+ \frac{1}{n} \int_0^t \int_{\mathbb{R}^N} |u|\,|v_h|\,\Big|\nabla\varphi\Big(\frac{x}{n}\Big)\Big|\,dx\,ds.
$$

Hence, we have

$$
\sup_{[0,T]} \int_{(|x|\leq n)} |v_h|\,dx \;\leq\; C + \frac{C}{n^2} \int_0^T \int_{(n\leq|x|\leq 2n)} |v_h|\,dx\,ds
$$
$$
+ \frac{C}{n}\, \|u\|_{L^\infty(0,T;L^2)}\, \|v_h\|_{L^\infty(0,T;L^2)}
$$

or

$$
\sup_{[0,T]} \int_{(|x|\leq n)} |v_h|\,dx \;\leq\; \Big(C + \frac{C(h)}{n}\Big) + \frac{C}{n^2} \int_0^T \int_{(n\leq|x|\leq 2n)} |v_h|\,dx\,ds. \tag{3.37}
$$

In particular, we have

$$
\sup_{[0,T]} \int_{(|x|\leq n)} |v_h|\,dx
$$
$$
\leq\; C + \frac{C(h)}{n} + \frac{C}{n^2}\Big(\int_0^T \int_{(n\leq|x|\leq 2n)} |v_h|^2\,dx\,ds\Big)^{1/2} n^{N/2}.
$$

If $N \leq 4$, we deduce that $\sup_{[0,T]} \int_{\mathbb{R}^N} |v_h|\,dx \leq C$. If $N \geq 5$, we deduce that

$$
\sup_{[0,T]} \int_{(|x|\leq n)} |v_h|\,dx \;\leq\; C + C(h)\,n^{\frac{N-4}{2}}
$$

and we insert this bound in the right-hand side of (3.37) to obtain

$$
\sup_{[0,T]} \int_{(|x|\leq n)} |v_h|\,dx \;\leq\; C + \frac{C(h)}{n} + \frac{C}{n^2} + C(h)\,n^{\frac{N-8}{2}}.
$$

If $N \leq 7$, we obtain the same bound as before. If $N \geq 8$, we go back to (3.37).

In conclusion, we have shown

$$\sup_{[0,T]} \int_{\mathbb{R}^N} |v_h|\, dx \leq C. \tag{3.38}$$

We deduce (3.32) from (3.38) letting h go to 0_+: indeed, since $u \in L^2(0,T; H^1(\mathbb{R}^N))^N$, v_h converges to $\partial_k u_i$ in $L^2(\mathbb{R}^N \times (0,T))$ as h goes to 0_+ and (3.32) follows.   □

3.3 Second derivative estimates

In this section we want to present various a priori estimates on weak solutions of Navier–Stokes equations and their second derivatives in x. This will yield similar estimates upon $\frac{\partial u}{\partial t}$. We first consider the case when $\Omega = \mathbb{R}^N$ (and the periodic case) and next we consider the case of Dirichlet boundary conditions which presents specific difficulties already mentioned in section 3.1.

We thus begin with $\Omega = \mathbb{R}^N$ and we mention without further detail that all the results we state and prove below are also valid mutatis mutandis in the periodic case. Also, as explained in the preceding section, we always normalize the force term in such a way that (3.28) (unless explicitly mentioned) holds. Then we observe that in view of the results shown in the preceding section, we have for any weak solution as built in Theorems 2.1–2.2

$$\left. \begin{array}{l} (u \cdot \nabla)u, \quad \nabla p \in L^\alpha(0,T; L^\beta(\mathbb{R}^N)), \\[2mm] L^2(0,T; \mathcal{H}^1(\mathbb{R}^N)), \quad L^1\left(0,T; L^{\frac{N}{N-1},1}(\mathbb{R}^N)\right) \end{array} \right\} \tag{3.39}$$

whenever $1 < \alpha < 2$, $\frac{1}{\beta} = 1 - \frac{2-\alpha}{N\alpha}$. In particular, we can choose $\alpha = \beta = \frac{N+2}{N+1}$, i.e. $\frac{5}{4}$ if $N = 3$.

Next, any weak solution u is the sum of u_1 and u_2 where u_1, u_2 are respectively solutions of

$$\frac{\partial u_1}{\partial t} - \nu \Delta u_1 = 0 \quad \text{in } \mathbb{R}^N \times (0,T),\; u_1|_{t=0} = u_0 \quad \text{in } \mathbb{R}^N \tag{3.40}$$

$$\frac{\partial u_2}{\partial t} - \nu \Delta u_2 = g \quad \text{in } \mathbb{R}^N \times (0,T),\; u_2|_{t=0} = 0 \quad \text{in } \mathbb{R}^N \tag{3.41}$$

where $g = f - (u \cdot \nabla)u - \nabla p$. Of course, u_1 is smooth for $t > 0$ and its global regularity on $\mathbb{R}^N \times [0,T]$ depends on the properties of u_0, and, in view of (3.39), $g \in L^\alpha(0,T; L^\beta(\mathbb{R}^N))$ if $f \in L^\alpha(0,T; L^\beta(\mathbb{R}^N))$. Notice that the reduction to (3.28) leaves invariant this property together with the $L^2(0,T;\mathcal{H}^1(\mathbb{R}^N))$ or the $L^1(0,T;L^{\frac{N}{N-1},1}(\mathbb{R}^N))$ regularity. Therefore, by classical L^p (L^q) estimates on heat equations (see also Appendix D),

$$\frac{\partial u_2}{\partial t}, D_x^2 u_2 \in L^\alpha(0,T;L^\beta(\mathbb{R}^N)) \quad \text{if} \quad f \in L^\alpha(0,T;L^\beta(\mathbb{R}^N)) \quad (3.42)$$

with $1 < \alpha < 2$, $\frac{1}{\beta} = 1 - \frac{2-\alpha}{N\alpha}$. In conclusion, we find

$$\left. \begin{array}{c} \dfrac{\partial u}{\partial t}, D_x^2 u \in L^\alpha(0,T;L^\beta(\mathbb{R}^N)) \quad \text{if} \quad f \in L^\alpha(0,T;L^\beta(\mathbb{R}^N)), \\[2mm] 1 < \alpha < 2, \quad \dfrac{1}{\beta} = 1 - \dfrac{2-\alpha}{N\alpha} \end{array} \right\} \quad (3.43)$$

if u_0 belongs to an appropriate (Besov) space and in particular to $W^{s,\beta}(\mathbb{R}^N)$ where $s = 2\frac{\alpha-1}{\alpha}$ (we still assume that $u_0 \in L^2(\mathbb{R}^N)$), while if $u_0 \in L^\beta(\mathbb{R}^N)$ ($\cap L^2(\mathbb{R}^N)$), we obtain for all $\varepsilon > 0$

$$\left. \begin{array}{c} \dfrac{\partial u}{\partial t}, D_x^2 u \in L^\alpha(\varepsilon,T;L^\beta(\mathbb{R}^N)) \quad \text{if} \quad f \in L^\alpha(0,T;L^\beta(\mathbb{R}^N)), \\[2mm] 1 < \alpha < 2, \quad \dfrac{1}{\beta} = 1 - \dfrac{2-\alpha}{N\alpha}. \end{array} \right\} \quad (3.44)$$

Of course, these results really mean that we have a priori estimates in these norms for weak solutions in addition to such regularity information.

The borderline cases are more delicate: in particular, we do not know if (3.42) holds with $L^\alpha(0,T;L^\beta(\mathbb{R}^N))$ replaced by $L^2(0,T;\mathcal{H}^1(\mathbb{R}^N))$— this does not seem to be known for the heat equation! As we shall see later on, we can prove that $\frac{\partial u}{\partial t}, D_x^2 u \in L^p(0,T;L^1(\mathbb{R}^N))$ for all $p \in [1,2)$. A similar difficulty occurs with $L^1(0,T;L^{\frac{N}{N-2},1}(\mathbb{R}^N))$: however, in that case, using the results of Appendix D, we can deduce conclusions similar to (3.43)–(3.44) replacing $\frac{\partial u}{\partial t}, D_x^2 u \in L^1(0,T;L^{\frac{N}{N-1},1}(\mathbb{R}^N))$ by $\frac{\partial u}{\partial t}, D_x^2 u \in L^{\frac{N}{N-1},1}(\mathbb{R}^N)$ a.e. $t \in (0,T)$ (we already know they belong to $L^p(\mathbb{R}^N)$ for $p < \frac{N}{N-1}$) and $\frac{\partial u}{\partial t}, D_x^2 u \in L^{1,\infty}(0,T;L^{\frac{N}{N-1},1}(\mathbb{R}^N))$ where $a \in L^{1,\infty}(0,T;X)$ means meas $\{\|a(t)\|_X > \lambda\} \leq \frac{C}{\lambda}$ for all $\lambda > 0$, for some $C \in [0,\infty)$.

At this stage, we obtain the conclusions (3.43)–(3.44) and if we insist upon having $\alpha = \beta$ we find $\alpha = \beta = \frac{N+2}{N+1}$. However, as shown by P. Constantin [98], one can obtain a better integrability on $D_x^2 u$ by a different argument: in [98] it was shown that $D_x^2 u \in L^p(\mathbb{R}^N \times (0,T))$ if $p < 4/3$, $N = 3$ (under appropriate conditions on f, u_0). We shall show below that $D_x^2 u \in L^{\frac{4}{3},\infty}(\mathbb{R}^N \times (0,T))$ (for all $N \geq 2$) by a variant of the argument in [89]. Before we even state a precise result, we would like to explain formally the origin of this exponent $\frac{4}{3}$. Differentiating (3.1) (and taking $f = 0$ to simplify) we obtain

$$\frac{\partial}{\partial t}(\partial_k u_i) + (u \cdot \nabla)\,\partial_k u_i - \nu\Delta(\partial_k u_i) \;=\; -(\partial_k u_j\,\partial_j u_i + \partial_{ik} p)$$

and we remark that the right-hand side belongs to $L^1(\mathbb{R}^N \times (0,T))$ since $u \in L^2(0,T;H^1(\mathbb{R}^N))^N$ by the definition of a weak solution and by (3.29)— we could, as in [98], avoid the use of (3.29) by taking the curl of (3.1). The maximal regularity we can deduce from this fact and the preceding equation is $D_x^2(\partial_k u_i) \in L^1(\mathbb{R}^N \times (0,T))$. In fact, even if the convection term $(u \cdot \nabla)$ were not creating additional difficulties, this is not correct since the L^1 maximal regularity is not true for the heat equation, but let us ignore this borderline problem for the sake of the argument. Then we recall that $\partial_k u_i \in L^2(\mathbb{R}^N \times (0,T))$ and thus, by interpolation, $D_x(\partial_k u_i) \in L^p(\mathbb{R}^N \times (0,T))$ where $\frac{1}{p} = \frac{1}{2}\left(\frac{1}{2}\right) + \frac{1}{2}1 = \frac{3}{4}$ and we recover the claim $D_x^2 u \in L^{4/3}(\mathbb{R}^N \times (0,T))$. As we shall see below, this formal (and false!) argument can be almost justified and the only price to be paid is the replacement of $L^{\frac{4}{3}}$ by $L^{\frac{4}{3},\infty}$! Let us also observe by the way that "$D_x^3 u \in L^1$" is a good guess since it also yields "$D_x^2 u \in L^1(0,T;L^{\frac{N}{N-1},1}(\mathbb{R}^N))$)" (Sobolev's embeddings) which is the borderline case of (3.43) (and (3.39)) taking $\alpha = 0$. Also, $\frac{\partial D_x u}{\partial t},\,D_x^3 u \in L^1(\mathbb{R}^N \times (0,T))$ implies that $D_x^2 u \in L^2(0,T;L^1(\mathbb{R}^N))$ which is the other borderline case of (3.43) taking $\alpha = 2$. Notice however that these two (equally false!) deductions do not use the L^2 integrability of Du which seems to be the key to the improvement of the $L^p(\mathbb{R}^N \times (0,T))$ integrability of $D_x^2 u$ from $p = \frac{N+2}{N+1}$ to $p = \frac{4}{3}$ (or almost). Notice finally that only in the case $N = 2$ (the nice case), $\frac{N+2}{N+1}$ coincides with $\frac{4}{3}$.

We now state precisely this integrability result; as in the preceding section and above, u is any weak solution as built in Theorems 3.1–3.2 and f has been normalized to have zero divergence.

Theorem 3.7. *We assume that Du_0, Df are bounded measures respectively on $\mathbb{R}^N$ and $\mathbb{R}^N \times [0,T]$. Then, we have*

$$\left.\begin{aligned}
&D_x^2 u \,\in\, L^{\frac{4}{3},\infty}(\mathbb{R}^N \times (0,T)) \quad \text{and} \\[2mm]
&\sup_{R>0} \frac{1}{R}\int_{\mathbb{R}^N}\!\!\int_0^T \sum_{i,k=1}^N |\nabla\partial_k u_i|^2\,1_{(|\partial_k u_i|<R)}\,dx\,dt \;<\; \infty.
\end{aligned}\right\} \tag{3.45}$$

Remarks 3.4. 1) Taking the curl instead of an arbitrary first derivative of (3.2) in the proof below, we obtain (3.45) with $|D^2 u|^2\,1_{(|Du|<R)}$ replaced by $|\nabla\,\mathrm{curl}\,u|^2\,1_{(|\mathrm{curl}\,u|<R)}$ assuming only that $\mathrm{curl}\,u_0$ and $\mathrm{curl}\,f$ (before or after normalization) are bounded measures.

2) As we shall see below, the proof of Theorem 3.7 also shows that

$$\Delta v \in \mathcal{M}_b(\mathbb{R}^N),\ v \in L^2(\mathbb{R}^N)\ \Rightarrow\ Dv \in L^{\frac{4}{3},\infty}(\mathbb{R}^N) \qquad (3.46)$$

$$\left.\begin{array}{l} \dfrac{\partial v}{\partial t} - \dfrac{1}{2}\Delta v \in \mathcal{M}_b(\mathbb{R}^N \times [0,T]),\ v|_{t=0} \in \mathcal{M}_b(\mathbb{R}^N), \\[2mm] v \in L^2(\mathbb{R}^N \times (0,T))\ \Rightarrow\ Dv \in L^{\frac{4}{3},\infty}(\mathbb{R}^N \times (0,T)). \end{array}\right\} \qquad (3.47)$$

Here and below $\mathcal{M}_b$ denotes the space of bounded measures (with a norm denoted by $\|\cdot\|_m$). Even though we do not know if in Theorem 3.7 $L^{\frac{4}{3},\infty}$ can be replaced by $L^{4/3}$, we can check that such an extension is not true for (3.46) or (3.47). Indeed, we claim that $Dv \in L^{4/3}(\mathbb{R}^N)$ is not true in general since, if it were, we would have estimates of Dv in $L^{4/3}$ in terms of $\|\Delta v\|_m + \|v\|_{L^2}$ or $\|\frac{\partial v}{\partial t} - \frac{1}{2}\Delta v\|_m + \|v(0)\|_m + \|v\|_{L^2}$, and, if we choose for (3.46), $N = 4$, $v_\varepsilon(x) = \min\left(\frac{1}{|x|^2}, \frac{1}{\varepsilon^2}\right)\zeta(x)$ where $\zeta \in C_0^\infty(\mathbb{R}^N)$, $\zeta(x) \equiv 1$ if $|x| \le 1$, $\varepsilon \in (0,1)$, we check easily that $\|\Delta v_\varepsilon\|_m$ is bounded while $\|v_\varepsilon\|_{L^2} = \left(|S^3|\,|\log\varepsilon| + C\right)^{1/2}$ and $\|\nabla v_\varepsilon\|_{L^{4/3}} = \left(|S^3|\,2^{4/3}|\log\varepsilon| + C\right)^{3/4}$, and we reach a contradiction. In the case of (3.47), we choose $v_\varepsilon = (2\pi(t+\varepsilon))^{-N/2}\exp-\frac{|x|^2}{2(t+\varepsilon)}$ for $t \in [0,T]$, $x \in \mathbb{R}^N$, $N = 2$, $T \in (0,\infty)$, $\varepsilon \in (0,1)$. Obviously, $\frac{\partial v_\varepsilon}{\partial t} - \frac{1}{2}\Delta v_\varepsilon \equiv 0$, $\|v_\varepsilon(0)\|_m = 1$ and

$$\|v_\varepsilon\|_{L^2(\mathbb{R}^2 \times (0,T))} = \left(\int_0^T (4\pi)^{-1}(t+\varepsilon)^{-2}(t+\varepsilon)dt\right)^{1/2} = C\left(\log\left(1+\frac{T}{\varepsilon}\right)\right)^{1/2}.$$

On the other hand, we have

$$\|Dv_\varepsilon\|_{L^{\frac{4}{3}}(\mathbb{R}^2 \times (0,T))} = C\left(\int_0^T (t+\varepsilon)^{-1}\,dt\right)^{3/4} = C\left(\log\left(1+\frac{T}{\varepsilon}\right)\right)^{3/4}$$

and we reach a contradiction. $\square$

Proof of Theorem 3.7

Step 1. We prove

$$\left.\begin{array}{c} \displaystyle\sup_{R>0} \frac{1}{R}\int_{\mathbb{R}^N}\int_0^T |\nabla\partial_k u_i|^2\,1_{(|\partial_k u_i|<R)}\,dt\,dx < \infty, \\[3mm] \text{for all }\ 1 \le i,k \le N. \end{array}\right\} \qquad (3.48)$$

Let us first explain formally the proof of this estimate (directly related to the idea of renormalized solutions for elliptic and parabolic equations—see for instance R.J. DiPerna and P.L. Lions [128], P.L. Lions and F. Murat

[**308**] and Appendix E). First of all, we differentiate (3.1) and we find for all $1 \leq i, k \leq N$

$$\frac{\partial}{\partial t}(\partial_k u_i) + u \cdot \nabla(\partial_k u_i) - \nu\Delta(\partial_k u_i) = \partial_k f_i - \partial_{ki}^2 p - \partial_k u_j \, \partial_j u_i. \quad (3.49)$$

In view of Theorem 3.6, of the properties of weak solutions and of the assumption made upon f, we see that the right-hand side is a bounded measure on $\mathbb{R}^N \times (0, T)$. Then, we multiply (3.49) by $T_R(\partial_k u_i)$ where $T_R(z) = z$ if $|z| \leq R$, $= R$ if $z \geq R$, $= -R$ if $z \leq -R$, and we find, integrating by parts over $\mathbb{R}^N$,

$$\frac{d}{dt}\int_{\mathbb{R}^N} S_R(\partial_k u_i)\,dx + \int_{\mathbb{R}^N}(u \cdot \nabla)\,S_R(\partial_k u_i)\,dx$$

$$+ \nu \int_{\mathbb{R}^N} |\nabla\partial_k u_i|^2 \, 1_{(|\partial_k u_i| < R)}\,dx \;\leq\; C\,R$$

since $|T_R(z)| \leq R$ for all $z \in \mathbb{R}$, where $S_R(z) = \frac{|z|^2}{2}$ if $|z| \leq R$, $R|z| - \frac{R^2}{2}$ if $|z| \geq R$. Hence,

$$\int_0^T \int_{\mathbb{R}^N} |\nabla\partial_k u_i|^2 \, 1_{(|\partial_k u_i| \leq R)}\,dx\,dt \;\leq\; CR + \int_{\mathbb{R}^N} S_R(\partial_k u_i(0))\,dx \;\leq\; CR,$$

since $0 \leq S_R(z) \leq R|z|$ for all $z \in \mathbb{R}$.

In order to justify this computation, we argue as in the proof of Theorem 3.6 and we find, defining $v_h(x, t) = \frac{1}{h}\left(u_i(x+he_k, t) - u_i(x, t)\right)$ (i, k fixed in $\{1, \dots, N\}$, $h \in (0, 1)$),

$$\left.\begin{aligned}
\frac{\partial v_h}{\partial t} + u \cdot \nabla v_h - \nu\Delta v_h &= g_h \qquad \text{bounded in} \quad L^1(\mathbb{R}^N \times (0, T)), \\
v_h|_{t=0} &= v_h^0 \qquad \text{bounded in} \quad L^1(\mathbb{R}^N).
\end{aligned}\right\}$$

We then multiply by $T_R(v_h)$ justifying the computations exactly as in the proof of Theorem 3.6 and we obtain finally

$$\int_0^T \int_{\mathbb{R}^N} |\nabla T_R(v_h)|^2 \, dx\,dt \;=\; \int_0^T \int_{\mathbb{R}^N} |\nabla v_h|^2 \, 1_{(|v_h| < R)}\,dx\,dt \;\leq\; C\,R$$

$$(3.50)$$

for some $C \geq 0$ independent of $R \in (0, \infty)$ and $h \in (0, 1)$. The only new point to check is the fact that we have for each h, R

$$\int_0^T \int_{\mathbb{R}^N} |u|\,|S_R(v_h)|\,1_{(n \leq |x| \leq 2n)}\,dx\,dt \;\to\; 0 \qquad \text{as} \quad n \to +\infty,$$

and this is immediate since $u, v_h \in L^2(\mathbb{R}^N \times (0, T))$.

Once (3.50) is established, we conclude easily upon letting h go to 0. Indeed, we then deduce that $T_R(\partial_k u_i) \in L^2(0, T; H^1(\mathbb{R}^N))$ and

$$\left.\begin{array}{l} \displaystyle\int_0^T \int_{\mathbb{R}^N} |\nabla T_R(\partial_k u_i)|^2 \, dx \, dt \leq C\,R, \\[4mm] \text{for all} \quad R \in (0, \infty),\ 1 \leq i, k \leq N. \end{array}\right\}$$

Since, as shown above, $D^2 u \in L^q_{loc}(\mathbb{R}^N \times (0, T))$ for some $q > 1$ (observe indeed that since $f \in \mathcal{M}_b(\mathbb{R}^N \times (0, T))$ and $f \in L^2(0, T; H^{-1}(\mathbb{R}^N))$, $f \in L^p(0, T; L^{4/3}(\mathbb{R}^N))$ for all $p < 4/3$), we conclude since $\nabla T_R(\partial_k u_i) = \nabla \partial_k u_i\, 1_{(|\partial_k u_i| < R)}$ a.e. on $\mathbb{R}^N \times (0, T)$.

Step 2. We prove that $D^2_x u \in L^{\frac{4}{3}, \infty}(\mathbb{R}^N \times (0, T))$. For each $(i, k) \in \{1, \dots, N\}^2$ fixed, we set $v = \partial_k u_i$ and we recall that $v \in L^2((\mathbb{R}^N \times (0, T))$ while, by step 1,

$$\sup_{R > 0} \frac{1}{R} \int_{\mathbb{R}^N} dx \int_0^T dt\, |\nabla v|^2\, 1_{(|v| < R)} < \infty.$$

We then decompose ∇v as follows for each $R \in (0, \infty)$

$$\left.\begin{array}{l} |\nabla v| = |\nabla v|\, 1_{(|v| < R)} + |\nabla v|\, 1_{(|v| \geq R)}, \\[2mm] \big\|\nabla v\, 1_{(|v| < R)}\big\|_{L^2(\mathbb{R}^N \times (0, T))} \leq C\, R^{1/2}. \end{array}\right\} \qquad (3.51)$$

Next, we wish to estimate $|\nabla v|\, 1_{(|v| > R)}$. In order to do so, we write

$$\int_{\mathbb{R}^N} \int_0^T |\nabla v|\, 1_{(|v| \geq R)} \, dx \, dt$$

$$= \sum_{j=0}^\infty \int_{\mathbb{R}^N} \int_0^T |\nabla v|\, 1_{(R2^j \leq |v| < R2^{j+1})} \, dx \, dt$$

$$\leq \sum_{j=0}^\infty C(R2^j)^{1/2} \operatorname{meas}\{(x, t) \in \mathbb{R}^N \times (0, T) \,/$$

$$R2^j \leq |v(x, t)| < R2^{j+1}\}^{1/2}$$

$$\leq C\left(\sum_{j=0}^\infty (R2^j)^2 \operatorname{meas}\{(x, t) \in \mathbb{R}^N \times (0, T) \,/\right.$$

$$\left. R2^j \leq |v(x, t)| < R2^{j+1}\}\right)^{1/2} \left(\sum_{j=0}^\infty (R2^j)^{-1}\right)^{1/2}$$

$$\leq C\, R^{-1/2}\left(\int_{\mathbb{R}^N} \int_0^T 1_{(R \leq |v|)}\, |v|^2 \, dx \, dt\right)^{1/2} \leq C\, R^{-1/2}.$$

This estimate, combined with (3.51), yields

$$|\nabla v| = d_1 + d_2, \quad \|d_1\|_{L^2(\mathbb{R}^N \times (0,T))} \leq CR^{1/2}, \atop \|d_2\|_{L^1(\mathbb{R}^N \times (0,T))} \leq CR^{-1/2}. \Bigg\}$$

This is enough to complete the proof of (3.45) and of Theorem 3.7. Indeed, we have for all $\lambda > 0$

$$\text{meas}\{|\nabla v| > \lambda\} \leq \text{meas}\left\{d_1 > \frac{\lambda}{2}\right\} + \text{meas}\left\{d_2 > \frac{\lambda}{2}\right\}$$

$$\leq C\left(\frac{R}{\lambda^2} + \frac{1}{R^{1/2}\lambda}\right) = C\lambda^{-4/3} \quad \text{if we choose } R = \lambda^{2/3}. \quad \square$$

The estimate (3.45) and the argument used in step 2 above can be used to derive some regularity information on $\frac{\partial u}{\partial t}$ and $D_x^2 u$ in $L^p(0,T;L^1(\mathbb{R}^N))$ for all $p \in [1,2)$ which correspond, roughly speaking, to the regularity of $(u \cdot \nabla)u$ and ∇p contained in (3.39), namely $(u \cdot \nabla)u$, $\nabla p \in L^2(0,T; \mathcal{H}^1(\mathbb{R}^N))$. More precisely, we have

Theorem 3.8. *We assume that Du_0, Df are bounded measures on $\mathbb{R}^N$, $\mathbb{R}^N \times [0,T]$ respectively and that $f \in L^p(0,T;L^1(\mathbb{R}^N))$ for all $p \in [1,2)$. Then, we have*

$$D_x^2 u, \frac{\partial u}{\partial t} \in L^p(0,T;L^1(\mathbb{R}^N)) \qquad \text{for all} \quad p \in [1,2). \tag{3.52}$$

Remark 3.5. If we do not normalize f (to have zero divergence), we need to assume that $\nabla(-\Delta)^{-1} \operatorname{div} f$ is a bounded measure on $\mathbb{R}^N \times [0,T]$ in addition to the assumptions made in the above theorem. $\quad \square$

Proof of Theorem 3.8. We use the estimate shown in step 1 of the proof of Theorem 3.7 and the fact (shown in Theorem 3.6) that $Du \in L^\infty(0,T;L^1(\mathbb{R}^N))$ while, of course, $Du \in L^2(\mathbb{R}^N \times (0,T))$. Therefore, $Du \in L^q(0,T;L^{\frac{q}{q-1}}(\mathbb{R}^N))$ for $2 \leq q \leq \infty$. Next, we write with the notation of the proof of step 2 of Theorem 3.6

$$\int_{\mathbb{R}^N} |\nabla v|\, dx = \sum_{j=0}^\infty \int_{\mathbb{R}^N} |\nabla v|\, 1_{(2^j \leq |v| < 2^{j+1})}\, dx + \int_{(|v|<1)} |\nabla v|\, dx$$

$$\leq \sum_{j=0}^\infty f_j\, \alpha_j^{1/2} + \underline{f}$$

where

$$f_j(t) = \left(\int_{\mathbb{R}^N} |\nabla v|^2 \, 1_{(2^j \le |v| < 2^{j+1})} dx \right)^{1/2},$$

$$\underline{f}(t) = \left(\int_{\mathbb{R}^N} |\nabla v|^2 \, 1_{(|v|<1)} \, dx \right)^{1/2},$$

$$\alpha_j(t) = \text{meas} \left\{ x \in \mathbb{R}^N \, / \, 2^j \le |v(x,t)| < 2^{j+1} \right\}$$

for all $j \ge 0$, a.e. $t \in (0,T)$.

In view of (3.45), we have for some $C \ge 0$ independent of j, t

$$\|f_j\|_{L^2(0,T)} \le C \, 2^{j/2}, \quad \|\underline{f}\|_{L^2(0,T)} \le C. \tag{3.53}$$

On the other hand, we recalled above that $v \in L^q(0, T; L^{\frac{q}{q-1}}(\mathbb{R}^N))$ for $2 \le q \le \infty$. Hence, choosing $q < \infty$ ($q \ge 2$), we deduce

$$\sum_{j=0}^{\infty} \alpha_j \, 2^{j \frac{q}{q-1}} \in L^{q-1}(0,T). \tag{3.54}$$

Using (3.53) and (3.54), we may then write

$$\int_{\mathbb{R}^N} |\nabla v| \, dx \le \underline{f} + \left(\sum_{j=0}^{\infty} f_j^2 \, 2^{-j \frac{q}{q-1}} \right)^{1/2} \left(\sum_{j=0}^{\infty} \alpha_j \, 2^{j \frac{q}{q-1}} \right)^{1/2}$$

and $\underline{f} \in L^2(0,T)$, $\left(\sum_{j=0}^{\infty} \alpha_j \, 2^{j \frac{q}{q-1}} \right)^{1/2} \in L^{2(q-1)}(0,T)$, while $\left(\sum_{j=0}^{\infty} f_j^2 \right.$
$\left. 2^{-j \frac{q}{q-1}} \right)^{1/2} \in L^2(0,T)$ since we obviously have, because of (3.53),

$$\int_0^T \sum_{j=0}^{\infty} f_j^2 \, 2^{-j \frac{q}{q-1}} \, dt \le C \sum_{j=0}^{\infty} 2^j \, 2^{-j \frac{q}{q-1}}$$

$$= C \sum_{j=0}^{\infty} 2^{-\frac{j}{q-1}} = C \left(1 - 2^{-\frac{1}{q-1}} \right)^{-1}.$$

We have thus shown that ∇v, and thus $D^2 u$, $\in L^{2\frac{q-1}{q}}(0, T; L^1(\mathbb{R}^N))$. Since q is arbitrary in $[2, \infty)$, we have proven Theorem 3.8 for $D_x^2 u$. The claim for $\frac{\partial u}{\partial t}$ then follows from equation (3.1) and (3.30). $\square$

Remark 3.6. The method introduced above can be used (and modified slightly) to build global weak solutions for two-dimensional Navier–Stokes equations ($N = 2$, $\Omega = \mathbb{R}^N$ for example or the periodic case) when the initial data u_0 satisfies: $u_0 \in L^{2,\infty}(\mathbb{R}^2)$, $\omega_0 = \text{curl} \, u_0 \in \mathcal{M}_b(\mathbb{R}^2)$. Indeed, one only needs to obtain appropriate a priori estimates and the existence follows upon regularizing u_0 and passing to the limit. These a priori estimates can be obtained as follows: first of all, $\omega = \operatorname{curl} u$ should solve

$$\frac{\partial \omega}{\partial t} + u \cdot \nabla \omega - \nu \Delta \omega = 0 \quad \text{in } \mathbb{R}^2 \times (0, \infty), \quad \omega|_{t=0} = \omega_0 \quad \text{in } \mathbb{R}^2 \quad (3.55)$$

(we take $f = 0$ to simplify the presentation). Indeed, if $N = 2$, $\operatorname{curl}((u \cdot \nabla)u) = (u \cdot \nabla)\operatorname{curl} u$ since $\operatorname{div} u = 0$. Next, using symmetrization results due to A. Alvino, P.L. Lions and G. Trombetti [8],[9], we deduce that for each $t > 0$, $\omega(t)$ is "dominated" by $\overline{\omega}(t) = \omega_0^\sharp * e^{-\frac{|x|^2}{4\pi\nu t}}(4\pi\nu t)^{-1}$ where $\omega_0^\sharp$ is the (Schwarz) spherically symmetric decreasing rearrangement of ω_0 and domination means in particular that all L^p and Orlicz norms of $\omega(t)$ are less than $\overline{\omega}(t)$. Therefore, $\omega \in L^{2,\infty}(0, \infty; L^2(\mathbb{R}^2)) \cap L^\infty(0, \infty; L^1(\mathbb{R}^2))$ and $\|\omega(t)\|_{L^\infty(\mathbb{R}^2)} \leq \frac{1}{4\pi\nu t}\|\omega_0\|_{M_b}$. In particular, we deduce that $u \in L^\infty(0, \infty; L^{2,\infty}(\mathbb{R}^2))$. In addition, (3.55) yields as in the proof of Theorem 3.7

$$\frac{1}{R}\int_0^\infty dt \int_{\mathbb{R}^2} dx \, |\nabla\omega|^2 \, 1_{(|\omega|<R)} \leq \frac{1}{\nu}\|m_0\|_m, \quad \text{for all } R > 0. \quad (3.56)$$

Next, we estimate, in a manner similar to the proof of Theorem 3.7, for all $\lambda > 0$, $R > 0$

$$\text{meas} \{|D\omega| > \lambda, \, |\omega| \geq R\}$$

$$= \sum_{j=0}^\infty \text{meas} \{|D\omega| > \lambda, \, 2^j R \leq |\omega| < 2^{j+1}R\}$$

$$\leq \sum_{j=0}^\infty \text{meas} \{|D\omega| > \lambda, \, |\omega| < 2^{j+1}R\}^{1/2} \, \text{meas} \{|\omega| \geq 2^j R\}^{1/2}$$

$$\leq C \sum_{j=0}^\infty \frac{(R2^{j+1})^{1/2}}{\lambda} \, (R2^j)^{-1} \leq C R^{-1/2}\lambda^{-1}$$

in view of (3.56) and the fact that $\omega \in L^{2,\infty}(\mathbb{R}^2 \times (0, \infty))$. Therefore, we deduce for all $\lambda > 0$

$$\text{meas} \{|D\omega| > \lambda\}$$

$$= \text{meas} \{|D\omega| > \lambda, \, |\omega| < R\} + \text{meas} \{|D\omega| > \lambda, \, |\omega| \geq R\}$$

$$\leq C\left\{\frac{R}{\lambda^2} + \frac{1}{\lambda R^{1/2}}\right\} = C\lambda^{-4/3} \quad \text{choosing } R = \lambda^{1/2}.$$

Hence, $\nabla\omega \in L^{\frac{4}{3},\infty}(\mathbb{R}^2 \times (0, \infty))$ and thus $D^2 u \in L^{\frac{4}{3},\infty}(\mathbb{R}^2 \times (0, \infty))$. Even if more estimates can be derived for u, Du, $D^2 u$, $\frac{\partial u}{\partial t}$, p, the estimates listed above are already enough to pass to the limit in regularized problems and build a solution. □

We now conclude this section by considering in more detail the case of Dirichlet boundary conditions. First of all, we take $\alpha \in (1,2)$ and set $\frac{1}{\beta} = 1 - \frac{2-\alpha}{N\alpha}$. If we assume that $f \in L^\alpha(0,T;L^\beta(\Omega))$—and we always normalize it to satisfy div $f = 0$ in $\mathcal{D}'(\Omega \times (0,T))$—then by L^p-regularity results for linear Stokes equations (see V.A. Solonnikov [444],[445] for instance) and extensions due to Y. Giga and H. Sohr [183] about $L^p_t(L^q_x)$ regularity, we deduce immediately

$$\frac{\partial u}{\partial t}, \, D^2_x u, \, \nabla p \; \in \; L^\alpha(\varepsilon,T;L^\beta(\Omega)) \tag{3.57}$$

for all $\varepsilon > 0$ and we can take $\varepsilon = 0$ if $u_0 \in W^{2(1-\frac{1}{\alpha}),\beta}(\Omega)$. We next decompose p as follows: $p = p_0 + p_1$ where p_0, p_1 satisfy

$$\left. \begin{array}{l} -\Delta p_0 = 0 \quad \text{in } \Omega, \quad \int_\Omega p_0 \, dx = 0 \quad \text{a.e. } t \in (0,T), \\[2mm] -\Delta p_1 = \partial_i u_j \, \partial_j u_i \quad \text{in } \Omega, \quad p_1 = 0 \quad \text{on } \partial\Omega. \end{array} \right\} \tag{3.58}$$

Observe that since $(u \cdot \nabla)u \in L^\alpha(0,T;L^\beta(\Omega))$ (and $\partial_i u_j \, \partial_j u_i = \text{div}\,((u \cdot \nabla)u))$, $\nabla p_1 \in L^\alpha(0,T;L^\beta(\Omega))$ and thus $\nabla p_0 \in L^\alpha(\varepsilon,T;L^\beta(\Omega))$. From now on, in order to simplify the presentation we take $\varepsilon = 0$. Therefore, we deduce from (3.58) and using the arguments developed in the preceding section

$$\nabla p_1 \in L^\alpha(0,T;L^\beta(\Omega)), \quad \nabla p_0 \in L^\alpha(0,T;L^\beta(\Omega)), \tag{3.59}$$

$$\left. \begin{array}{l} D^2 p_1 \in L^1(0,T;\mathcal{H}^1(\Omega)), \\[2mm] \nabla p_1 \in L^1\left(0,T;L^{\frac{N}{N-1},1}(\Omega)\right) \cap L^2(0,T;\mathcal{H}^1(\Omega)), \\[2mm] p_1 \in L^1\left(0,T;L^{\frac{N}{N-2},1}(\Omega)\right), \\[2mm] (u \cdot \nabla)u \in L^2(0,T;\mathcal{H}^1(\Omega)) \cap L^1\left(0,T;L^{\frac{N}{N-1},1}(\mathbb{R}^N)\right), \end{array} \right\} \tag{3.60}$$

(if $N = 2$, we replace $L^{\frac{N}{N-2},1}(\Omega)$ by $C(\overline{\Omega})$)

$$p_0 \in L^\alpha(0,T;H^k_{\text{loc}}(\Omega)) \qquad \text{for all} \quad k \geq 0. \tag{3.61}$$

Here and everywhere below, whenever we write $L^p(X_{\text{loc}}(\Omega))$ for some function space X we mean $L^p(X(K))$ for any relatively compact subdomain K of Ω such that $\overline{K} \subset \Omega$.

Next, we claim that if Df, Du_0 are bounded measures respectively on $\Omega \times [0, T)$, Ω, then we have

$$Du \in L^\infty(0, T; L^1_{\text{loc}}(\Omega)), \tag{3.62}$$

$$D^2u \in L^p(0, T; L^1_{\text{loc}}(\Omega)) \qquad \text{for all} \quad p < 2, \tag{3.63}$$

$$\left. \begin{aligned} &\sup_{R>0} \frac{1}{R} \int_0^T dt \int_K dx \sum_{i,k=1}^{N} |\nabla \partial_k u_i|^2 \, 1_{(|\partial_k u_i| < R)} \ < \ \infty, \\ &\text{for all compact sets} \quad K \subset \Omega, \end{aligned} \right\} \tag{3.64}$$

$$D^2u \in L^{\frac{4}{3}, \infty}\left(0, T; L^{\frac{4}{3}, \infty}_{\text{loc}}(\Omega)\right). \tag{3.65}$$

Let us recall before explaining the proof of all these estimates that (3.62) is the analogue of the estimate (3.32) (Theorem 3.6), while (3.64)–(3.65) correspond to Theorem 3.7. Finally, (3.63) is the analogue of (3.52); we do not know if a similar estimate holds for $\frac{\partial u}{\partial t}$. If we go back to the proofs made in the case when $\Omega = \mathbb{R}^N$, we see that we considered in each case $v_h = \frac{1}{h}(u_i(x + he_k, t) - u_i(x, t))$ for $h \in (0, h_0)$ (where $x \in \Omega_{h_0} = \{x \in \Omega, \ \text{dist}\,(x, \partial\Omega) > h_0\}$, $t \in [0, T]$) which satisfies

$$\left. \begin{aligned} &\frac{\partial v_h}{\partial t} + (u \cdot \nabla)v_h - \nu \Delta v_h \ = \ m_h, \\ &m_h \quad \text{is bounded in} \quad L^1(0, T; L^1_{\text{loc}}(\Omega)) \end{aligned} \right\} \tag{3.66}$$

and as in the proof of Theorem 3.6

$$\frac{\partial |v_h|}{\partial t} + (u \cdot \nabla)|v_h| - \nu \Delta |v_h| \ \le \ |m_h|. \tag{3.67}$$

Then, if we fix a compact set $K \subset \Omega$, we choose $\varphi \in C_0^\infty(\Omega)$ and h_0 in such a way that $0 \le \varphi \le 1$ in Ω, $\varphi \equiv 1$ on K, $\text{Supp}\,\varphi \subset \Omega_{h_0}$. We may next multiply (3.67) by φ and we find

$$\int_\Omega |v_h| \varphi \, dx(t) \ \le \ C + \int_0^T \int_\Omega |u|\,|v_h|\,|\nabla\varphi| \, dx \, dt$$

$$\le \ C + C \int_0^T \int_\Omega |u|^2 + |\nabla u|^2 \, dx \, dt \ \le \ C$$

and (3.62) follows.

Similarly, (3.64) follows upon multiplying (3.66) by $\varphi\, T_R(v_h)$. We then obtain as in the proof of Theorem 3.7

$$\int_0^T \int_\Omega \varphi |\nabla v_h|^2\, 1_{(|v_h|<R)}\, dx\, dt$$

$$\leq CR + \left| \int_0^T \int_\Omega \nabla v_h T_R(v_h) \cdot \nabla \varphi\, dx\, dt \right| + \int_0^T \int_\Omega |u|\, |\nabla\varphi|\, |S_R(v_h)|\, dx\, dt$$

$$\leq CR + \left| \int_0^T \int_\Omega S_R(v_h)\Delta\varphi\, dx\, dt \right| + R \int_0^T \int_\Omega |u|\, |\nabla\varphi|\, |v_h|\, dx\, dt$$

$$\leq CR.$$

Once (3.64) is established, (3.65) follows exactly as in the proof of Theorem 3.7.

Finally, we claim that if $f \in L^1(0,T; L_{\text{loc}}^{\frac{N}{N-1},1}(\Omega))$, then

$$D^2 u, \frac{\partial u}{\partial t} \in L^{1,\infty}\left(0,T; L_{\text{loc}}^{\frac{N}{N-1},1}(\Omega)\right), \tag{3.68}$$

$$u \in L^1(0,T; C_{\text{loc}}(\Omega)) \qquad \text{if } N = 3. \tag{3.69}$$

Indeed, we observe that we have, in view of (3.60),

$$\frac{\partial u}{\partial t} - \nu\Delta u = g \in L^1\left(0,T; L_{\text{loc}}^{\frac{N}{N-1},1}(\Omega)\right).$$

In particular, we deduce

$$\left(\frac{\partial}{\partial t} - \nu\Delta\right)(u\varphi) = g\varphi - 2\nu\nabla u \cdot \nabla\varphi - u\Delta\varphi \qquad \text{in } \mathbb{R}^N \times (0,T)$$

and $g\varphi - 2\nu\nabla u \cdot \nabla\varphi - u\Delta\varphi \in L^1(0,T; L^{\frac{N}{N-1},1}(\mathbb{R}^N))$. Therefore, (3.68) follows from the result shown in Appendix D while (3.69) is a consequence of Lemma 3.1.

Let us summarize the regularity information we have obtained with the

Theorem 3.9. *Let* $f \in L^2(0,T; H^{-1}(\Omega)) \cap L^\alpha(0,T; L^\beta(\Omega)) \cap L^1(0,T; L^{\frac{N}{N-1},1}(\mathbb{R}^N))$ *for* $1 < \alpha < 2$ *(with* $\frac{1}{\beta} = 1 - \frac{2-\alpha}{N\alpha}$*) satisfy:* div $f = 0$ *in* $\mathcal{D}'(\Omega \times (0,T))$, Df *is a bounded measure on* $\Omega \times [0,T)$. *Let* $u_0 \in W^{2(1-\frac{1}{\alpha}),\beta}(\Omega) \cap L^2(\Omega)$ *for* $1 < \alpha < 2$ *satisfy:* div $u_0 = 0$ *in* $\mathcal{D}'(\Omega)$, $u_0 \cdot n = 0$ *on* $\partial\Omega$, Du_0 *is a bounded measure on* Ω. *Then, there exists a weak solution* u *(or* (u,p)*) as in Theorems 3.3–3.4 satisfying in addition (3.57) (with* $\varepsilon = 0$*), (3.59)–(3.61), (3.62)–(3.65), (3.68)–(3.69) and the local energy inequality (3.10) if* $N = 3$.

The only new information is the local inequality (3.10) if $N = 3$ which was first obtained by L. Caffarelli, R. Kohn and L. Nirenberg [77]. Let us emphasize that contrary to *all* the other information listed in Theorem 3.9, we do not know if *any* weak solution satisfies (3.10). But we can build at least one that satisfies (3.10). In fact, the existence procedure taken from chapter 2 and recalled in section 3.1 (see (3.16)–(3.17) in the case $\Omega = \mathbb{R}^N$) yields such a solution. Indeed, one can check without any difficulty that all the estimates derived above hold uniformly: in particular, ∇p_ε is bounded in $L^{\frac{5}{4}}(\Omega \times (0,T))$ (take $\alpha = \frac{5}{4}$ and thus $\beta = \frac{5}{4}$) and this is enough to derive (3.10). Indeed, if we compare with the proof of Theorem 3.2 (step 3), we have only to explain how to pass to the limit in $u_\varepsilon p_\varepsilon$. The preceding bound shows that—normalizing p_ε such that $\int_\Omega p_\varepsilon \, dx = 0$ a.e. $t \in (0,T)$—p_ε converges weakly to some p in $L^{\frac{5}{4}}(0,T; L^{15/7}(\Omega))$ while we already know that u_ε converges to u (a weak solution) in $L^5(0,T; L^q(\Omega))$ for $q < \frac{30}{11}$ and we conclude since $\frac{11}{30} + \frac{7}{15} = \frac{5}{6} < 1$.

3.4 Temperature and the Rayleigh–Benard equations

In this section, we study the Navier–Stokes equations ((3.1)–(3.2)) complemented with an equation for the internal energy e (or equivalently for the temperature T), namely (1.41) written under the assumption that ρ is constant, namely

$$\frac{\partial e}{\partial t} + \operatorname{div}(ue) - \operatorname{div}(k\nabla\theta) = \frac{\nu}{2}(\partial_i u_j + \partial_j u_i)^2 \quad \text{in } \Omega \times (0,T). \quad (3.70)$$

In order to simplify the presentation, we shall consider here only the case when we have

$$e = C_0\theta, \quad C_0 > 0, \quad k \in (0,\infty). \quad (3.71)$$

In that case, (3.70) reduces to

$$\frac{\partial \theta}{\partial t} + \operatorname{div}(u\theta) - \alpha\Delta\theta = \frac{\nu}{2C_0}(\partial_i u_j + \partial_j u_i)^2 \quad \text{in } \Omega \times (0,T) \quad (3.72)$$

where $\alpha = \frac{k}{C_0} \in (0,\infty)$. In the case of Dirichlet boundary conditions (for the Navier–Stokes part of the system of equations), various boundary conditions are possible like, for instance, Neumann boundary conditions

$$\frac{\partial \theta}{\partial n} = 0 \quad \text{on } \partial\Omega \times (0,T). \quad (3.73)$$

Recall that n denotes the unit outward normal to $\partial\Omega$. In the periodic case, we simply require θ to be periodic.

At this stage, let us mention that if most of this section is devoted to the model described above (and a variant of it), we also discuss at the end of the section an interesting (both physically and mathematically) variant of the classical homogeneous, incompressible Navier–Stokes equations, namely a model for a homogeneous incompressible flow with internal degrees of freedom taken from S.N. Antontsev, A.V. Kazhikov and V.N. Monakhov [17].

Let us first observe that, at least formally, that is when solutions are smooth, (3.1) and (3.72) are equivalent to (3.1) and

$$\frac{\partial}{\partial t}\left\{\frac{|u|^2}{2} + C_0\theta\right\} + \operatorname{div}\left(u\left\{\frac{|u|^2}{2} + C_0\theta + p\right\}\right) - \nu\Delta\frac{|u|^2}{2} - k\Delta\theta$$
$$= f \cdot u + \nu\,\partial_i u_j\,\partial_j u_i \quad \text{in} \quad \Omega \times (0,T), \tag{3.74}$$

which is nothing but the "total energy" equation. Recall also that we have

$$-\Delta p = \partial_i u_j\,\partial_j u_i = \operatorname{div}\left((u \cdot \nabla)u\right) \quad \text{in} \quad \Omega \times (0,T) \tag{3.75}$$

at least if we normalize f to satisfy, as we can always do as explained in the preceding sections,

$$\operatorname{div} f = 0 \quad \text{in} \quad \Omega \times (0,T), \tag{3.76}$$

an assumption we always make from now on without recalling it.

Observe that we have, at least formally, $\frac{\partial}{\partial n}|u|^2 = 0$ on $\partial\Omega \times (0,T)$ in the case of Dirichlet boundary conditions and thus we deduce from (3.74)

$$\int_\Omega \frac{|u(t)|^2}{2} + C_0\theta(t)\,dx = \int_\Omega \frac{|u_0|^2}{2} + C_0\theta_0\,dx + \int_0^t ds \int_\Omega dx\, f \cdot u \tag{3.77}$$

denoting by θ_0 the initial condition for θ, i.e.

$$\theta|_{t=0} = \theta_0 \quad \text{in} \quad \Omega. \tag{3.78}$$

Of course, in the periodic case, we assume that θ_0 is given on $\mathbb{R}^N$ and periodic, and we always assume from now on that $\theta_0 \in L^1(\Omega)$.

From the above considerations, we see that there are two ways to look at this system of equations (often called the Rayleigh–Benard equations). Either we decouple the two parts and solve first (3.1)–(3.2), then, given a weak solution u of (3.1)–(3.2), we attempt to solve (3.72)–(3.73) (with the initial condition (3.78)). The other possible approach is to build simultaneously (u,T) solving (3.1)–(3.2), (3.74) and (3.78). The reason why these two approaches might not yield the same solutions is the fact that we do

not know if any (or even some) weak solution of (3.1)–(3.2) satisfies the
energy identity

$$\frac{\partial}{\partial t}\left(\frac{|u|^2}{2}\right) + \text{div}\left(u\,\frac{|u|^2}{2}\right) - \nu\Delta\frac{|u|^2}{2} + \nu|\nabla u|^2 \;=\; f\,u \qquad \text{in} \quad \Omega \times (0,T).$$

If it were the case, then both approaches could be reconciled but this is
an open problem that can be solved only if $N = 2$. Indeed, in the two-
dimensional case, the regularity and uniqueness of weak solutions allow us
to compare the two approaches presented below but we shall not discuss
this point further here.

We begin with the decoupled approach in which u is a given weak solution
of (3.1)–(3.2) as available from Theorems 3.1–3.4. Then, we wish to solve
(3.72)–(3.73) and (3.78). Let us write $D = \frac{\nu}{2C_0}\,(\partial_i u_j + \partial_j u_i)^2$; obviously,
$D \in L^1(\Omega \times (0,T))$. Therefore, from heat equation considerations, we
cannot expect a better integrability for θ than: $\theta \in L^\infty(0,T;L^1(\Omega)) \cap
L^1(0,T;L^q(\Omega))$ for all $q < \frac{N}{N-2}$. Hence, $u\theta \in L^1_{\text{loc}}$ if $N = 2$ or $N = 3$. This
explains why solving (3.72) in the sense of distributions is not adapted to
the problem in hand. Instead, we use the notion of renormalized solutions
which is more flexible (and also more precise)—see R.J. DiPerna and P.L.
Lions [128], P.L. Lions and F. Murat [308].

We recall the *definition*: we shall say that T is a renormalized solution
of (3.72), (3.73) and (3.78) if $\theta \in C([0,T];L^1(\Omega)) \cap L^1(0,T;L^q(\Omega))$ for all
$q < \frac{N}{N-2}$ satisfies

$$\left.\begin{array}{l} T_R(\theta) \in L^2(0,T;H^1(\Omega)) \qquad \text{for all} \quad R > 0 \quad \text{and} \\[2mm] \displaystyle\lim_{R\to\infty}\frac{1}{R}\int_\Omega dx \int_0^T dt\,|\nabla T_R(\theta)|^2 \;=\; 0 \end{array}\right\} \tag{3.79}$$

$$\left.\begin{array}{l} \displaystyle\int_0^T dx \int_\Omega dx\left[\beta(\theta)\left\{\frac{\partial\varphi}{\partial t} + u\cdot\nabla\varphi + \alpha\Delta\varphi\right\} - \beta''(\theta)|\nabla\theta|^2\varphi + D\beta'(\theta)\varphi\right] \\[4mm] \displaystyle + \int_\Omega dx\,\beta(\theta_0)\,\varphi(0) \;=\; \int_\Omega dx\,\beta(\theta(T))\,\varphi(T) \end{array}\right\} \tag{3.80}$$

for all $\beta \in C^2(\mathbb{R})$ such that β' has compact support and for all $\varphi \in C^\infty(\overline{\Omega}\times
[0,T])$ (periodic in the periodic case, with compact support in $\mathbb{R}^N \times [0,T]$
if $\Omega = \mathbb{R}^N$). Some explanations are necessary: indeed—see [308] for more
details and the preceding section for a related argument—(3.79) yields the
fact that $\nabla\theta \in L^s(\Omega \times (0,T))$ for all $s < \frac{N+1}{N}$. In particular, $\nabla\theta \in L^1_{\text{loc}}$
and $\nabla T_R(\theta) = \nabla\theta\,1_{(|\theta|<R)}$ a.e. This explains why $\beta''(\theta)|\nabla\theta|^2 \in L^1_{\text{loc}}$ since
$\beta \in C^2(\mathbb{R})$ and β' has compact support. Next, we remark that (3.80) is a

weak formulation of the following equation and conditions

$$\left(\frac{\partial}{\partial t} + u \cdot \nabla - \alpha \Delta\right)\beta(\theta) + \alpha\beta''(\theta)|\nabla\theta|^2 = D\beta'(\theta) \qquad \text{in } \Omega \times (0,T) \left.\vphantom{\frac{\partial}{\partial n}}\right\}$$
$$\frac{\partial}{\partial n}\beta(\theta) = 0 \quad \text{on } \partial\Omega \times (0,T), \quad \beta(\theta)|_{t=0} = \beta(\theta_0) \quad \text{in } \Omega.$$

The equation for $\beta(\theta)$ follows formally from (3.72) and the notion recalled above simply consists in requesting that these natural changes of variables are indeed possible. Finally, the condition on $T_R(\theta)$ follows, at least formally, from (3.72) and the integrability of D and θ_0 since we deduce from (3.72) upon multiplying by $T_R(\theta)$

$$\alpha \int_0^T dt \int_\Omega dx \, |\nabla T_R(\theta)|^2 = \int_0^T dt \int_\Omega dx \, D T_R(\theta) + \int_\Omega dx \, S_R(\theta_0).$$

Next, we observe that $\frac{1}{R}T_R(\theta)$ is bounded by 1 and converges a.e. to 0 while $\frac{1}{R}S_R(\theta_0)$ is bounded by θ_0 and converges a.e. to 0, and (3.79) follows. Let us finally recall a few facts from [308]: if $|u|\theta \in L^1_{\text{loc}}$ and θ is a renormalized solution of (3.72) then θ satisfies (3.72) in the sense of distributions. On the other hand, using the fact that $u \in L^2(0,T;H^1(\Omega))$, it is not difficult to check that if θ satisfies (3.72) in the sense of distributions and $\theta \in L^2(\Omega \times (0,T))$ then θ is a renormalized solution.

With this notion, the following result proved in Appendix E holds.

Theorem 3.10. *There exists a unique renormalized solution of* (3.72), (3.73) *and* (3.78).

Remarks 3.7. 1) Recall that u is any weak solution of (3.1)–(3.2) as given by Theorems 3.1–3.4.

2) Using the results of R.J. DiPerna and P.L. Lions [128], we immediately see that this result also holds if $k = 0$ ($\alpha = 0$). Then, we only know that $\theta \in C([0,T];L^1(\mathbb{R}^3))$.

3) If $\Omega = \mathbb{R}^N$, we can take θ_0 to be in $L^1(\mathbb{R}^N) + L^\infty(\mathbb{R}^N)$ (changing appropriately the spaces to which θ belongs).

4) One can show that $\text{infess}_{x\in\Omega}\,\theta(x,t)$ ($\in [-\infty,+\infty)$) is a nondecreasing function of t. This is a simple consequence of the fact that $D \geq 0$ a.e.

5) If u satisfies (3.10) and $|u|\theta \in L^1_{\text{loc}}(\Omega \times (0,T))$—these two facts hold if $N = 3$—then we see that we have

$$\frac{\partial}{\partial t}\left(\frac{|u|^2}{2} + C_0\theta\right) + \text{div}\left(u\left\{\frac{|u|^2}{2} + C_0\theta + p\right\}\right) - \nu\Delta\frac{|u|^2}{2} - k\Delta T \left.\vphantom{\frac{\partial}{\partial t}}\right\} \quad (3.81)$$
$$\leq f\cdot u + \nu\partial_i u_j\,\partial_j u_i \qquad \text{in } \mathcal{D}'(\Omega \times (0,T)),$$

and in all cases, we only obtain

$$\left. \begin{array}{c} \displaystyle\int_\Omega \frac{|u(x,t)|^2}{2} + C_0\theta(x,t)\,dx \\[3mm] \displaystyle\leq \int_\Omega \frac{|u_0|^2}{2} + C_0\theta_0\,dx + \int_0^t ds \; <f,u>_{H^{-1}\times H^1} \end{array} \right\} \qquad (3.82)$$

and

$$\frac{d}{dt}\int_\Omega \frac{|u|^2}{2} + C_0\theta\,dx \;\leq<f,u>_{H^{-1}\times H^1} \qquad \text{in} \quad \mathcal{D}'(0,T). \qquad (3.83)$$

In particular if $f = 0$ the total energy is known to be conserved if and only if u satisfies some energy identity instead of an energy inequality, an open problem if $N \geq 3$ as we saw in section 3.1. $\quad\square$

We now turn to the second approach where we solve simultaneously (3.1) and (3.74). This will lead to weak solutions (u,θ) for which the total energy is conserved (when $f = 0$), a fact which is physically expected of course. The price to be paid for this "improvement" is the requirement that $N = 2$ or $N = 3$ because of integrability requirements for $|u|^3$ or $|u|\theta$. The case when $N = 2$ being straightforward in view of the regularity of weak solutions, we consider only the case when $N = 3$, and we begin with the case when $\Omega = \mathbb{R}^3$, the periodic case being completely similar.

Theorem 3.11. *Let $\theta_0 \in L^1(\mathbb{R}^3)$. Then there exists (u,θ) such that u is a weak solution of the Navier–Stokes equations (3.1)–(3.2) (as in Theorem 3.2) satisfying (3.8)–(3.10), $\theta \in L^\infty(0,T;L^1(\mathbb{R}^3)) \cap L^{\frac{5}{3},\infty}(\mathbb{R}^3 \times (0,T)) \cap L^1(0,T;L^q(\mathbb{R}^3))$ for all $q < 3$, $\sup_{R>0} \frac{1}{R}\int_{\mathbb{R}^3} dx \cdot \int_0^T dt\,|\nabla\theta|^2\,1_{|\theta|<R} < \infty$, and (u,θ) solves (3.74), (3.78) in a weak form, namely we have for all $\varphi \in C_0^\infty(\mathbb{R}^3 \times [0,T))$*

$$\left. \begin{array}{c} \displaystyle\int_{\mathbb{R}^3} dx \int_0^T dt \left\{ \left(\frac{|u|^2}{2}+C_0\theta\right)\frac{\partial\varphi}{\partial t} + u\left(\frac{|u|^2}{2}+C_0\theta+p\right)\cdot\nabla\varphi \right. \\[3mm] \displaystyle\left. + \nu\frac{|u|^2}{2}\Delta\varphi + k\theta\Delta\varphi + f\cdot u\varphi + \nu\partial_i u_j\,\partial_j u_i\varphi \right\} \\[3mm] \displaystyle+ \int_{\mathbb{R}^3} dx \left\{ \frac{|u|^2}{2} + C_0\theta_0 \right\}\varphi(x,0) \; = \; 0. \quad\square \end{array} \right\} \qquad (3.84)$$

In the case of Dirichlet boundary conditions, we have the

Theorem 3.12. *Let $\theta_0 \in L^1(\Omega)$, let $u_0 \in W^{2/5,5/4}(\Omega) \cap L^2(\Omega)$ be such that $\operatorname{div} u_0 = 0$ in Ω and $u_0 \cdot n = 0$ on $\partial\Omega$, let $f \in L^{5/4}(\Omega \times (0,T))$ be* such that div $f = 0$ in $\mathcal{D}'(\Omega \times (0,T))$. Then there exists (u, θ) such that $u \in L^{5/4}(0,T;W^{2,\frac{5}{4}}(\Omega))$, $\frac{\partial u}{\partial t}$, $\nabla p \in L^{5/4}(\Omega \times (0,T))$, u is a weak solution of (3.1)–(3.2) as in Theorem 3.4 satisfying (3.8)–(3.10), $\theta \in L^{\infty}(0,T;L^1(\Omega)) \cap L^1(0,T;L^q(\Omega))$ for all $q < \frac{N}{N-2}$, $\sup_{R>0} \frac{1}{R} \int_{\Omega} dx \int_0^T dt \; |\nabla\theta|^2 \, 1_{(|\theta|<R)} < \infty$ and (u, θ) solves (3.74), (3.73), (3.78) in a weak form, namely (3.84) holds (with $\mathbb{R}^3$ replaced by Ω and $C_0^{\infty}(\mathbb{R}^3 \times [0,T))$ replaced by $C_0^{\infty}(\overline{\Omega} \times [0,T))$).

Remarks 3.8. 1) Remark 3.7 (3) also holds for Theorems 3.11 and 3.12.

2) Remark 3.7 (4) holds in the contexts of Theorems 3.11 and 3.12.

3) We deduce from the above results the following facts

$$\frac{d}{dt} \int_{\Omega} \frac{|u|^2}{2} + C_0\theta \, dx = \int_{\Omega} f \cdot u \, dx \qquad \left(= <f, u>_{H^{-1} \times H^1} \right)$$

$$\frac{\partial \theta}{\partial t} + \mathrm{div}\,(u\theta) - \alpha\Delta\theta \geq \frac{\nu}{2C_0}(\partial_i u_j + \partial_j u_i)^2 \quad \text{in} \quad \mathcal{D}'(\Omega \times (0,T)).$$

4) From the weak formulation, one deduces easily that $\frac{|u|^2}{2} + C_0\theta$ is continuous in t with values into $\mathcal{M}_b(\overline{\Omega})$ endowed with the weak $*$ topology (weak topology of measures).

5) In Theorem 3.12, the weak formulation incorporates the Neumann boundary condition (3.73) together with the observation already made above, namely: $\frac{\partial |u|^2}{\partial n} = 0$ on $\partial\Omega \times (0,T)$ since $u = 0$ on $\partial\Omega \times (0,T)$ (at least formally).

6) The only (new) term in (3.84) whose meaning has to be explained is the term $u\theta$. Since $u \in L^{\infty}(0,T;L^2) \cap L^2(0,T;L^6)$ (Sobolev embeddings), $u \in L^{\frac{10}{3}}$ while $\theta \in L^q$ for all $q < \frac{5}{3}$. Hence $\theta u \in L^1(\Omega \times (0,T))$. $\square$

The proof of Theorem 3.11 being similar (and in fact simpler) to the one of Theorem 3.12, we only present the latter.

Proof of Theorem 3.12. With the notation of section 2.4, we consider the solution u^{ε} of

$$\left.\begin{array}{l} \dfrac{\partial u^{\varepsilon}}{\partial t} + u_{\varepsilon} \cdot \nabla u^{\varepsilon} - \nu\Delta u^{\varepsilon} + \nabla p^{\varepsilon} = f^{\varepsilon} \quad \text{in } \Omega \times (0,T), \\[2mm] u^{\varepsilon}|_{t=0} = u_0^{\varepsilon} \quad \text{in } \Omega, \quad u^{\varepsilon} = 0 \quad \text{on } \partial\Omega \times (0,T), \\[2mm] u^{\varepsilon} \in C^2(\overline{\Omega} \times [0,T]), \quad \mathrm{div}\, u^{\varepsilon} = 0 \quad \text{in } \Omega \times (0,T). \end{array}\right\} \quad (3.85)$$

We already know, extracting subsequences if necessary, that, as ε goes to 0_+, $u^{\varepsilon}, u_{\varepsilon}$ converge weakly in $L^2(0,T;H_0^1(\Omega)) \cap L^{\infty}(0,T;L^2(\Omega))$ (weak $*$) to a weak solution u of (3.1)–(3.2) satisfying (3.8), (3.9). In addition, f^{ε} and $u_{\varepsilon} \cdot \nabla u^{\varepsilon}$ (same proof as for $u \cdot \nabla u$) being bounded in $L^{5/4}(\Omega \times$

$(0, T))$, we deduce as in the proof of Theorem 3.9 that u^ε is bounded in $L^{5/4}\left(0, T; W^{2, \frac{5}{4}}(\Omega)\right)$, $\frac{\partial u^\varepsilon}{\partial t}$ and ∇p^ε are bounded in $L^{\frac{5}{4}}(\Omega \times (0, T))$: in particular, $D_x^2 u^\varepsilon$, $\frac{\partial u^\varepsilon}{\partial t}$, ∇p^ε are bounded in $L^{5/4}(\Omega \times (0, T))$ and (3.10) holds. Finally, let us recall that u^ε converges to u as ε goes to 0_+ in $L^p(0, T; L^2(\Omega))$ for all $p \in [1, \infty)$, and in $L^2(0, T; L^q(\Omega))$ for all $1 \le q < 6$.

Next, we introduce the solution θ^ε of

$$
\left.
\begin{aligned}
&\frac{\partial \theta^\varepsilon}{\partial t} + u_\varepsilon \cdot \nabla \theta^\varepsilon - \alpha \Delta \theta^\varepsilon = \frac{\nu}{2C_0} (\partial_i u_j^\varepsilon + \partial_j u_i^\varepsilon)^2 \quad \text{in } \Omega \times (0, T) \\
&\frac{\partial \theta^\varepsilon}{\partial n} = 0 \quad \text{on } \partial\Omega \times (0, T), \quad \theta^\varepsilon|_{t=0} = \theta_0^\varepsilon \quad \text{in } \Omega
\end{aligned}
\right\}
\tag{3.86}
$$

where $\theta_0^\varepsilon \in C_0^\infty(\Omega)$, θ_0^ε converges to θ_0 in $L^1(\Omega)$ as ε goes to 0_+.

Since u_ε is smooth and $(\partial_i u_j^\varepsilon + \partial_j u_i^\varepsilon)^2 \in C^1(\overline{\Omega} \times (0, T))$ (for example), this is nothing but a standard linear parabolic problem and we know there exists θ_ε in, say, $C^{2,1}(\overline{\Omega} \times (0, T))$, i.e. $u, D_x u, D_x^2 u, \frac{\partial u}{\partial t} \in C(\overline{\Omega} \times (0, T))$.

Since $(\partial_i u_j^\varepsilon + \partial_j u_i^\varepsilon)^2$ is bounded in $L^1(\Omega \times (0, T))$, we deduce when $\Omega = \mathbb{R}^N$ in the periodic case from estimates on solutions of heat equations (via, for instance, symmetrization results due to [8],[9]) that θ^ε is bounded in $C([0, T]; L^1(\Omega)) \cap L^{\frac{5}{3}, \infty}(\Omega \times (0, T)) \cap L^1(0, T; L^q(\Omega))$ for all $q < 3$. In the case of Dirichlet boundary conditions ($\Omega \ne \mathbb{R}^N$), we deduce from the results shown in Appendix E that θ^ε is bounded in $C([0, T]; L^1(\Omega)) \cap L^1(0, T; L^q(\Omega))$ for all $q < 3$. Finally, as in the proof of Theorem 3.10, we deduce that $T_R(\theta^\varepsilon) R^{-1/2}$ is bounded in $L^2(0, T; H^1(\Omega))$ uniformly in R, ε. Therefore, without loss of generality we may assume that θ^ε converges weakly in $L^a(\Omega \times (0, T))$ for all $a \in (1, \frac{5}{3})$ to some $\theta \in L^\infty(0, T; L^1(\Omega)) \cap L^1(0, T; L^q(\Omega))$ ($\forall q < 3$). Also, as in the proof of Remark 3.6 (see also P.L. Lions and F. Murat [308]), $\nabla \theta^\varepsilon$ is bounded in $L^r(\Omega \times (0, T))$ and thus $\nabla \theta \in L^r(\Omega \times (0, T))$ for all $r < \frac{5}{4}$.

Next, we deduce from (3.85) and (3.86) that we have

$$
\left.
\begin{aligned}
&\frac{\partial}{\partial t}\left(\frac{|u^\varepsilon|^2}{2} + C_0 \theta^\varepsilon\right) + \operatorname{div}\left\{u_\varepsilon\left(\frac{|u^\varepsilon|^2}{2} + C_0 \theta^\varepsilon\right) + u^\varepsilon p^\varepsilon\right\} \\
&- \nu \Delta \frac{|u^\varepsilon|^2}{2} - k \Delta \theta^\varepsilon = f^\varepsilon \cdot u^\varepsilon + \nu \, \partial_i u_j^\varepsilon \, \partial_j u_i^\varepsilon \quad \text{in } \Omega \times (0, T), \\
&\frac{\partial}{\partial n}\left(\frac{|u^\varepsilon|^2}{2} + C_0 \theta^\varepsilon\right) = 0 \quad \text{on } \partial\Omega.
\end{aligned}
\right\}
\tag{3.87}
$$

In addition, we know that $\frac{|u^\varepsilon|^2}{2}$, θ^ε are bounded in $C([0, T]; L^1(\Omega)) \cap L^1(0, T; L^q(\Omega))$ ($\forall q < 3$) and that $\nabla \frac{|u^\varepsilon|^2}{2}$, $\nabla \theta^\varepsilon$ are bounded in $L^r(\Omega \times (0, T))$ ($\forall r < \frac{5}{4}$). From these bounds and equation (3.85), observing that

$\partial_i u_j^\varepsilon \, \partial_j u_i^\varepsilon = \partial_i((u^\varepsilon \cdot \nabla) u_i^\varepsilon)$ and using classical compactness theorems, we deduce easily that $\frac{|u^\varepsilon|^2}{2} + C_0 \theta^\varepsilon$ converges to $\frac{|u|^2}{2} + C_0 \theta$ in $L^r(0,T;L^1(\Omega)) \cap L^1(0,T;L^q(\Omega))$ (for all $1 \le r < \infty$, $1 \le q < 3$). Therefore, θ^ε converges to θ in $L^r(0,T;L^1(\Omega)) \cap L^1(0,T;L^q(\Omega))$ (for all $1 \le r < \infty$, $1 \le q < 3$). Then, deducing (3.84) from (3.87) is an easy exercise using these convergences and the bounds collected above.

Remark 3.9. Combining the methods developed in this chapter and those introduced in chapter 2, it is possible to study density-dependent models with temperature such as

$$\left.\begin{array}{l} \dfrac{\partial \rho}{\partial t} + \mathrm{div}\,(\rho u) = 0, \quad \mathrm{div}\, u = 0 \\[2mm] \dfrac{\partial(\rho u_i)}{\partial t} + \mathrm{div}\,(\rho u\, u_i) - \dfrac{1}{2}\,\partial_j\big(\mu(\rho,\theta)(\partial_i u_j + \partial_j u_i)\big) + \nabla p = \rho f \\[2mm] C_0 \dfrac{\partial(\rho\theta)}{\partial t} + C_0\, \mathrm{div}\,(\rho u\theta) - \mathrm{div}\,(k(\rho,\theta)\nabla\theta) = \dfrac{\mu(\rho,\theta)}{2}(\partial_i u_j + \partial_j u_i)^2 \end{array}\right\}$$

(3.88)

where $\mu, k \in C([0,\infty) \times \mathbb{R})$, $\inf\{\mu(t,s)\,/\,|t| \le R,\ s \in \mathbb{R}\}$, $\inf\{k(t,s)\,/\,|t| \le R,\ s \in \mathbb{R}\} > 0$ for all $R > 0$ (for instance).

However, we shall not attempt to present here precise results on such a system of equations. $\square$

Finally, we conclude this section and this chapter with a model for an incompressible, homogeneous, newtonian fluid taking into account internal degrees of freedom (for more details see S.N. Antontsev, A.V. Kazhikov and V.M. Monakhov [**17**]). We only describe the three-dimensional situation with Dirichlet boundary conditions and we look for $u(x,t), \omega(x,t) \in \mathbb{R}^3$ solutions of

$$\left.\begin{array}{l} \dfrac{\partial u}{\partial t} + (u \cdot \nabla)u - \nu\Delta u + \nabla p = f + (\omega \times u), \\[2mm] \mathrm{div}\, u = 0 \quad \text{in } \Omega \times (0,T), \quad u = 0 \quad \text{on } \partial\Omega \times (0,T), \\[2mm] \dfrac{\partial \omega}{\partial t} + \mathrm{div}\,(u\omega) + F(p)\omega = m \quad \text{in } \Omega \times (0,T), \\[2mm] \displaystyle\int_\Omega p\,dx = 0 \quad \text{in } (0,T) \end{array}\right\}$$

(3.89)

where F is a continuous, non-negative function on $\mathbb{R}$ satisfying

$$|F(t)| \le C(1+|t|^\alpha) \quad \text{on } \mathbb{R}, \text{ for some } \alpha \in \left[0, \frac{5}{3}\right)$$

(3.90)

and for some non-negative constant C. Finally, we keep the initial condition (3.2) where $u_0 \in L^2(\Omega)$ satisfies (3.3) and $u_0 \cdot n = 0$ on $\partial\Omega$, and we add an initial condition for ω

$$\omega|_{t=0} = \omega_0 \quad \text{in} \quad \Omega, \tag{3.91}$$

and we assume that $m \in L^\infty(\Omega \times (0,T))^3$, $\omega_0 \in L^\infty(\Omega)^3$, $u_0 \in W^{\frac{4}{5}, \frac{15}{14}}(\Omega) \cap L^2(\Omega)$ (div $u_0 = 0$ in Ω, $u_0 \cdot n = 0$ on $\partial\Omega$), $f \in L^2(0,T; H^{-1}(\Omega))^3 \cap L^{\frac{5}{4}}(\Omega \times (0,T))^3$. Then, we can prove

Theorem 3.13. *There exists a solution* (u,p,ω) *of* (3.89) *(in the sense of distributions)*, (3.2) *and* (3.91) *such that* $u \in L^2(0,T; H_0^1(\Omega)) \cap C([0,T]; L_w^2(\Omega)) \cap C([0,T]; L^1(\Omega)) \cap L^{\frac{5}{3}}(0,T; W^{2,\frac{15}{14}}(\Omega))$, $\frac{\partial u}{\partial t} \in L^{\frac{5}{3}}(0,T; L^{\frac{15}{14}}(\Omega))$, $p \in L^{\frac{5}{3}}(0,T; W^{1,\frac{15}{14}}(\Omega))$, $\omega \in L^\infty(\Omega \times (0,T))$, $\omega \in C([0,T]; L^q(\Omega))$ *for all* $1 \leq q < \infty$.

Remark 3.10. *The proof below also shows that* u *satisfies the energy inequalities* (3.8)–(3.10) *and that* ω *satisfies for all* $\beta \in C^1(\mathbb{R}^3; \mathbb{R})$

$$\int_\Omega \beta(\omega(x,t))dx + \int_0^t ds \int_\Omega dx\, F(p)\omega \cdot \nabla\beta(\omega)$$

$$= \int_\Omega \beta(\omega^0)dx + \int_0^t ds \int_\Omega dx\, m \cdot \nabla\beta(\omega) \quad \text{for all} \quad t \in [0,T]. \quad \square$$

Proof of Theorem 3.13. Following the arguments developed in chapter 2 (the situation being somewhat easier here), we introduce the following approximated system of equations

$$\left.\begin{aligned}
\frac{\partial u^\varepsilon}{\partial t} + u_\varepsilon \cdot \nabla u^\varepsilon - \nu\Delta u^\varepsilon + \nabla p^\varepsilon &= f^\varepsilon + (\omega^\varepsilon \times u^\varepsilon), \\
\text{div}\, u^\varepsilon &= 0 \quad \text{in } \Omega \times (0,T) \\
\frac{\partial \omega^\varepsilon}{\partial t} + \text{div}\,(u_\varepsilon\omega^\varepsilon) + F(p^\varepsilon)\omega^\varepsilon &= m^\varepsilon \quad \text{in } \Omega \times (0,T), \\
\int_\Omega p^\varepsilon\, dx &= 0 \quad \text{in } (0,T)
\end{aligned}\right\} \tag{3.92}$$

with $u^\varepsilon = 0$ on $\partial\Omega \times (0,T)$, $u^\varepsilon|_{t=0} = u_0^\varepsilon$, $\omega^\varepsilon|_{t=0} = \omega_0^\varepsilon$, where $u_\varepsilon, f^\varepsilon, u_0^\varepsilon$ have been defined previously (see chapter 2 in particular) and $\omega_0^\varepsilon \in C_0^\infty(\Omega)$, $m^\varepsilon \in C_0^\infty(\Omega \times (0,T))$, $\omega_0^\varepsilon, m^\varepsilon$ are bounded uniformly in ε respectively in $L^\infty(\Omega)$, $L^\infty(\Omega \times (0,T))$ and $\omega_0^\varepsilon, m^\varepsilon$ converge respectively to ω_0, m a.e. and in $L^q(\Omega)$, $L^q(\Omega \times (0,T))$ for all $1 \leq q < \infty$.

The existence of smooth (on $\overline{\Omega} \times [0,T]$) solutions $(u^\varepsilon, p^\varepsilon, \omega^\varepsilon)$ of (3.92) is an easy adaptation of the argument introduced in section 2.4 and of the bounds we obtain now. First of all, multiplying the equation satisfied by u^ε by itself, we find the usual energy identity valid for all $t \in [0, T]$

$$\left.\begin{aligned} \int_\Omega \frac{1}{2} |u^\varepsilon(x,t)|^2 \, dx + \int_0^t ds \int_\Omega dx \, \nu |\nabla u^\varepsilon|^2(x,s) \\ = \frac{1}{2} \int_\Omega |u_0^\varepsilon|^2 \, dx + \int_0^t ds \int_\Omega dx \, f^\varepsilon \cdot u^\varepsilon \end{aligned}\right\} \qquad (3.93)$$

which yields a bound (uniform in ε) on u^ε in $C([0,T]; L^2(\Omega)) \cap L^2(0,T; H_0^1(\Omega))$. Next, using the equation satisfied by u^ε, we obtain easily for all $t \in [0,T]$, for all $\beta \in C^1(\mathbb{R}^3; \mathbb{R})$

$$\left.\begin{aligned} \int_\Omega \beta(\omega^\varepsilon(x,t)) \, dx + \int_\Omega \int_0^t F(p^\varepsilon) \omega^\varepsilon \cdot \nabla \beta(\omega^\varepsilon) \, dx \, ds \\ = \int_\Omega \beta(\omega_0^\varepsilon) \, dx + \int_0^t \int_\Omega m^\varepsilon \cdot \nabla \beta(\omega^\varepsilon) \, dx \, ds. \end{aligned}\right\} \qquad (3.94)$$

In particular choosing $\beta(x) = |x_i|^m x_i$ for $m > 0$, $i = 1,2,3$, we obtain

$$\sup_{\varepsilon>0} \sup_{t\in[0,T]} \int_\Omega |\omega^\varepsilon|^q \, dx < \infty \qquad \text{for all} \quad 1 < q < \infty \qquad (3.95)$$

and keeping track of the precise bounds as $q \to +\infty$ (or applying directly the maximum principle), we deduce

$$\sup_{\varepsilon>0} \sup \left\{ |\omega^\varepsilon| \, / \, x \in \overline{\Omega}, \, t \in [0,T] \right\} < \infty. \qquad (3.96)$$

Then, going back to the equation satisfied by u^ε, we deduce using the preceding bounds (and Sobolev embeddings) that $f^\varepsilon + \omega^\varepsilon \times u^\varepsilon - (u_\varepsilon \cdot \nabla) u^\varepsilon$ is bounded in $L^{\frac{5}{3}}(0,T; L^{\frac{15}{14}}(\Omega))$ uniformly in ε. Therefore, $u^\varepsilon, \frac{\partial u^\varepsilon}{\partial t}, p^\varepsilon$ are bounded uniformly in ε respectively in $L^{\frac{5}{3}}(0,T; W^{2,\frac{15}{14}}(\Omega))$, $L^{\frac{5}{3}}(0,T; L^{\frac{15}{14}}(\Omega))$, $L^{\frac{5}{3}}(0,T; W^{1,\frac{15}{14}}(\Omega))$. From these bounds, we deduce easily that, extracting subsequences if necessary, u^ε converges to some $u \in C([0,T]; L_w^2(\Omega)) \cap L^2(0,T; H_0^1(\Omega)) \cap C([0,T]; L^1(\Omega)) \cap L^{\frac{5}{3}}(0,T; W^{2,\frac{15}{14}}(\Omega))$ and the convergence is a weak convergence in $L^\infty(0,T; L^2(\Omega))$ (weak $*$) $\cap L^2(0,T; H_0^1(\Omega)) \cap L^{\frac{5}{3}}(0,T; W^{2,\frac{15}{14}}(\Omega))$ and a strong convergence in $C([0,T]; L^p(\Omega))$ ($\forall \, 1 \le p < 2$), in $L^q(0,T; W^{1,q}(\Omega))$ ($\forall \, 1 \le q < 2$) and in $L^2(0,T; L^q(\Omega))$ ($\forall \, 1 \le q < 6$). Similarly, $\frac{\partial u^\varepsilon}{\partial t}, \nabla p^\varepsilon$ converge weakly respectively in $L^{\frac{5}{3}}(0,T; L^{\frac{15}{14}}(\Omega))$, $L^{\frac{5}{3}}(0,T; W^{1,\frac{15}{14}}(\Omega))$ to $\frac{\partial u}{\partial t}$ and ∇p for some p which satisfies: $\int_\Omega p \, dx = 0$ in $(0,T)$. In addition, we may assume that ω^ε converges weakly in $L^\infty(\Omega \times (0,T))$ (weak $*$) and strongly in $C([0,T]; W^{-s,p}(\Omega))$

($\forall\, s > 0$, $\forall\, 1 \leq p < \infty$) to some $\omega \in L^\infty(\Omega \times (0,T)) \cap C([0,T]; L^q_w(\Omega))$ ($\forall\, q < \infty$) which satisfies $\omega|_{t=0} = \omega_0$ on Ω. Observe indeed that, because of (3.90), $F(p^\varepsilon)$ is bounded in $L^r(\Omega \times (0,T))$ where $r = \frac{5}{3\alpha}$. Finally, we assume without loss of generality that $F(p^\varepsilon)$ converges weakly in $L^r(\Omega \times (0,T))$ to some $\overline{F} \geq 0$.

Obviously, we can pass to the limit in the equation satisfied by u^ε. We also recover the energy inequalities (3.8)–(3.10) mentioned in Remark 3.10 from (3.93) and its local variant, namely $\frac{\partial}{\partial t} \frac{|u^\varepsilon|^2}{2} + \mathrm{div}\left(u_\varepsilon \frac{|u^\varepsilon|^2}{2} + u^\varepsilon p^\varepsilon\right) - \nu\Delta \frac{|u^\varepsilon|^2}{2} + \nu|\nabla u^\varepsilon|^2 = f^\varepsilon \cdot u^\varepsilon$. In order to complete the proof of Theorem 3.13, it only remains to pass into the limit in the equation satisfied by ω^ε and to show that $\omega \in C([0,T]; L^q(\Omega))$ for all $1 \leq q < \infty$.

In fact, we are first going to show that ω^ε converges to ω in $C([0,T]; L^q(\Omega))$ for all $1 \leq q < \infty$ and that ω satisfies the desired equation with, however, $F(p)$ replaced by $\overline{F}$. Then we shall show that $\overline{F} = F(p)$.

The second step is easy: indeed, once we know that ω^ε converges to ω in, say, $L^q(\Omega \times (0,T))$ for all $1 \leq q < \infty$, then we deduce from the convergences of $u^\varepsilon, \nabla u^\varepsilon, f^\varepsilon$ listed above that $f^\varepsilon + \omega^\varepsilon \times n^\varepsilon - (u_\varepsilon \cdot \nabla)u^\varepsilon$ converges to $f + \omega \times u - (u \cdot \nabla)u$ in $L^{q_1}(0,T; L^{q_2}(\Omega))$ for all $1 \leq q_1 < \frac{5}{3}$, $1 \leq q_2 < \frac{15}{14}$. Hence, using the results of Y. Giga and H. Sohr [**183**] on Stokes equations, we deduce that u^ε converges to u in $L^q(0,T; W^{2,q}(\Omega))$, $\frac{\partial u^\varepsilon}{\partial t}$ converges to $\frac{\partial u}{\partial t}$ in $L^{q_1}(0,T; L^{q_2}(\Omega))$ and ∇p^ε converges to ∇p in $L^{q_1}(0,T; L^{q_2}(\Omega))$ for all $q_1 < \frac{5}{3}$, $1 \leq q_2 < \frac{15}{14}$. Since we normalize p^ε and p to satisfy $\int_\Omega p^\varepsilon\, dx = \int_\Omega p\, dx = 0$ on $(0,T)$, we deduce that p^ε converges to p in $L^{q_1}(0,T; W^{1,q_2}(\Omega))$ for all $q_1 < \frac{5}{3}$, $1 \leq q_2 < \frac{15}{14}$ and in particular in $L^q(\Omega \times (0,T))$ for all $1 \leq q < 5/3$. Since F satisfies (3.90) and F is continuous, we deduce easily that $F(p^\varepsilon)$ converges in $L^{r'}(\Omega \times (0,T))$ to $F(p)$ for $1 \leq r' < r = \frac{5}{3\alpha}$. Hence $\overline{F} = F(p)$ and we conclude.

Finally, the above claim on ω^ε is proven by a convenient adaptation of the method introduced in steps 1–3 of the proof of part 1 of Theorem 2.4 (chapter 2, section 2.3). More precisely, we claim that if $F(p^\varepsilon)\omega^\varepsilon$, $F(p^\varepsilon)|\omega^\varepsilon|^2$ converge weakly in $L^r(\Omega \times (0,T))$ respectively to $\overline{F}\omega$ and $\overline{F}\omega_2$—where ω_2 is the weak $* L^\infty(\Omega \times (0,T))$ limit of $|\omega^\varepsilon|^2$—then the convergence of ω^ε to ω in $C([0,T]; L^2(\Omega))$ and thus in $C([0,T]; L^q(\Omega))$ for all $1 \leq q < \infty$ follows easily. Indeed, if this claim is shown, then ω and ω_2 solve respectively: $\omega, \omega_2 \in L^\infty(\Omega \times (0,T)) \cap C([0,T]; L^q_w(\Omega))$ ($\forall\, 1 < q < \infty$)

$$\frac{\partial \omega}{\partial t} + \mathrm{div}\,(u\omega) + \overline{F}\omega = m \quad \text{in } \Omega \times (0,T), \quad \omega|_{t=0} = \omega_0 \quad \text{in } \Omega \qquad (3.97)$$

$$\left.\begin{aligned} &\frac{\partial \omega_2}{\partial t} + \mathrm{div}\,(u\omega_2) + 2\overline{F}\omega_2 = 2m \cdot \omega \quad \text{in } \Omega \times (0,T), \\ &\omega_2|_{t=0} = |\omega_0|^2 \quad \text{in } \Omega. \end{aligned}\right\} \qquad (3.98)$$

In addition, the proof of Theorem 2.4 mentioned above adapts easily to show that $|\omega|^2$ also solves (3.98) and that $\omega_2 = |\omega|^2$ (uniqueness of transport equations, recall that $\operatorname{div} u = 0$ and $F \geq 0$). Hence, ω_ε converges (strongly) to ω in $L^2(\Omega \times (0,T))$. Finally, the convergence in $C([0,T]; L^2(\Omega))$ of ω^ε also follows from the adaptation of the arguments of section 2.3: indeed, we deduce from (3.97) (and the uniqueness) that $\omega \in C([0,T]; L^2(\Omega))$ and from (3.98) that we have for all $t \in [0,T]$

$$\int_\Omega |\omega(x,t)|^2 \, dx + \int_0^t \int_\Omega 2\overline{F}|\overline{\omega}|^2 \, dx \, ds \ = \ \int_\Omega |\omega_0(x)|^2 \, dx,$$

while we deduce from (3.94) taking $\beta(\omega) = |\omega|^2$ for all $s \in [0,T]$

$$\int_\Omega |\omega^\varepsilon(x,s)|^2 \, dx + \int_0^s \int_\Omega 2F(p^\varepsilon)|\omega^\varepsilon|^2 \, dx \, d\sigma \ = \ \int_\Omega |\omega_0^\varepsilon(x)|^2 \, dx.$$

Then, if $s_n \underset{n}{\to} t$ ($s_n \in (0,T]$) and $\varepsilon_n \underset{n}{\to} 0$, we already know that $\omega^{\varepsilon_n}(s_n)$ converges weakly in $L^2(\Omega)$ to $\omega(t)$. The above equalities together with the fact that, as claimed above, $\int_0^{s_n} \int_\Omega 2F(p^{\varepsilon_n})|\omega^{\varepsilon_n}|^2 \, dx \, dv \underset{n}{\to} \int_0^t \int_\Omega 2\overline{F}\omega_2 \, dx \, d\sigma = \int_0^t \int_\Omega 2\overline{F}|\omega|^2 \, dx \, d\sigma$, show that $\omega^{\varepsilon_n}(s_n)$ converges (strongly) in $L^2(\Omega)$ to $\omega(t)$, and we conclude.

The only claim remaining to prove is the weak convergence of $F(p^\varepsilon)\omega^\varepsilon$, $F(p^\varepsilon)|\omega^\varepsilon|^2$ respectively to $\overline{F}\omega, \overline{F}\omega_2$. Since the proofs are entirely similar, we only detail them in the case of $F(p^\varepsilon)\omega^\varepsilon$. First of all, since p^ε is bounded in $L^{\frac{5}{3}}\left(0,T; W^{1,\frac{15}{14}}(\Omega)\right)$, there exists, for all $\delta > 0$, $p_\delta^\varepsilon \in L^{\frac{5}{3}}(0,T; C^1(\overline{\Omega}))$ (for instance) such that p_δ^ε is bounded in $L^{\frac{5}{3}}\left(0,T; W^{1,\frac{15}{14}}(\Omega)\right)$ uniformly in $\varepsilon, \delta > 0$ and

$$\|p^\varepsilon - p_\delta^\varepsilon\|_{L^{5/3}(\Omega \times (0,T))} \ \leq \ \delta. \tag{3.99}$$

Then, we introduce $F_n \in C_b^1(\mathbb{R}, \mathbb{R})$ (F_n and F_n' are bounded and continuous on $\mathbb{R}$) such that (3.90) holds uniformly in n with F replaced by F_n and F_n converges to F uniformly on compact sets of $\mathbb{R}$. Obviously, $F^n(p_\delta^\varepsilon)$ is bounded in $L^r(\Omega \times (0,T))$ uniformly in ε, δ, n and without loss of generality, we may assume that $F^n(p_\delta^\varepsilon)$ converges weakly to $\overline{F}_\delta^n$ as ε goes to 0. Next, we estimate $F^n(p_\delta^\varepsilon) - F(p^\varepsilon)$ in $L^1(\Omega \times (0,T))$ and we have for all $R \in (0, \infty)$, $\gamma \in (0, 1)$,

$$
\int_0^T dt \int_\Omega dx \, |F^n(p_\delta^\varepsilon) - F(p^\varepsilon)|
$$

$$
\leq \int_0^T dt \int_\Omega dx \, \{F^n(p_\delta^\varepsilon) + F(p_\delta^\varepsilon)\} \, 1_{|p_\delta^\varepsilon| \geq R}
$$

$$
+ C \Big\{ \sup_{|s| \leq R} |F^n(s) - F(s)| \Big\} + \int_0^T dt \int_\Omega dx \, |F(p_\delta^\varepsilon) - F(p^\varepsilon)|
$$

$$
\leq C \Big\{ \sup_{|s| \leq R} |F^n(s) - F(s)| \Big\}
$$

$$
+ C \int_0^T dt \int_\Omega dx \, (|p_\delta^\varepsilon|^\alpha + 1) \cdot \left(1_{|p_\delta^\varepsilon| \geq R} + 1_{|p_\delta^\varepsilon - p^\varepsilon| \geq \gamma} \right)
$$

$$
+ C \int_0^t dt \int_\Omega dx \, (|p|^\alpha + 1) \left(1_{|p| \geq R} + 1_{|p_\delta^\varepsilon - p^\varepsilon| \geq \gamma} \right)
$$

$$
+ C \, \sup \{ |F(x) - F(y)| \, / \, |x - y| \leq \gamma, \, |x|, |y| \leq R \}
$$

$$
\leq \varepsilon_n(R) + \omega_R(\gamma) + CR^{-\left(\frac{5}{4} - \alpha\right)} + C\gamma^{-\left(\frac{5}{4} - \alpha\right)} \|p_\delta^\varepsilon - p^\varepsilon\|_{L^{\frac{5}{4}}(\Omega \times (0,T))}^{\frac{5}{4} - \alpha}
$$

$$
\leq \varepsilon_n(R) + \omega_R(\gamma) + CR^{-\left(\frac{5}{4} - \alpha\right)} + C\gamma^{-\left(\frac{5}{4} - \alpha\right)} \delta^{\frac{5}{4} - \alpha}
$$

where we used (3.90) and (3.99) and $\varepsilon_n(R) \underset{n}{\to} 0$ for $R > 0$ fixed, $\omega_R(\gamma) \to 0$ as $\gamma \to 0_+$ for $R > 0$ fixed. Hence, letting first n go to $+\infty$ and δ go to 0, then γ go to 0 and finally R go to $+\infty$, we deduce

$$
\lim_{\substack{n \to \infty \\ \delta \to 0}} \sup_{\varepsilon > 0} \|F^n(p_\delta^\varepsilon) - F(p^\varepsilon)\|_{L^1(\Omega \times (0,T))} = 0.
$$

In particular, we deduce that $\overline{F}_\delta^n$ converges in $L^1(\Omega \times (0,T))$ to $\overline{F}$ as n goes to $+\infty$ and δ goes to 0. Therefore, we have only to show that $F^n(p_\delta^\varepsilon)\omega^\varepsilon$ converges weakly in $L^{\frac{5}{3}}(\Omega \times (0,T))$ (say) to $\overline{F}_\delta^n \omega$. But this means we can now assume without loss of generality that $F(p^\varepsilon)$ is replaced by $F^n(p_\delta^\varepsilon)$ which is bounded on $\Omega \times (0,T)$ and satisfies: $F^n(p_\delta^\varepsilon)$ is bounded in $L^{\frac{5}{3}}(0, T; C^1(\overline{\Omega}))$. In other words, we may assume that $F(p^\varepsilon) = F^\varepsilon$ is bounded in $L^\infty(\Omega \times (0,T)) \cap L^{\frac{5}{3}}(0, T; C^1(\overline{\Omega}))$. Repeating the above argument, i.e. approximating F^ε (in $L^{\frac{5}{3}}(0, T; C^1(\overline{\Omega}))$), we may in fact assume that F^ε is bounded in $L^\infty(\Omega \times (0,T)) \cap L^{\frac{5}{3}}(0, T; C^k(\overline{\Omega}))$ for an arbitrary $k \geq 0$ and thus in particular F^ε is bounded in $L^q(0, T; C^2(\overline{\Omega}))$ for any $q \in [1, \infty)$.

Next, from the equation satisfied by ω^ε, we deduce that

$$
\frac{\partial \omega^\varepsilon}{\partial t} \quad \text{is bounded in} \quad L^r(\Omega \times (0,T)) + L^\infty(0, T; H^{-1}(\Omega)). \tag{3.100}
$$

We then write (in the sense of distributions)

$$F^\varepsilon \omega^\varepsilon = \frac{\partial}{\partial t}\left(\omega^\varepsilon \int_0^t F^\varepsilon \, ds\right) - \frac{\partial \omega^\varepsilon}{\partial t}\left(\int_0^t F^\varepsilon \, ds\right)$$

and we conclude easily since we can use the above bounds on $\frac{\partial \omega^\varepsilon}{\partial t}$ and on F^ε to deduce that $\left(\int_0^t F^\varepsilon \, ds\right)$ converges uniformly on $\overline{\Omega} \times [0, T]$ and in $L^q(0, T; C^1(\overline{\Omega}))$ to $\left(\int_0^t \overline{F} \, ds\right)$. $\quad\square$

4

EULER EQUATIONS AND OTHER
INCOMPRESSIBLE MODELS

This chapter is essentially devoted to the study of incompressible (homogeneous) Euler equations, namely

$$\frac{\partial u}{\partial t} + (u \cdot \nabla)u + \nabla p = 0, \quad \operatorname{div} u = 0 \quad \text{in } \Omega \times (0, \infty); \qquad (4.1)$$

$$u|_{t=0} = u_0 \qquad \text{in} \quad \Omega \qquad (4.2)$$

with u_0 given on Ω satisfying

$$\operatorname{div} u_0 = 0 \qquad \text{in} \quad \Omega. \qquad (4.3)$$

(Substracting a gradient term from u_0, we can always make such an assumption.) Of course, we have to prescribe boundary conditions (unless $\Omega = \mathbb{R}^N$ or in the periodic case) which take here the following form

$$u \cdot n = 0 \qquad \text{on} \quad \partial\Omega \times (0, \infty) \qquad (4.4)$$

and we assume that u_0 satisfies

$$u_0 \cdot n = 0 \quad \text{on} \quad \partial\Omega. \qquad (4.5)$$

Recall that we assume that Ω, in the case of "Dirichlet boundary conditions" (4.4), is a bounded, smooth, connected open set of $\mathbb{R}^N$ ($N \geq 2$) and n denotes the unit outward normal to $\partial\Omega$. Let us also mention that we could as well consider extensions of (4.1) with a right-hand side (a force term) but we shall not do so here to simplify the presentation.

In fact, sections 4.1–4.4 are devoted to the above system of equations while two variants are considered in the final two sections of this chapter (sections 4.5–4.6).

4.1 A brief review of known results

The situation is completely different in two dimensions (i.e. $N = 2$) and in dimensions $N \geq 3$. This is due to the following fact: if $N = 2$ (and only if $N = 2$), $\omega = \operatorname{curl} u$ (a scalar if $N = 2$, $\omega = \frac{\partial u_2}{\partial x_1} - \frac{\partial u_1}{\partial x_2}$) satisfies the following equation, deduced from (4.1) by taking the curl of the equation and observing that if $N = 2$, $\operatorname{curl}[(u \cdot \nabla)u] = u \cdot \nabla(\operatorname{curl} u)$ when $\operatorname{div} u = 0$:

$$\frac{\partial \omega}{\partial t} + (u \cdot \nabla)\omega = 0. \tag{4.6}$$

(This fact was also used in chapter 3 in the context of Navier–Stokes equations.)

When $N \geq 3$, the only results which are available concern the existence and uniqueness of smooth solutions (say continuous in t with values in H^s for $s > \frac{N}{2} + 1$, or in $C^{1,\alpha}$ for $\alpha \in (0,1)$ in the case of a bounded domain) on a maximal time interval $[0, T_0)$ where $T_0 \in (0, +\infty]$ and if $T_0 < \infty$ the solution's norm blows up as t goes to T_{0_-}. In fact, it is even known—see J.T. Beale, T. Kato and A. Majda [28], G. Ponce [391]—that $\|\omega(t)\|_{L^\infty}$ has to blow up (at a "certain integral rate") when t goes to T_0. It is not known whether T_0 can be finite or in other words if smooth solutions become singular in finite time. We shall come back to this fundamental issue in sections 4.3 and 4.4.

If $N = 2$, the Cauchy problem for incompressible Euler equations is much better understood and we refer the reader to various existing surveys on the question: see A. Majda [316], J.Y. Chemin [90].

Before we state results on the above problem, let us first define precisely what we mean by solution of (4.1)–(4.2) ((4.4) in the case of Dirichlet boundary conditions): we consider $u \in L^\infty(0, \infty; L^2(\Omega))^N$, satisfying $\operatorname{div} u = 0$ in $\mathcal{D}'(\Omega \times (0, \infty))$ and $u \cdot n = 0$ on $\partial\Omega \times (0, \infty)$, such that we have for all $\varphi \in C^\infty(\overline{\Omega} \times [0, \infty))^N$ (for instance) vanishing on $\overline{\Omega}$ for t large

$$\int_0^\infty dt \int_\Omega dx \, u \cdot \left\{ \frac{\partial \varphi}{\partial t} + (u \cdot \nabla)P\varphi \right\} + \int_\Omega dx \, u_0 \cdot \varphi(x, 0) = 0. \tag{4.7}$$

Let us recall that we denote by P the projection on divergence-free vector fields ($\operatorname{div}(P\varphi) = 0$ in Ω, $(P\varphi) \cdot n = 0$ on $\partial\Omega$ in the case of Dirichlet boundary conditions, $\operatorname{curl} P\varphi = \operatorname{curl} \varphi$ in Ω) that we used several times in chapters 2 and 3.

If $\Omega = \mathbb{R}^N$, the above formulation is replaced by the (equivalent) usual weak formulation of (4.1), namely

$$\left. \begin{array}{l} \int_0^\infty dt \int_\Omega dx \, u \cdot \left\{ \frac{\partial \varphi}{\partial t} + (u \cdot \nabla)\varphi \right\} + \int_\Omega dx \, u_0 \cdot \varphi(x, 0) = 0, \\ \text{for all } \varphi \in C_0^\infty(\Omega \times [0, \infty)), \quad \operatorname{div} \varphi = 0 \quad \text{in } \Omega \times (0, \infty), \end{array} \right\} \tag{4.8}$$

and in the periodic case, we impose (4.7) for all $\varphi \in C^\infty(\mathbb{R}^N \times [0, \infty))$, periodic in x, vanishing on $\mathbb{R}^N$ for t large and satisfying $\operatorname{div} \varphi = 0$ in $\mathbb{R}^N \times [0, \infty)$. In general, (4.8) is contained in (4.7) but the converse might not be always true (in the case of Dirichlet boundary conditions).

We may now state a few typical results that are available when $N = 2$.

Theorem 4.1. *Let $u_0 \in L^2(\Omega)^2$ satisfy (4.3) (and (4.5) in the case of Dirichlet boundary conditions). We assume that $\operatorname{curl} u_0 \in L^r(\Omega)$ for some $r \in (1, +\infty]$. Then, there exists $u \in C([0, \infty); W^{1,r})$ in the periodic or in the case of Dirichlet boundary conditions, $u \in u_0 + C([0, \infty); L^q \cap W^{1,r})$ with $q = \max\left(1, \frac{2r}{r+2}\right)$ if $\Omega = \mathbb{R}^2$, $u \in C([0, \infty); L^2 \cap W^{1,s})$ for all $s \in (1, +\infty)$ if $r = +\infty$ and Ω is bounded, $u \in C([0, \infty); L^2 \cap W^{1,s}_{\text{loc}})$ for all $s \in (1, +\infty)$ if $r = +\infty$ and $\Omega = \mathbb{R}^2$, $\operatorname{curl} u \in L^\infty(\Omega \times (0, \infty))$ if $r = +\infty$.*

Furthermore, such a solution is unique when $r = +\infty$, and if $u_0 \in W^{k,p}(\Omega)$ where $k \in \mathbb{N}$, $1 < p < \infty$, $k > 1 + 2/p$, resp. $u_0 \in C^{k,\alpha}(\overline{\Omega})$ where $k \in \mathbb{N}$, $k \geq 1$, $\alpha \in (0, 1)$, then $u \in C([0, \infty); W^{k,p})$, $u_t \in C([0, \infty); W^{k-1,p})$, resp. $u \in C([0, \infty); C^{k,\alpha})$, $u_t \in C([0, \infty); C^{k-1,\alpha})$.

Remarks 4.1. 1) We shall see below (Corollary 4.1) additional properties of solutions when $r \in (1, \infty)$.

2) It is possible to consider cases when $\Omega = \mathbb{R}^2$ and $\operatorname{curl} u_0 \in L^r(\mathbb{R}^2)$ $(1 < r < \infty)$ but we shall not do so here. We shall come back to the specific case $\Omega = \mathbb{R}^2$ in the next section.

3) Also, in the next section, we shall discuss the important borderline case $r = 1$, different (and more precise) formulations of the equation (including the vorticity equation (4.6)).

4) When $1 < r < \infty$, the uniqueness of the above solutions is not known. We shall see in the next section that, for "generic" $u_0 \in L^2$, there exists a unique solution $u \in C([0, \infty); L^2)$.

5) The existence and regularity properties of the pressure are discussed below in the case when $r \in (1, \infty)$. If $u_0 \in W^{k,p}$ (resp. $C^{k,\alpha}$) then the pressure lies in $C([0, \infty); W^{k,p})$ (resp. $C[0, \infty); C^{k,\alpha})$).

6) In the above result, one could add in the existence of solutions the conservation of energy, namely the fact that $\int_\Omega |u(t)|^2 \, dx$ is independent of t.

7) The growth of high order estimates of solutions as t goes to $+\infty$ is an interesting open problem: for instance, if $u_0 \in H^3(\mathbb{R}^2)$, how does the H^3 norm of $u(t)$ behaves as t goes to $+\infty$? Only an upper bound of the form $e^{e^{Ct}}$ is known.

8) We shall briefly sketch below parts of the above result leaving aside the regularity results which follow in a direct way from the L^∞ bound on $\operatorname{curl} u$ and the uniqueness in the case $r = +\infty$ originally shown in V.I. Yudovich [494] and extensively studied (among other topics) in J.Y. Chemin [90]. Again, the crucial bound is the L^∞ bound on $\operatorname{curl} u$ which then implies

that u is, uniformly in t, "almost" Lipschitz (i.e. admits a $t|\log t|$ modulus of continuity). □

Corollary 4.1. *Under the assumptions of Theorem 4.1 and if $p < +\infty$, for any solution satisfying the properties listed in Theorem 4.1, there exists $p \in C([0,\infty); L^q)$ with $1 < q \leq \frac{r}{2-r}$ if $r < 2$, $1 < q < \infty$ if $r = 2$, $1 < q \leq \infty$ if $r > 2$ (replacing L^∞ by C_0 if $r = +\infty$) such that (4.1) holds (in the sense of distributions). In addition, in the periodic case or in the case of Dirichlet boundary conditions, we have:* $\frac{\partial u}{\partial t} \in C([0,\infty); L^q)$, $p \in C([0,\infty); W^{1,q})$ *where $q = r$ if $r > 2$, $q \in [1,r)$ if $r = 2$, $q = \frac{2r}{4-r}$ if $\frac{4}{3} \leq r < 2$; in the periodic case, $p \in C([0,\infty); W^{2,\frac{r}{2}})$ if $r \geq 2$; $\frac{\partial u}{\partial t} \in C([0,\infty); W^{-1,q})$, $p \in C([0,\infty); L^q)$ where $q = \frac{r}{2-r}$ if $1 < r < \frac{4}{3}$.*

Finally, if $\Omega = \mathbb{R}^2$, we have: $\frac{\partial u}{\partial t} \in C([0,\infty); L^q)$, $p \in C([0,\infty); W^{1,q})$ *where $\frac{2r}{r+2} \leq q \leq r$ if $r > 2$, $1 \leq q < r$ if $r = 2$, $1 \leq q \leq \frac{2r}{4-r}$ if $\frac{4}{3} \leq r < 2$, $D^2p \in C([0,\infty); L^1)$ if $r = 2$, $p \in C([0,\infty); W^{2,r/2})$ if $r > 2$; $\frac{\partial u}{\partial t} \in C([0,\infty); \dot{W}^{-1,q})$, $p \in C([0,\infty); L^q)$ where $1 < q \leq \frac{r}{2-r}$ if $1 < r < \frac{4}{3}$ and $\dot{W}^{-1,q} = \{T \in \mathcal{S}' \mid (-\Delta)^{-1/2}T \in L^q\}$.* □

Remarks 4.2. 1) The proof of Corollary 4.1, given below after the proof of Theorem 4.1, shows in fact a bit more. When $r = \frac{4}{3}$, $q = \frac{2r}{4-r}$ becomes 1 and we may replace L^1 by the Hardy space $\mathcal{H}^1$ (see section 3.2 in chapter 3); similarly, we may replace $W^{1,1}$ by $\{p \in L^1, \nabla p \in \mathcal{H}^1\}$. The same remark holds when $r = 2$ or when $q = 1$, $\Omega = \mathbb{R}^2$ replacing L^1 by $\mathcal{H}^1$, $W^{1,1}$ by $\{f \in L^1, \nabla f \in \mathcal{H}^1\}$ or $W^{2,1}$ by $\{f \in W^{1,1}, D^2f \in \mathcal{H}^1\}$. Finally, using the results of R. Coifman, P.L. Lions, Y. Meyer and S. Semmes [95] in the proofs below, we shall see that, in the case when $\Omega = \mathbb{R}^2$ (or in the periodic case), $\frac{\partial u}{\partial t}$ (and thus $u - u_0) \in C([0,\infty); \mathcal{H}^q)$ where $\frac{2r}{r+2} \leq q \leq \frac{2r}{4-r}$ if $r \leq 2$ where $\mathcal{H}^q = L^q$ if $q > 1$—indeed observe that $\frac{\partial u}{\partial t} = -P((u \cdot \nabla)u)$, $(u \cdot \nabla)u \in C([0,\infty); \mathcal{H}^q)$ and P maps $\mathcal{H}^q$ into $\mathcal{H}^q$ (if $q > \frac{2}{3}$)—and $D^2p \in C([0,\infty); \mathcal{H}^{r/2})$, $Dp \in C([0,\infty); \mathcal{H}^q)$ where $\frac{2r}{r+2} \leq q \leq \frac{2r}{r-4}$ if $r \leq 2$. This last observation on D^2p follows from the fact that we have

$$-\Delta p = \det(D^2\phi) \quad \text{where} \quad u = \nabla^\perp \phi = \begin{pmatrix} \frac{\partial \phi}{\partial x_2} \\ -\frac{\partial \phi}{\partial x_1} \end{pmatrix} \tag{4.9}$$

and $D^2\phi \in C([0,\infty); L^r)$. Indeed, by the results of [95], $\det(D^2\phi) \in C([0,\infty); \mathcal{H}^{r/2})$ and thus $\frac{\partial^2 p}{\partial x_i \partial x_j} = (-\Delta)^{-1} \frac{\partial^2 \det(D^2\phi)}{\partial x_i \partial x_j} \in C([0,\infty); \mathcal{H}^{r/2})$, since $r/2 > 1/2$.

2) The proof of Corollary 4.1 also shows that $\frac{\partial u}{\partial t} \in C([0,\infty); \dot{W}^{-1,q})$ when $\Omega = \mathbb{R}^2$ and $1 < q \leq \frac{r}{2-r}$ if $1 < r < 2$, $1 < q < \infty$ if $r = 2$, $1 < q \leq +\infty$ if $r > 2$. We can even replace $\dot{W}^{-1,q}$ by $\{f \in \mathcal{S}' \mid (-\Delta)^{-1/2}f \in C^\alpha\}$ where $\alpha = 1 - \frac{2}{r}$ in the case when $r > 2$. □

Proof of Theorem 4.1. The proof is divided into several steps. Let us recall that, for the reasons mentioned in Remark 4.1 (8)), we only prove here the case when $1 < r < \infty$.

Step 1. Fundamental a priori bounds. We first wish to explain the heart of the matter. First of all, multiplying (4.1) by u, we expect u to satisfy the following local form of the energy identity (or energy conservation), namely

$$\frac{\partial}{\partial t}\left(\frac{|u|^2}{2}\right) + \operatorname{div}\left\{u\left[\frac{|u|^2}{2} + p\right]\right\} = 0 \qquad (4.10)$$

which implies, at least formally, the conservation of energy, i.e. the fact that $\int_\Omega |u|^2\,dx$ should be independent of t and thus a bound on u in $C([0,\infty); L^2)$.

Next, using (4.6), we deduce formally

$$\frac{\partial}{\partial t}\{|\omega|^r\} + \operatorname{div}\{u|\omega|^r\} = 0 \qquad (4.11)$$

hence integrating in x

$$\int_\Omega |\omega|^r\,dx \qquad \text{is independent of } t. \qquad (4.12)$$

In particular, this yields a bound on ω in $C([0,\infty); L^r)$ hence a bound on ∇u in $C([0,\infty); L^2)$: indeed observe that $\operatorname{div} u = 0$ and $u \cdot n = 0$ in the case of Dirichlet boundary conditions. This bound, combined with the $C([0,\infty); L^2)$ bound on u, yields a bound on u in $C([0,\infty); W^{1,r})$ in the periodic case, in the case of Dirichlet boundary conditions or if $\Omega = \mathbb{R}^2$, $r \geq 2$. Finally, if $\Omega = \mathbb{R}^2$, we observe that (4.1) yields

$$\frac{\partial u}{\partial t} + P[(u \cdot \nabla)u] = 0. \qquad (4.13)$$

Since $(u \cdot \nabla)u$ is bounded in $C\big([0,\infty); L^{\frac{2r}{r+2}}\big)$ in view of the preceding bounds, $\frac{\partial u}{\partial t}$ is also bounded in $C\big([0,\infty); L^{\frac{2r}{r+2}}\big)$ and thus $u - u_0$ is bounded in $C\big([0,T]; L^{\frac{2r}{r+2}}\big)$ for all $T \in (0,\infty)$.

Step 2. Navier–Stokes approximation. In the periodic case or in the case when $\Omega = \mathbb{R}^2$, there is no difficulty and we simply approximate (4.1) by the Navier–Stokes equation

$$\frac{\partial u}{\partial t} + (u \cdot \nabla)u - \nu \Delta u + \nabla p = 0, \quad \operatorname{div} u = 0 \qquad (4.14)$$

keeping the same initial condition (4.2), where $\nu > 0$.

However, in the case of Dirichlet boundary conditions, a fundamental difficulty arises with boundary conditions. It is of course tempting to use again (4.14) with Dirichlet boundary conditions, namely: $u = 0$ on $\partial\Omega \times (0, \infty)$. The difficulty which then appears is emphasized in the following remark.

Remark 4.3. *(2D Navier–Stokes $\rightarrow$ 2D Euler, with Dirichlet boundary conditions?).* We consider (4.14) in the case of Dirichlet boundary conditions: $u = 0$ on $\partial\Omega \times (0, \infty)$. It is quite clear that a boundary layer will form since solutions of Euler equations only satisfy $u \cdot n = 0$ on $\partial\Omega$ and there is no reason why the tangential part of u should vanish (and they do not in general!). It is not known if the "L^2 strength" of this "layer" goes to 0 and, more precisely, the limit as r goes to 0 of solutions of Navier–Stokes equations with Dirichlet boundary conditions to solutions of Euler equations is an important open problem. Equivalent formulations of this problem can be found in T. Kato [**238**]. $\square$

To circumvent that difficulty, we modify the Dirichlet boundary conditions associated to (4.14) in the following way

$$u \cdot n = 0, \quad \omega = 0 \quad \text{on} \quad \partial\Omega \times (0, \infty). \tag{4.15}$$

Then (4.14)–(4.15) is equivalent to

$$\left.\begin{array}{l} \dfrac{\partial\omega}{\partial t} + (u \cdot \nabla)\omega - \nu\Delta\omega = 0 \quad \text{in } \Omega \times (0, \infty), \\[2mm] \omega = 0 \quad \text{on } \partial\Omega \times (0, \infty), \quad \operatorname{curl} u = \omega, \\[2mm] \operatorname{div} u = 0 \quad \text{in } \Omega \times (0, \infty), \quad u \cdot n = 0 \quad \text{on } \partial\Omega \times (0, \infty). \end{array}\right\} \tag{4.16}$$

Notice also that since we are in two dimensions

$$\Delta u = \nabla \operatorname{div} u + \nabla^\perp \operatorname{curl} u \quad \text{where} \quad \nabla^\perp = \begin{pmatrix} \frac{\partial}{\partial x_2} \\ -\frac{\partial}{\partial x_1} \end{pmatrix}$$

hence, if $\operatorname{div} u = 0$, $\Delta u \cdot n = (n \cdot \nabla^\perp) \operatorname{curl} u$ and we deduce from (4.15)

$$\Delta u \cdot n = 0 \quad \text{on} \quad \partial\Omega \times (0, \infty), \tag{4.17}$$

and we may rewrite (4.14) as

$$\frac{\partial u}{\partial t} + P((u \cdot \nabla)u) - \nu\Delta u = 0 \quad \text{in} \quad \Omega \times (0, \infty). \tag{4.18}$$

We then claim that the system (4.14)–(4.15) (or equivalently (4.16)) can be solved exactly as the usual Navier–Stokes equations (i.e. with the usual full Dirichlet boundary conditions)—see section 3.1—and thus that we obtain (in all cases) for each $\nu > 0$ a unique solution $u_\nu \in L^2(0, T; H^1) \cap C([0, \infty); L^2)$, $\frac{\partial u_\nu}{\partial t} \in L^2(0, T; H^{-1})$ (for all $T \in (0, \infty)$). All these facts follow from a modified energy identity that we briefly sketch (and from (4.18) to obtain the regularity information on $\frac{\partial u_\nu}{\partial t}$). Multiplying by u, integrating by parts over $\partial\Omega$ and using the fact that $u \cdot n = 0$ on $\partial\Omega \times (0, \infty)$, we obtain

$$\frac{1}{2}\frac{d}{dt}\int_\Omega |u|^2\,dx + \nu\int_\Omega |\nabla u|^2\,dx = \nu\int_{\partial\Omega} \frac{\partial u}{\partial n} \cdot u\,dS.$$

Using the boundary conditions, namely $u \cdot n = 0$, $\omega = 0$ on $\partial\Omega$, it is easy to check that $\frac{\partial u}{\partial n} \cdot u = \kappa|u|^2$ on $\partial\Omega$, where κ is the curvature of $\partial\Omega$. Hence, we obtain for some $C_0 > 0$ that depends only on Ω

$$\frac{1}{2}\frac{d}{dt}\int_\Omega |u|^2\,dx + \nu\int_\Omega |\nabla u|^2\,dx \leq C_0\nu\int_{\partial\Omega} |u|^2\,dS.$$

(Recall that in the periodic case, or if $\Omega = \mathbb{R}^2$, or in the usual Dirichlet case we obtain an equality with a right-hand side that vanishes.) Then, using a classical trace inequality, we deduce from the preceding inequality

$$\frac{1}{2}\frac{d}{dt}\int_\Omega |u|^2\,dx + \nu\int_\Omega |\nabla u|^2\,dx \leq C\nu\left(\int_\Omega |\nabla u|^2\,dx\right)^{1/2}\left(\int_\Omega |u|^2\,dx\right)^{1/2}$$

where, here and below, C denotes various positive constants that depend only on Ω. Hence

$$\frac{d}{dt}\int_\Omega |u|^2\,dx + \nu\int_\Omega |\nabla u|^2\,dx \leq C\nu\int_\Omega |u|^2\,dx, \qquad (4.19)$$

and our claim is shown. We have obtained in fact the following bound

$$\int_\Omega |u(x, t)|^2\,dx \leq e^{C\nu t}\int_\Omega |u_0|^2\,dx \qquad \text{for all } t \geq 0. \qquad (4.20)$$

Let us finally observe that, for each $\nu > 0$, u_ν is in fact smooth for $t > 0$: for instance, multiplying (4.16) by $t\omega_\nu$ and integrating by parts, we immediately obtain for all $T \in (0, \infty)$:

$$t|\omega_\nu(t)|^2_{L^2} + \int_0^t s|\nabla\omega_\nu(s)|^2_{L^2}\,ds \leq C(T, \nu) \qquad \text{for } 0 \leq t \leq T$$

or

$$t\|u_\nu\|^2_{H^1} + \int_0^t s\|u_\nu\|^2_{H^2}\,ds \leq C(T, \nu) \qquad \text{for } 0 \leq t \leq T \quad \square$$

Step 3. Existence when $r \geq 2$. We treat here the case when $r \geq 2$ and we first consider either periodic or Dirichlet boundary conditions. Then, we explain how to modify the proof when $\Omega = \mathbb{R}^2$. First of all, we obtain an estimate on ω_ν in $C([0, \infty); L^r)$. We simply multiply (4.16) by $|\omega_\nu|^{r-2} \omega_\nu$ and obtain in all cases (either periodic or Dirichlet boundary conditions) for all $t \geq 0$

$$
\left.
\begin{aligned}
\frac{1}{r} \int_\Omega |\omega_\nu(t)|^r \, dx + \nu(r-1) \int_0^t ds \int_\Omega |\nabla \omega_\nu|^2 \, |\omega_\nu|^{r-2} \, dx \\
= \frac{1}{r} \int_\Omega |\text{curl } u_0|^r \, dx.
\end{aligned}
\right\}
\tag{4.21}
$$

This identity yields the desired bound. Of course, the preceding calculation has to be justified. This is not difficult and we leave it to the reader (multiply (4.16) by $|T_R(\omega_\nu)|^{r-2} T_R(\omega_\nu)$ where $T_R(\omega_\nu) = \max(\min(\omega_\nu, R), -R)$ for $R > 0$ and let R go to $+\infty$, observing that for instance $\omega_\nu \in L^2(0, \infty; H^1) \cap C([0, \infty); L^1))$. Observe also that $\sqrt{\nu}\,\omega_\nu$ is bounded in $L^2(0, \infty; H^1)$ (take $r = 2$ in (3.21)).

The bound on ω_ν yields a bound on u_ν in $C([0, \infty); W^{1,r})$ exactly as in step 1. Then, from the equation (4.16), we deduce that $\frac{\partial \omega_\nu}{\partial t}$ is bounded in $L^2(0, T; W^{-1,q})$ for all $T \in (0, \infty)$, $1 < q < 2$ for instance. This is enough to ensure that ω_ν is relatively compact in $C([0, T]; L^r - w)$ (recall that $L^r - w$ means L^r endowed with the weak topology) by the observation detailed in Appendix C. Hence, u_ν is relatively compact in $C([0, T]; W^{1,r} - w)$ by the same argument as in step 1 and u_ν is relatively compact in $C([0, T]; L^q)$ for all $1 \leq q < \infty$, $T \in (0, \infty)$ by the Rellich–Kondrakov theorem. Extracting subsequences if necessary, we may thus assume that, as ν goes to 0, u_ν converges to some u in $C([0, T]; W^{1,r} - w) \cap C([0, T]; L^q)$ for all $1 \leq q < \infty$ while ω_ν converges to some ω in $C([0, T]; L^r - w)$ (for all $T \in (0, \infty)$). In particular, we have on passing to the limit: $u(0) = u_0$, $\omega(0) = \text{curl } u_0$ in Ω, $\int_\Omega |u(t)|^2 \, dx \leq \int_\Omega |u_0|^2 \, dx$ for all $t \geq 0$ in view of (4.20),

$$u \in C([0, T]; L^2), \quad \text{div } u = 0, \quad \text{curl } u = \omega \quad \text{in } \Omega \times (0, \infty) \tag{4.22}$$

$$\frac{\partial \omega}{\partial t} + \text{div}(u\omega) = 0 \quad \text{in } \Omega \times (0, \infty) \tag{4.23}$$

and $u \cdot n = 0$ on $\partial\Omega \times (0, \infty)$ in the case of Dirichlet boundary conditions. Furthermore, we may pass to the limit in (4.18) and we obtain

$$\frac{\partial u}{\partial t} + P((u \cdot \nabla)u) = 0 \quad \text{in } \Omega \times (0, \infty). \tag{4.24}$$

We claim that this yields (4.7) and we explain why (4.7) holds in the case of Dirichlet boundary conditions: indeed, $\nabla u \in L^\infty(0, \infty; L^r)$, $u \in$

$L^\infty(0,\infty; W^{1,r})$ and thus we have for all $\varphi \in C^\infty(\overline{\Omega} \times [0,\infty))$ (vanishing for t large)

$$
\begin{aligned}
0 &= \int_0^\infty dt \int_\Omega dx \left\{ -u \cdot \frac{\partial \varphi}{\partial t} + (u \cdot \nabla)u \cdot P\varphi \right\} - \int_\Omega dx\, u_0 \cdot \varphi(x,0) \\
&= \int_0^\infty dt \int_\Omega dx \left\{ -u \cdot \frac{\partial \varphi}{\partial t} + \operatorname{div}(u \otimes u) \cdot P\varphi \right\} - \int_\Omega dx\, u_0 \cdot \varphi(x,0) \\
&= -\int_0^\infty dt \int_\Omega dx \left\{ u \cdot \frac{\partial \varphi}{\partial t} + u \cdot (u \cdot \nabla)[P\varphi] \right\} - \int_\Omega dx\, u_0 \cdot \varphi(x,0)
\end{aligned}
$$

since $\operatorname{div} u = 0$ and $u \cdot n = 0$ on $\partial\Omega \times (0,\infty)$.

In order to complete the proof of Theorem 4.1 in this case, it only remains to show that $\omega \in C([0,\infty); L^r)$ and thus $u \in C([0,\infty); W^{1,2})$. This fact is a consequence of (4.23). Indeed, by general results due to R.J. DiPerna and P.L. Lions [128] and recalled in section 2.3 on transport equations with divergence-free vector fields, we see that ω is necessarily equal to the unique renormalized solution of (4.23) which satisfies $\omega \in C([0,\infty); L^r)$ (and $\int_\Omega |\omega(t)|^r\, dx$ is independent of $t \geq 0$). We have used here the fact that $r \geq 2$ and thus, in particular, $u \in C([0,\infty); W^{1,2})$ while $\omega \in C([0,\infty); L^2)$.

Remarks 4.4. 1) Let us observe that since $\omega \in C([0,\infty); L^r)$ and $\int_\Omega |\omega(t)|^r\, dx$ is independent of t, the identity (4.21) immediately yields the convergence of ω_ν to ω in $C([0,T]; L^r)$ and thus of u_ν to u in $C([0,T]; W^{1,r})$ for all $T \in (0,\infty)$.

2) (4.24) shows that $\frac{\partial u}{\partial t} \in C([0,\infty); L^q)$ for all $1 \leq q < r$, and, multiplying (4.24) by u, one can easily justify the energy conservation since we have for all $t \geq 0$

$$
\begin{aligned}
\int_\Omega |u(t)|^2\, dx - \int_\Omega |u_0|^2\, dx &= -2 \int_0^t ds \int_\Omega dx\, [(u \cdot \nabla)u] \cdot Pu \\
&= -2 \int_0^t ds \int_\Omega dx\, [(u \cdot \nabla)u] \cdot u \\
&= -\int_0^t ds \int_\Omega dx\, (u \cdot \nabla)|u|^2 = 0.
\end{aligned}
$$

This verification only requires $|u|^2|\nabla u|$ to be integrable and this is the case as soon as $u \in L^\infty(0,\infty; W^{1,r})$ with $r \geq \frac{3}{2}$ since, in this case, $|u|^2 \in L^\infty(0,\infty; L^3)$ by Sobolev inequalities. $\square$

We now briefly explain how to modify the above proof in the case when $\Omega = \mathbb{R}^2$. All the steps of the proof are easily adapted, replacing $C([0,T]; L^q)$ or $C([0,T]; L^2)$ by $C([0,T]; L^q_{\text{loc}})$ or $C([0,T]; L^2_{\text{loc}})$: in particular we obtain $u \in L^\infty(0,\infty; L^2) \cap C([0,\infty); L^2_{\text{loc}})$, $\nabla u \in L^\infty(0,\infty; L^r)$, $u \in C([0,\infty);$

$W^{1,r}_{loc}$) satisfying $u(0) = u_0$ in $\mathbb{R}^2$, $\operatorname{div} u = 0$ in $\mathbb{R}^2$ and

$$\frac{\partial u}{\partial t} + P((u \cdot \nabla)u) = 0 \qquad \text{in} \quad \mathbb{R}^2 \times (0, \infty). \qquad (4.25)$$

In particular, $\frac{\partial u}{\partial t} \in L^\infty(0, \infty; L^{\frac{2r}{r+2}})$ hence $u - u_0 \in C([0, \infty); L^{\frac{2r}{r+2}})$. This yields the fact that $u \in C([0, \infty); L^2 \cap W^{1,r})$ since $\frac{2r}{r+2} \leq 2 \leq r$. $\quad\square$

Remark 4.5. The proofs "given" in Remark 4.4 are easily adapted to the case when $\Omega = \mathbb{R}^2$ and yield the conservation of energy together with the convergence (up to the extraction of subsequences) of u_ν in $C([0, T];$ $L^2 \cap W^{1,r})$ for all $T \in (0, \infty)$. $\quad\square$

Step 4. Existence when $1 < r < 2$. We now treat the case when $1 < r < 2$ and we begin again by excluding the case when $\Omega = \mathbb{R}^2$ and thus considering periodic or Dirichlet boundary conditions. In fact, we shall give the proof in the case of Dirichlet boundary conditions, the periodic case being similar and even somewhat simpler.

We shall deduce the existence of a solution when $1 < r < 2$ from the case we just treated. We then introduce $u_0^\varepsilon \in C^1(\overline{\Omega})$ satisfying $u_0^\varepsilon \cdot n = 0$ in Ω, $\operatorname{div} u_0^\varepsilon = 0$ in Ω and such that u_0^ε converges in $W^{1,r}$ to u_0 as ε goes to 0_+: the existence of u_0^ε can be obtained by considering $P\tilde{u}_0^\varepsilon$ where $\tilde{u}_0^\varepsilon \in C^\infty(\overline{\Omega})$ converges to u_0 in $W^{1,r}$. Next, we denote by u^ε a solution of the Euler equation provided by the preceding steps (it is in fact unique) and by $\omega^\varepsilon = \operatorname{curl} u^\varepsilon$. We have the following information: $u^\varepsilon \in C([0, \infty); W^{1,q})$, $\omega^\varepsilon \in C([0, \infty); L^q)$ for all $1 \leq q < \infty$. (4.7), (4.13) and (4.6) hold with u, ω replaced respectively by u, ω. Also, we have

$$\int_\Omega |\omega^\varepsilon(t)|^r \, dx = \int_\Omega |\omega_0^\varepsilon|^r \, dx \xrightarrow{\varepsilon} \int_\Omega |\omega_0|^r \, dx$$

defining $\omega_0^\varepsilon = \operatorname{curl} u_0^\varepsilon$, $\omega_0 = \operatorname{curl} u_0$. In particular, as in step 1, we see that $u^\varepsilon \in L^\infty(0, \infty; W^{1,r})$ and without loss of generality we may assume that u^ε converges weakly to some $u \in L^\infty(0, \infty; W^{1,r})$ which satisfies $u \cdot n = 0$ on $\partial\Omega \times (0, \infty)$, $\operatorname{div} u = 0$ in $\Omega \times (0, \infty)$.

Then we invoke the compactness results shown in R.J. DiPerna and P.L. Lions [**128**] essentially recalled and proved in section 2.3—only the L^∞ case was established there but the general L^r case follows as well by considering $T_R(\omega^\varepsilon)$ instead of ω^ε—and we deduce that ω^ε converges in $C([0, T]; L^r)$ (for all $T \in (0, \infty)$) to the unique renormalized solution ω of (4.6) satisfying $\omega|_{t=0} = \omega_0$ in Ω. Here, as in [**128**], renormalized solution means that we have for all $\beta \in C_b^\infty(\mathbb{R}; \mathbb{R})$

$$\frac{\partial \beta(\omega)}{\partial t} + \operatorname{div}(u\beta(\omega)) = 0 \qquad \text{in} \quad \mathcal{D}'(\Omega \times (0, \infty)). \qquad (4.26)$$

Therefore, by the same argument as in step 1, u^ε converges in $C([0,T];$ $W^{1,r})$ (for all $T \in (0,\infty)$) to u and we can pass to the limit in (4.7) completing the proof of Theorem 4.1 in this case. Notice, by the way, that, in particular, u^ε converges to u in $C([0,T];L^2)$ and thus Remark 4.1 (6) is deduced in that case from Remark 4.4 (2).

In the case when $\Omega = \mathbb{R}^2$, the same proof as above applies (we now simply regularize u_0 by convolution) and shows that ω^ε converges to $\omega = \operatorname{curl} u$ in $C([0,T];L^r)$ (for all $T \in (0,\infty)$) which is the unique renormalized solution of (4.6) in $\mathbb{R}^2 \times (0,\infty)$ satisfying $\omega|_{t=0} = \omega_0$ in $\mathbb{R}^2$. Since u^ε is bounded in $L^\infty(0,\infty;L^2)$, we deduce that ∇u^ε converges to ∇u in $C([0,T];L^r)$, u^ε converges to u in $C([0,T];W_{\text{loc}}^{1,r})$ for all $T \in (0,\infty)$ where $u \in C([0,\infty);W_{\text{loc}}^{1,r}) \cap L^\infty(0,\infty;L^2)$, $\nabla u \in C([0,\infty);L^r)$. It only remains to show that u^ε converges to u in $C([0,T];L^2)$ (for all $T \in (0,\infty)$) and that $u-u_0 \in C([0,\infty);L^1)$ when $1 < r < 2$. In order to do so, we follow Remark 4.2 (1): $(u^\varepsilon \cdot \nabla)u^\varepsilon$ is bounded, in view of [95], in $C([0,\infty),\mathcal{H}^{\frac{2r}{r+2}})$. Since $\frac{\partial u^\varepsilon}{\partial t} = -P((u^\varepsilon \cdot \nabla)u^\varepsilon)$ and $\frac{2r}{r+2} > \frac{2}{3}$ if $r > 1$, we deduce that $\frac{\partial u^\varepsilon}{\partial t}$ is bounded in $C([0,\infty);\mathcal{H}^{\frac{2r}{r+2}})$ and thus $u^\varepsilon - u_0^\varepsilon$ is bounded in $C([0,T];\mathcal{H}^{\frac{2r}{r+2}})$ for all $T \in (0,\infty)$. On the other hand, since $\nabla(u^\varepsilon - u_0^\varepsilon)$ converges to $\nabla(u-u_0)$ in $C([0,T];L^r)$ and $1 < r < 2$, we deduce from Sobolev embeddings that $u^\varepsilon - u_0^\varepsilon$ converges to $u-u_0$ in $C([0,T];L^{\frac{2r}{2-r}})$ for all $T \in (0,\infty)$. Hence, by interpolation, $u^\varepsilon - u_0^\varepsilon$ converges to $u-u_0$ in $C([0,T];L^q)$ for $1 \le q \le \frac{2r}{2-r}$ and our claims are shown since u_0^ε converges to u_0 in L^2 by construction. $\square$

Remark 4.6. It is possible to give a much more elementary proof of the convergence of u^ε to u in $C([0,T];L^2)$ ($\forall\, T \in (0,\infty)$) which also yields the fact that $u \in u_0 + C([0,\infty);L^q \cap W^{1,r})$ for $\frac{2r}{r+1} < q < r$ (but does not reach L^1!) when $1 < r < 2$. One simply observes that

$$\frac{\partial u^\varepsilon}{\partial t} = -\sum_{j=1}^{2} \frac{\partial}{\partial x_j} P(u_j^\varepsilon u^\varepsilon)$$

and $u_j^\varepsilon u^\varepsilon$ is bounded in $C([0,\infty);L^1 \cap L^{\frac{r}{2-r}})$. Hence, we have

$$u^\varepsilon - u_0^\varepsilon = \sum_{j=1}^{2} \frac{\partial}{\partial x_j}(f_j^\varepsilon), \quad \nabla(u^\varepsilon - u_0^\varepsilon) = g^\varepsilon$$

where f^ε is bounded in $C([0,T];L^q)$ for all $T \in (0,\infty)$, $1 < q \le \frac{r}{2-r}$, while g^ε is bounded in $C([0,\infty);L^r)$. These two facts imply easily that $u^\varepsilon - u_0^\varepsilon$ is bounded in $C([0,T];L^s)$ for $\frac{2r}{r+1} < s \le \frac{2r}{2-r}$, $0 < T < \infty$. Observe that $U^\varepsilon = (-\Delta)^{-1/2}(u^\varepsilon - u_0^\varepsilon)$ is bounded in $C([0,T];L^q)$ while $-\Delta U^\varepsilon$ is bounded in $C([0,\infty); L^r)$, and this is enough to show our claims following the end of the proof of Theorem 4.1. □

We now turn to the

Proof of Corollary 4.1. We begin with the existence of p which is a straightforward consequence of (4.7). Indeed, we have $< \frac{\partial u}{\partial t} + \operatorname{div}(u \otimes u), \varphi >_{\mathcal{D}' \times \mathcal{D}} = 0$ for all $\varphi \in C_0^\infty(\Omega \times (0,\infty))$ such that $\operatorname{div} \varphi = 0$. This implies the existence of a distribution p such that (4.1) holds in the sense of distributions. Next, we need to show the integrability requirements indicated in Theorem 4.1. Since P is bounded from $W_0^{1,q}$ into $W^{1,q}$ for all $1 < q < \infty$, we deduce from the weak formulation (4.7) (using $P\varphi$ for arbitrary smooth φ) that $\frac{\partial u}{\partial t} \in C([0,\infty); W^{-1,q})$ for $1 < q \leq \frac{r}{2-r}$ if $r < 2$, $1 < q < \infty$ if $r = 2$, $1 < q \leq \infty$ if $r > 2$ since $u \in C([0,\infty); L^2 \cap L^{\frac{2r}{2-r}})$ if $r \leq 2$ ($L^2 \cap L^q$ if $r = 2$ for $2 < q < \infty$, $L^2 \cap L^\infty$ if $r > 2$) by Sobolev embeddings. The integrability of p then follows since $\nabla p = -\frac{\partial u}{\partial t} - \operatorname{div}(u \otimes u)$.

In the periodic case or when $\Omega = \mathbb{R}^2$, the argument is a bit simpler: we just write

$$\frac{\partial u}{\partial t} = -\sum_{j=1}^2 \frac{\partial}{\partial x_j} P(u_j u), \quad \nabla p = -\frac{\partial u}{\partial t} - \operatorname{div}(u \otimes u).$$

All the claims listed in Corollary 4.1 follow easily from the bounds on u, Sobolev embeddings, the fact that $\frac{\partial u}{\partial t} = -P((u \cdot \nabla)u)$ when $r \geq \frac{4}{3}$ and that we have

$$-\Delta p = \operatorname{div}\{(u \cdot \nabla)u\} = \sum_{i,j=1}^2 \frac{\partial u_i}{\partial x_j} \frac{\partial u_j}{\partial x_i} = \sum_{i,j=1}^2 \frac{\partial^2}{\partial x_i \partial x_j}(u_i u_j). \quad \Box$$

$$(4.27)$$

Remarks 4.7. 1) The proof we gave of Theorem 4.1 shows a few additional properties of at least one weak solution. The first one is the fact that $\omega = \operatorname{curl} u$ is a renormalized solution of (4.6), that is it satisfies (4.26). This fact can be recovered a posteriori when $r \geq 2$ using a regularization technique as in Lemma 2.3 (section 2.3, chapter 2) for the equation (4.6) which is satisfied in the sense of distributions, but it is not clear that this can be done when $1 < r < 2$.

The second property we want to mention is the local form of the conservation of energy, namely equation (4.10): our existence proof shows it holds for at least one weak solution if $r \geq \frac{6}{5}$. Indeed, in that case we obtain the convergence of $u^\varepsilon, p^\varepsilon$ in $C([0,T]; L^3)$, $C([0,T]; L^{3/2})$ respectively for all $T \in (0,\infty)$ and we can recover (4.10).

2) Concerning the convergence of solutions of Navier–Stokes equations to solutions of Euler equations (with the boundary conditions modification introduced in step 2 of the proof of Theorem 4.1 in the case of Dirichlet boundary conditions), let us mention that when $1 < r < 2$ it is possible to show that ω_ν converges to ω in $C([0,T];L^r)$ and thus ∇u_ν converges to ∇u in $C([0,T];L^r)$ for all $T \in (0,\infty)$ using the (duality) method of the last section of R.J. DiPerna and P.L. Lions [128].

3) In the case of Dirichlet boundary conditions, it is possible to say a bit more about the regularity of p when $r \geq 2$. Indeed, we have (4.27) and thus $\Delta p \in C([0,\infty);L^{r/2})$ (L^1 being replaced by $\mathcal{H}^1$ when $r = 2$). In addition, one can show by the proof of Theorem 4.1 that there exists a weak solution (u,p) such that, denoting by κ the curvature of $\partial\Omega$, we have

$$\frac{\partial p}{\partial n} = -\kappa |u|^2 \quad \text{on} \quad \partial\Omega \times (0,\infty). \tag{4.28}$$

This is what we expect from (4.1) since $(u \cdot \nabla)u \cdot n = -(u \cdot \nabla)n \cdot u$ ($u \cdot n = 0$ on $\partial\Omega$ and thus $(u \cdot \nabla)(u \cdot n) = 0$ on $\partial\Omega$). Notice that if $\Omega \subset \mathbb{R}^N$, $N \geq 2$, $\kappa |u|^2$ is replaced by the "curvature quadratic form" applied to u. Combining (4.27) and (4.28), it is not difficult to show by elliptic regularity that $D^2 p \in C([0,\infty);L^{r/2}(\Omega))$. □

4.2 Remarks on Euler equations in two dimensions

This section is essentially devoted to a discussion of the Euler equation in two dimensions when the initial condition u_0 only lies in $L^2(\Omega)$ or belongs to L^2 and is such that $\operatorname{curl} u_0$ is a bounded measure. Roughly speaking, this corresponds to the case when $r = 1$ in the preceding section, a case which was of course excluded from our analysis. This borderline situation is not only very interesting mathematically but also corresponds to various relevant physical situations. We refer the reader to the fundamental series of works by R.J. DiPerna and A. Majda [129],[130] on this subject for a more complete discussion of the background of this issue (and of "vortex sheets"). Only at the end of this section shall we leave this issue to mention a few other questions of interest on Euler equations in two dimensions.

Let us now describe what we discuss below. First of all, if $u_0 \in L^2$, $\operatorname{div} u_0 = 0$ (and $u_0 \cdot n = 0$ on $\partial\Omega$ in the case of Dirichlet boundary conditions), the existence, uniqueness and stability of solutions are completely open. However, using the regularity which is available for smooth u_0 (and a few simple tricks), we shall see that there exists a G_δ set of initial conditions in L^2 (that is a countable intersection of dense open sets in L^2) for which there exists a unique solution of (4.1)–(4.4) in $C([0,\infty);L^2)$ with a conserved energy (i.e. $\int_\Omega |u(t)|^2 dx$ is independent of t). As we shall see this is a "cheap" result whose only merit is to indicate that the problem is well posed for most initial conditions in L^2.

The other angle of attack that we discuss in this section consists in pushing as much as we can towards L^1 the arguments developed in the previous section, which are obviously based upon the transport equation (4.6). Since (4.6) involves a divergence-free vector field we expect solutions of (4.6) to preserve the initial distributions function (or in other words, we expect the decreasing rearrangement of solutions to be independent of t)—and this is precisely the case with renormalized solutions. This will lead to two different kinds of results which are essentially optimal for this type of approach. However, we shall remain rather "far" from L^1 or bounded measures. We shall not address here in detail the problem of vortex sheets ($u_0 \in L^2$, curl u_0 is a bounded measure) and we refer instead to R.J. DiPerna and A. Majda [**129**],[**130**] for a discussion of the possible phenomena involved—see also the presentation of their results in L.C. Evans [**141**]. Let us also mention the existence of global weak solutions in the case when $u_0 \in L^2$, curl u_0 is a bounded measure such that $(\text{curl } u_0)^+$ (or $(\text{curl } u_0)^-$) $\in L^1$ which was obtained by J.M. Delort [**118**]; a simpler proof was proposed by A. Majda [**318**].

We now begin with our generic result. We introduce the Hilbert space H (for the L^2 scalar product) defined by: $H = \{u_0 \in L^2(\Omega)^2, \text{div } u_0 = 0 \text{ in } \mathcal{D}'(\Omega), u_0 \cdot n = 0 \text{ on } \partial\Omega\}$. In the case of Dirichlet boundary conditions, $H = \{u_0 \in L^2(\mathbb{R}^2)^2, \text{ div } u_0 = 0 \text{ in } \mathcal{D}'(\mathbb{R}^2)\}$ if $\Omega = \mathbb{R}^2$, $H = \{u_0 \in L^2_{\text{loc}}(\mathbb{R}^2)^2, u_0 \text{ is periodic}, \text{ div } u_0 = 0 \text{ in } \mathcal{D}'(\mathbb{R}^2)\}$ in the periodic case.

Theorem 4.2. *There exists a decreasing sequence of dense open sets $\mathcal{O}_n$ in H such that, for any $u_0 \in \bigcap_{n\geq 1} \mathcal{O}_n$, there exists a unique solution $u \in C([0,\infty); L^2)^2$ of (4.1)–(4.2) (and (4.4) in the case of Dirichlet boundary conditions) such that $\int_\Omega |u(t)|^2 dx$ is independent of $t \geq 0$. Furthermore, for any weak solution $\tilde{u} \in L^\infty(0,\infty; L^2)^2 \cap C([0,\infty); L^2 - w)^2$ of (4.1)–(4.2) (and (4.4) in the case of Dirichlet boundary conditions) such that $\int_\Omega |\tilde{u}(t)|^2 dx \leq \int_\Omega |u_0|^2 dx$ for all $t \geq 0$, then $\tilde{u} \equiv u$ in $\Omega \times (0,\infty)$.*

Proof of Theorem 4.2. The proof is based upon the fact that if $\overline{u}_0 \in L^2 \cap C^{1,\alpha}$ for some fixed $\alpha \in (0,1)$, then, see Theorem 4.1, there exists a unique solution $\overline{u} \in C([0,\infty); L^2 \cap C^{1,\alpha})$ of (4.1),(4.2) with u_0 replaced by $\overline{u}_0$ (and (4.4)) such that, in particular, $\|\nabla \overline{u}\|_{L^\infty(\Omega \times (0,T))} \leq C(\overline{u}_0, T)$ for all $T \in (0,\infty)$, and we can always assume that $C(\overline{u}_0, T) \geq 0$ is nondecreasing with respect to T.

The second ingredient which is basic for our proofs is the following (essentially classical) observation. If $\tilde{u}$ is any weak solution of the Euler equation as in Theorem 4.2 then we claim that we have for all $t \in [0, T]$

$$\|\tilde{u}(t) - \overline{u}(t)\|_{L^2} \leq e^{C(\overline{u}_0, T)t} \|u_0 - \overline{u}_0\|_{L^2}. \tag{4.29}$$

Indeed, on the one hand we have for all $t \geq 0$

$$\int_\Omega |\tilde{u}(t)|^2 \, dx \leq \int_\Omega |u_0|^2 \, dx, \quad \int_\Omega |\overline{u}(t)|^2 \, dx = \int_\Omega |\overline{u}_0|^2 \, dx \qquad (4.30)$$

and on the other hand we deduce from (4.7) using $\varphi = \overline{u}$ (a choice that can be justified by a simple approximation argument, take $\varphi_n = P(\tilde{\varphi}_n)$ where $\tilde{\varphi}_n$ converges in $C^{1,\alpha}$ to $\overline{u}$)

$$\int_\Omega \tilde{u}(t) \cdot \overline{u}(t) \, dx - \int_0^t ds \int_\Omega \tilde{u} \cdot \left\{ \frac{\partial \overline{u}}{\partial t} + (\tilde{u} \cdot \nabla)\overline{u} \right\} dx = \int_\Omega u_0 \cdot \overline{u}_0 \, dx$$

or equivalently using the equation satisfied by $\overline{u}$ (and the fact that div $\tilde{u} = 0$)

$$\int_\Omega \tilde{u}(t) \cdot \overline{u}(t) \, dx - \int_0^t ds \int_\Omega \tilde{u} \cdot \{[(\tilde{u}-\overline{u}) \cdot \nabla]\overline{u}\} \, dx = \int_\Omega u_0 \cdot \overline{u}_0 \, dx.$$

Next, we observe that $\int_0^t ds \int_\Omega \overline{u} \cdot \{[(\tilde{u}-\overline{u}) \cdot \nabla]\overline{u}\} \, dx = \int_0^t ds \int_\Omega (\tilde{u}-\overline{u}) \cdot \nabla\left(\frac{|\overline{u}|^2}{2}\right) dx = 0$ using the fact that div $\tilde{u} =$ div $\overline{u} = 0$ (and $(\tilde{u}-\overline{u}) \cdot n = 0$ in the case of Dirichlet boundary conditions), and thus we obtain for all $t \geq 0$

$$\int_\Omega \tilde{u}(t) \cdot \overline{u}(t) \, dx = \int_\Omega u_0 \cdot \overline{u}_0 \, dx + \int_0^t ds \int_\Omega dx \, (\tilde{u}-\overline{u}) \cdot \nabla \overline{u} \cdot (\tilde{u}-\overline{u}). \quad (4.31)$$

Combining (4.30) and (4.31), we deduce finally for $t \in [0, T]$

$$\int_\Omega |\tilde{u}(t)-\overline{u}(t)|^2 \, dx \leq \int_\Omega |u_0-\overline{u}_0|^2 \, dx + 2C(\overline{u}_0, T) \int_0^t ds \int_\Omega |\tilde{u}(s)-\overline{u}(s)|^2 \, dx \qquad (4.32)$$

and we deduce (4.29) from (4.32) using Grönwall's inequality.

We then introduce, for $n \geq 1$, the following open set $\mathcal{O}_n$

$$\mathcal{O}_n = \left\{ u_0 \in H \,/\, \exists \, \overline{u}_0 \in L^2 \cap C^{1,\alpha}, \, \|u_0-\overline{u}_0\|_{L^2} < \frac{1}{n} e^{-C(\overline{u}_0, n)} \right\} \quad (4.33)$$

and we wish to check the statements listed in Theorem 4.2 for this choice of $\mathcal{O}_n$. Since $\mathcal{O}_n$ contains $L^2 \cap C^{1,\alpha}(\cap H)$, it is not difficult to check that $\mathcal{O}_n$ is dense in H : indeed, $(L^2 \cap C^{1,\alpha})^2$ is dense in $(L^2)^2$ and $P((L^2 \cap W^{1,\alpha})^2) = (L^2 \cap C^{1,\alpha})^2 \cap H$.

Next, if $u_0 \in \bigcap_{n \geq 1} \mathcal{O}_n$, it is not difficult to show the uniqueness part of Theorem 4.2. Indeed, if $u_0 \in \bigcap_{n \geq 1} \mathcal{O}_n$, there exists, for each $n \geq 1$, some $\overline{u}_0^n \in L^2 \cap C^{1,\alpha} \cap H$ such that $\|u_0-\overline{u}_0^n\|_{L^2} < \frac{1}{n} e^{-C(\overline{u}_0^n, n)n}$. Then, if $\tilde{u}^1, \tilde{u}^2$

are two weak solutions as in Theorem 4.2, (4.29) implies $\sup_{t\in[0,n]}\|\tilde{u}^1(t)-\tilde{u}^2(t)\|_{L^2}<\frac{2}{n}$, and the uniqueness is proven.

The existence part also follows from (4.29). It is clearly enough to show that $\overline{u}^n$ is a Cauchy sequence in $C([0,T];L^2)$ (for all $T\in(0,\infty)$). Then, if $n,m\geq T$, we deduce from (4.29) that we have

$$\sup_{t\in[0,T]}\|\overline{u}^n(t)-\overline{u}^m(t)\|_{L^2}$$

$$\leq\min\left\{e^{C(\overline{u}_0^n,n)T},\,e^{C(\overline{u}_0^m,m)T}\right\}\|\overline{u}_0^n-\overline{u}_0^m\|_{L^2}$$

$$\leq\min\left\{e^{C(\overline{u}_0^n,n)T},\,e^{C(\overline{u}_0^m,m)T}\right\}\left[\|\overline{u}_0^n-u_0\|_{L^2}+\|u_0-\overline{u}_0^m\|_{L^2}\right]$$

$$\leq\min\left\{e^{C(\overline{u}_0^n,n)T},\,e^{C(\overline{u}_0^m,m)T}\right\}\left[\frac{1}{n}e^{-C(\overline{u}_0^n,n)n}+\frac{1}{m}e^{-C(\overline{u}_0^m,m)m}\right]$$

$$\leq\frac{1}{n}+\frac{1}{m}.$$

Our claim is shown, thus completing the proof of Theorem 4.2. □

We next discuss some other existence results based upon the fact that we expect $\omega=\operatorname{curl}u$ to satisfy (4.6), i.e. a transport equation with a divergence-free vector field (namely u). Thus, the distributions function of $\omega(t)$, namely the function $\mu_{\omega(t)}$ on $(0,\infty)$ defined by $\mu_{\omega(t)}(\lambda)=\operatorname{meas}\{x\,/\,|\omega(x,t)|>\lambda\}$ for $\lambda>0$, should be independent of t. Notice that this is precisely the case with renormalized solutions of (4.6). Indeed, (4.26) yields the fact (integrating in x) that $\int_\Omega\beta(\omega(t))\,dx$ is independent of t for all β.

From now on, in order to avoid unnecessary technicalities, we restrict ourselves to the case of Dirichlet boundary conditions (or to the periodic case).

The first type of result we wish to discuss consists in pushing the proof of Theorem 4.1 towards L^1. Looking carefully at the proof of Theorem 4.1, we see that we only need a bound on u in $W^{1,1}$. But, if we introduce the stream function, i.e. the solution of

$$-\Delta\psi=\omega\quad\text{in}\quad\Omega,\quad\psi=0\quad\text{on}\quad\partial\Omega\tag{4.34}$$

in the case of Dirichlet boundary conditions for instance assuming that Ω is simply connected (to simplify the presentation), then $u=\left(\begin{smallmatrix}-\frac{\partial}{\partial x_2}\psi\\\frac{\partial}{\partial x_1}\psi\end{smallmatrix}\right)$. Therefore, $Du\in L^1$ if and only if $D^2\psi\in L^1$, and by classical elliptic theory, this is the case in two dimensions as soon as $\omega\in L^1$, $\int_\Omega|\omega|\,|\log|\omega(t)||\,dx<\infty$. Observe in addition that $\int_\Omega|\omega(t)|\,|\log|\omega(t)||$ is, at least formally, independent of t. Once these observations are made, it is not hard to copy the proof of Theorem 4.1 and to show that if $u_0 \in L^2(\Omega)$, div $u_0 = 0$ in Ω, $u_0 \cdot n = 0$ on $\partial\Omega$, $\omega_0 = $ curl $u_0 \in L^1(\Omega)$ and $\int_\Omega |\omega_0| \, |\log \omega_0| \, dx < \infty$, then there exists a solution $u \in C([0,\infty); W^{1,1}(\Omega))$ of (4.1), (4.2), (4.4) such that $\int_\Omega |u(t)|^2 \, dx$ is independent of t, $\omega \in C([0,\infty); L^1(\Omega))$ (and even "$L^1 \log L^1$") is a renormalized solution of (4.6), i.e. satisfies (4.26). Indeed, regularizing ω_0, we obtain a sequence of solutions of (4.1) $(u^\varepsilon, \omega^\varepsilon)$ and, by the results of R.J. DiPerna and P.L. Lions [**128**], we check that ω^ε converges in $C([0,T]; L^1)$ ($\forall\, T \in (0,\infty)$) to some $\omega \in C([0,\infty); L^1(\Omega))$. In addition, $\omega^\varepsilon |\log |\omega^\varepsilon||$ also converges in $C([0,T]; L^1)$ ($\forall\, T \in (0,\infty)$) to $\omega |\log \omega|$. Hence, u^ε converges to some u in $C([0,T]; L^2)$ ($\forall\, T \in (0,\infty)$) and we conclude.

However, if we follow this argument, we can ask for less information on ω and we might simply try to deduce from the invariance in t of $\mu_{\omega^\varepsilon(t)}$ ($= \mu_{\omega_0^\varepsilon}$) some compactness in L^2 (or $C([0,T]; L^2)$) for all $T \in (0,\infty)$ of u^ε. This leads to the following question: find the optimal distributions (or rearrangement) invariant class for ω such that the corresponding velocity field u belongs to $L^2(\Omega)$ and then prove the existence of solutions of (4.1) with such initial conditions. This question can be solved completely using some symmetrization techniques. In order to do so, we need some notation. First of all, if $\omega \in L^1(\Omega)$, we denote by ω^* the decreasing rearrangement which is the inverse function of μ_ω: in other words, ω^* is the unique nondecreasing function in $L^1(0, |\Omega|)$ ($|\Omega| = \text{meas}(\Omega)$) such that $\omega^* \in L^1(0, |\Omega|)$ and $\mu_{\omega^*} = \mu_\omega$ a.e. In two dimensions, we denote by $\omega^\sharp$ the Schwarz symmetrization of ω (or spherically symmetric decreasing rearrangement), i.e. the unique spherically symmetric function in $L^1(\Omega^\sharp)$ which is nondecreasing with respect to $r = |x|$ such that $\mu_{\omega^\sharp} = \mu_\omega$ a.e., where $\Omega^\sharp$ is the ball centred at 0 with the same volume as Ω (or in other words with a radius R_0 given by $\left(\frac{|\Omega|}{\pi}\right)^{1/2}$). Obviously, $\omega^\sharp(x) = \omega^*(\pi|x|^2)$ a.e. in $\Omega^\sharp$.

From now on, we restrict ourselves (to simplify the presentation) to the case of Dirichlet boundary conditions, assuming in addition that Ω is simply connected (even though the results below hold in general), and we introduce the stream function, i.e. the solution of (4.34). We then recall the following general comparison result due to G. Talenti [**462**]

$$\psi^\sharp \leq \Psi \qquad \text{a.e. in} \quad \Omega^\sharp \tag{4.35}$$

$$\int_\Omega |\nabla\psi|^2 \, dx \leq \int_{\Omega^\sharp} |\nabla\Psi|^2 \, dx \tag{4.36}$$

where Ψ is the solution of (4.34) with Ω, ω replaced respectively by $\Omega^\sharp, \omega^\sharp$, namely

$$-\Delta\Psi = \omega^\sharp \quad \text{in} \quad \Omega^\sharp, \quad \Psi = 0 \quad \text{on} \quad \partial\Omega^\sharp. \tag{4.37}$$

Recall that Ψ is given by the explicit formula

$$\Psi(x) = \int_{|x|}^{R_0} \frac{1}{r} \int_0^r s\omega^\sharp(s)\,ds, \quad \Psi'(|x|) = -\frac{1}{|x|}\int_0^{|x|} s\omega^\sharp(s)\,ds. \quad (4.38)$$

Then, if we observe that $u = \begin{pmatrix} -\frac{\partial\psi}{\partial x_2} \\ \frac{\partial\psi}{\partial x_1} \end{pmatrix}$, we see that the optimal rearrangement invariant class ensuring that $u \in L^2(\Omega)$ is given by

$$\left\{ \omega \in L^1(\Omega) \;/\; \int_{\Omega^\sharp} |\nabla\Psi|^2\,dx < \infty \right\}.$$

But we have easily

$$\begin{aligned}
\int_{\Omega^\sharp} |\nabla\Psi|^2\,dx &= 2\pi \int_0^{R_0} r(\Psi'(r))^2\,dr = 2\pi \int_0^{R_0} \frac{1}{r}\left(\int_0^r s\omega^\sharp(s)\,ds\right)^2 dr \\
&= 2\pi \int_0^{R_0} \frac{1}{r}\left(\int_0^r s\omega^*(\pi s^2)\,ds\right)^2 dr \\
&= \frac{1}{2\pi} \int_0^{R_0} \frac{1}{r}\left(\int_0^{\pi r^2} \omega^*(s)\,ds\right)^2 dr \\
&= \frac{1}{4\pi} \int_0^{|\Omega|}\left(\int_0^t \omega^*(s)\,ds\right)^2 \frac{dt}{t}.
\end{aligned}$$

In conclusion, the optimal class is given by the space

$$L_2^1(\Omega) = \left\{ \omega \in L^1(\Omega) \;/\; \int_0^{|\Omega|}\left(\int_0^t \omega^*(s)\,ds\right)^2 \frac{dt}{t} < \infty \right\}. \quad (4.39)$$

For instance, in view of the observation made by A.B. Mergulis [341], it contains the Orlicz class of functions ω in $L^1(\Omega)$ such that $\int_\Omega |\omega| \cdot |\log|\omega||^{1/2}\,dx < \infty$.

We then have

Theorem 4.3. *Let $\omega_0 = \operatorname{curl} u_0 \in L_2^1(\Omega)$ (with $\operatorname{div} u_0 = 0$ in Ω, $u_0 \cdot n = 0$ on $\partial\Omega$). Then there exists a solution $u \in C([0,\infty); L^2(\Omega)^2)$ of (4.1), (4.2), (4.4) such that $\int_\Omega |u(t)|^2\,dx$ is independent of $t \geq 0$.*

Remarks 4.8. 1) The results due to J.M. Delort [118], that we mentioned above, show the existence of a weaker solution of (4.1), (4.2), (4.4) when $\omega_0 \in L^1(\Omega)$, $u_0 \in L^2(\Omega)$ since the conservation of energy is not known in that case and $u \in L^\infty(0,\infty; L^2(\Omega))$).

2) If $\Omega = \Omega^\sharp$ (is a ball!) and if we consider $\omega(x) = \frac{1}{|x|^2}|\log|x||^{-\alpha}$, then one can check that $\omega \in L^1$ if and only if $\alpha > 1$, $\omega \in L_2^1$ if and only if $\alpha > \frac{3}{2}$ while $D^2\psi\ (=D^2\Psi) \in L^1(\Omega)$ if and only if $\alpha > 2$. $\quad\square$

Proof of Theorem 4.3. Let us define $w_0^n = T_n(w_0)$ for $n \geq 1$ and let u_0^n be the unique element of $L^2(\Omega)^2$ such that: $\operatorname{curl} u_0^n = w_0^n$, $\operatorname{div} u_0^n = 0$ and $u_0^n \cdot n = 0$ on $\partial\Omega$. In view of Theorem 4.1, there exists a unique solution $u^n \in C([0,\infty); L^2(\Omega)^2)$ of (4.1), (4.2) and (4.4) with u_0 replaced by u_0^n, such that $\operatorname{curl} u^n \in L^\infty(\Omega \times (0,\infty))$. Of course, we wish to recover the existence result stated in Theorem 4.3 by passing to the limit as n goes to $+\infty$. To this end we recall that $\mu_{w^n(t)} = \mu_{w_0^n}$ for all $t \geq 0$, $n \geq 1$ and thus we have, writing by $w_t^n = w^n(t)$,

$$(w_t^n)^* = (w_0^n)^* = w_0^* \wedge n. \tag{4.40}$$

In particular, in view of the derivation above of $L_{\frac{1}{2}}^1(\Omega)$, u_0^n is bounded in $L^2(\Omega)^2$, and since $|u^n(t)|_{L^2} = |u_0^n|_{L^2}$, we finally deduce that u^n is bounded in $C([0,\infty); L^2(\Omega)^2)$.

Next, because of (4.7) and since P maps $\{\varphi \in C^{1,\alpha}(\overline{\Omega})^2 , \; \varphi = 0 \text{ on } \partial\Omega\}$ into $C^{1,\alpha}(\overline{\Omega})^2$ for any $0 < \alpha < 1$, $\frac{\partial u^n}{\partial t}$ is bounded in $C([0,\infty); X_\alpha')$ for any $0 < \alpha < 1$ where X_α is the closure of $C_0^\infty(\Omega)^2$ in $C^{1,\alpha}(\overline{\Omega})^2$ ($= \{v \in C^{1,\alpha}(\overline{\Omega})^2 , \; v = \nabla v = 0 \text{ on } \partial\Omega\}$). Then, we deduce from Appendix C that, extracting a subsequence if necessary, we may assume that u^n converges in $C([0,T]; L^2(\Omega)^2 - w)$ (for all $T \in (0,\infty)$) to some $u \in C([0,\infty); L^2(\Omega)^2 - w)$ satisfying (4.2) and (4.4). In addition, we have

$$\int_\Omega |u(t)|^2 \, dx \leq \varliminf_{n\to\infty} \int_\Omega |u^n(t)|^2 \, dx = \varliminf_{n\to\infty} \int_\Omega |u_0^n|^2 \, dx. \tag{4.41}$$

We are going to show below that we have

$$u^n(t_n) \underset{n}{\to} u(t) \quad \text{in} \quad L^2(\Omega)^2 \quad \text{if} \quad 0 \leq t_n \underset{n}{\to} t \geq 0. \tag{4.42}$$

If this claim were proved, we would deduce on the one hand that u_0^n converges in $L^2(\Omega)^2$ to u_0 and on the other hand that for all $t \geq 0$

$$\int_\Omega |u(t)|^2 \, dx = \liminf_n \int_\Omega |u^n(t)|^2 \, dx$$

$$= \liminf_n \int_\Omega |u_0^n|^2 \, dx = \int_\Omega |u_0|^2 \, dx.$$

Hence, in particular, $u \in C([0,\infty); L^2(\Omega)^2)$, and this fact combined with (4.42) would show that u^n converges to u in $C([0,T]; L^2(\Omega)^2)$ for all $T \in (0,\infty)$. Once this convergence is shown, Theorem 4.3 follows easily.

Therefore, we have only to prove (4.42). Let us first remark that we already know that $u^n(t_n)$ converges to $u(t)$ weakly in $L^2(\Omega^2)$. In addition, since $\Delta\psi^n(t_n)$ is bounded in $L^1(\Omega)$ (and $\nabla\psi^n(t_n)$ is bounded in $L^2(\Omega)^2$), we deduce from elliptic regularity that $\nabla \psi^n(t_n)$ and thus $u^n(t_n)$ converge in $L^p(\Omega)^2$ for all $1 \le p < 2$. Finally, let us observe that (4.40) yields for all $T \in (0, \infty)$

$$\sup_{n \ge 1} \sup_{t \in [0,T]} \int_0^\varepsilon \left(\int_0^\lambda \omega_t^n(s)^* \, ds \right)^2 \frac{d\lambda}{\lambda} \to 0 \quad \text{as} \quad \varepsilon \to 0_+. \qquad (4.43)$$

In view of these facts, (4.42) is deduced from the following

Lemma 4.1. *Let ω_n be bounded in $L^1(\Omega)$ and satisfy*

$$\left. \begin{array}{l} \displaystyle \int_0^{|\Omega|} \left(\int_0^t \omega_n^* \, ds \right)^2 \frac{dt}{t} < \infty, \\[3mm] \displaystyle \sup_{n \ge 1} \int_0^\varepsilon \left(\int_0^t \omega_n^* \, ds \right)^2 \frac{dt}{t} \to 0 \quad \text{as} \quad \varepsilon \to 0_+. \end{array} \right\} \qquad (4.44)$$

Then, ω_n is weakly relatively compact in $L^1(\Omega)$ and, extracting a subsequence if necessary, we may assume that ω_n converges weakly in $L^1(\Omega)$ to some $\omega \in L_2^1(\Omega)$. Denoting by ψ_n the solution of (4.34) (with ω replaced by ω_n), we then have

$$\psi_n \omega_n \underset{n}{\rightharpoonup} \psi \omega \quad \text{weakly in } L^1(\Omega), \quad \psi_n \underset{n}{\to} \psi \quad \text{in } H_0^1(\Omega). \qquad (4.45)$$

Proof of Lemma 4.1. The weak compactness in L^1 follows from the following observations

$$\int_\varepsilon^{|\Omega|} \left(\int_0^t \omega_n^* \, ds \right)^2 \frac{dt}{t} \ge \left(\int_0^\varepsilon \omega_n^* \, ds \right)^2 \log \frac{|\Omega|}{\varepsilon}, \quad \text{for all} \quad \varepsilon > 0,$$

and

$$\int_0^\varepsilon \omega_n^* \, ds = \sup_{|A| \le \varepsilon} \int_A |\omega_n| \, dx, \quad \text{for all} \quad \varepsilon > 0. \qquad (4.46)$$

Hence, we may assume that ω_n converges weakly in $L^1(\Omega)$ to some ω and, similarly, that $\omega_n^\sharp$ converges weakly in $L^1(\Omega^\sharp)$ to some $\tilde{\omega}$. Since $\omega_n^\sharp$ is radially increasing, we may assume without loss of generality that $\omega_n^\sharp$ converges to $\tilde{\omega}$ a.e. in $\Omega^\sharp$ and thus $\omega_n^\sharp$ converges to $\tilde{\omega}$ in $L^1(\Omega^\sharp)$. In particular, one deduces easily that $\tilde{\omega} \in L_2^1(\Omega)$ and, since $\int_0^t \omega^* \, ds \le \lim_n \int_0^t \omega_n^* \, ds = \int_0^t \tilde{\omega}^* \, ds$, $\omega \in L_2^1(\Omega)$. Therefore, as shown above when we introduced $L_2^1(\Omega)$, $\omega_n \psi_n$ is bounded in $L^1(\Omega)$, $\omega \psi \in L^1(\Omega)$ (observe indeed that $\int_\Omega |\omega \psi| \, dx \le \int_{\Omega^\sharp} \omega^\sharp \psi^\sharp \, dx \le \int_{\Omega^\sharp} \omega^\sharp \Psi^\sharp \, dx$)

$$\int_\Omega |\nabla \psi_n|^2 \, dx = \int_\Omega \omega_n \psi_n \, dx, \quad \int_\Omega |\nabla \psi|^2 \, dx = \int_\Omega \omega \psi \, dx \qquad (4.47)$$

$$\int_{\Omega^{\sharp}} |\nabla \Psi_n|^2 \, dx = \int_{\Omega^{\sharp}} \omega_n^{\sharp} \Psi_n \, dx, \quad \int_{\Omega^{\sharp}} |\nabla \Psi|^2 \, dx = \int_{\Omega^{\sharp}} \tilde{\omega} \, \Psi^{\sharp} \, dx \quad (4.48)$$

with the notation introduced before Theorem 4.3 (and obvious adaptations).

We first claim that $\omega_n^{\sharp} \Psi_n$, $\nabla \Psi_n$ converge, respectively, in $L^1(\Omega^{\sharp})$, $L^2(\Omega^{\sharp})^2$ to $\tilde{\omega}\Psi$, $\nabla \Psi$. Since $\omega_n^{\sharp}, \Psi_n$ are radially nonincreasing, non-negative and $\nabla \Psi_n = -\frac{x}{|x|^2} \int_0^{|x|} s \omega_n^*(s) \, ds$, we have only to show in view of (4.48)

$$\sup_{n \geq 1} \int_{|x| < \varepsilon} |\nabla \Psi_n|^2 \, dx \to 0 \quad \text{as} \quad \varepsilon \to 0_+ .$$

But this is immediate in view of (4.44) since we check as in the definition of $L_2^1(\Omega)$

$$\int_{|x| < \varepsilon} |\nabla \Psi_n|^2 \, dx = \frac{1}{4\pi} \int_0^{\pi \varepsilon^2} \left(\int_0^t \omega_n^* \, ds \right)^2 \frac{dt}{t}.$$

Next, we check that $\psi_n \omega_n$ is weakly relatively compact in $L^1(\Omega)$. Indeed, for any measurable set A

$$\int_A \psi_n \omega_n \, dx \leq \int_{A^{\sharp}} \psi_n^{\sharp} \omega_n^{\sharp} \, dx \leq \int_{A^{\sharp}} \Psi_n \omega_n^{\sharp} \, dx$$

and our claim is shown since $\Psi_n \omega_n^{\sharp}$ converges in $L^1(\Omega^{\sharp})$.

It only remains to show that $\psi_n \omega_n$ converges weakly in $L^1(\Omega)$ to $\psi\omega$. Recalling that ψ_n converges a.e. to ψ, we see that for all $R \in (0, \infty)$, $T_R(\psi_n)\omega_n$ converges weakly in $L^1(\Omega)$ to $T_R(\psi)\omega$ since ω_n converges weakly in $L^1(\Omega)$ to ω. We can then conclude observing that

$$\int_{\Omega} |\psi_n - T_R(\psi_n)| \, |\omega_n| \, dx = \int_{\Omega} (|\psi_n| - R)_+ |\omega_n| \, dx$$
$$\leq \int_{\Omega^{\sharp}} (\psi_n^{\sharp} - R)_+ \omega_n^{\sharp} \, dx \leq \int_{\Omega^{\sharp}} (\Psi_n - R)_+ \, \omega_n^{\sharp} \, dx$$
$$\leq \int_{\Omega^{\sharp}} \Psi_n \omega_n^{\sharp} \, 1_{(\Psi_n > R)} \, dx$$

and thus, since Ψ_n converges in $L^1(\Omega^{\sharp})$ (in $H^1(\Omega^{\sharp})$ in fact), $\Psi_n \omega_n^{\sharp}$ converges in $L^1(\Omega^{\sharp})$, these integrals can be made, uniformly in $n \geq 1$, arbitrarily small letting R go to $+\infty$. □

Remark 4.9. Let us mention in passing the so-called "vortex-patches" problem which was settled recently. One considers the case of an initial vorticity (in the case when $\Omega = \mathbb{R}^2$ for instance) ω_0 which is constant (say

A) and supported on a bounded, smooth domain D. Then, the corresponding solution of (4.1), (4.2), (4.4) satisfies for all $t \geq 0$: $\omega(t) = A \, 1_{D(t)}$ for some measurable set $D(t)$. It is proved in J.Y. Chemin [88] (a simplified proof can be found in A. Bertozzi and P. Constantin [58], see also Ph. Serfati [419]) that $D(t)$ is in fact a bounded, smooth domain for all $t \geq 0$. □

We conclude this (two-dimensional) section with a different topic which concerns the case when $\Omega = \mathbb{R}^2$ and when we no longer assume that $u_0 \in L^2(\mathbb{R}^2)^2$. The first systematic treatment of that question seems to be given in D. Benedetto, C. Marchioro and M. Pulvirenti [51] where the case of an initial condition u_0 satisfying ((4.3) and), for some $C \geq 0$,

$$\left. \begin{array}{l} |u_0(x)| \leq C(1+|x|^\alpha), \quad 0 \leq \alpha < \min\left(1, \dfrac{2}{p}\right), \\[2mm] \omega_0 = \operatorname{curl} u_0 \in L^p \cap L^\infty(\mathbb{R}^2) \quad \text{for some} \quad 1 \leq p < \infty, \end{array} \right\} \tag{4.49}$$

is treated. We wish to extend this analysis here to the case when $1 < p < \infty$ and

$$u_0 \in W^{1,p}_{\text{loc}}(\mathbb{R}^2), \quad \nabla u_0 \in L^p(\mathbb{R}^2), \quad \operatorname{div} u_0 = 0 \quad \text{a.e. in } \mathbb{R}^2. \tag{4.50}$$

We shall be mainly using the vorticity formulation of (4.1)

$$\frac{\partial \omega}{\partial t} + \operatorname{div}(u\omega) = 0 \quad \text{in} \quad \mathcal{D}'(\mathbb{R}^2 \times (0,\infty)) \tag{4.51}$$

requiring u to belong to $C([0,\infty); W^{1,p}_{\text{loc}}(\mathbb{R}^2))$ (i.e. $u \in C([0,\infty); W^{1,p}(B_R))$ for all $R \in (0,\infty)$), to satisfy (4.1) (in the weak sense defined in section 4.1) and

$$\left. \begin{array}{l} \nabla u \in C([0,\infty); L^p(\mathbb{R}^2)); \quad \operatorname{div} u = 0, \\[2mm] \operatorname{curl} u = \omega \quad \text{a.e. in} \quad \mathbb{R}^2 \times (0,\infty). \end{array} \right\} \tag{4.52}$$

Let us mention that (4.51), as in section 4.1, holds if $p \geq 2$ and, if $p < 2$, (4.51) holds in the renormalized sense, i.e. (4.51) holds with ω replaced by $\beta(\omega)$ for all $\beta \in C_b(\mathbb{R}; \mathbb{R})$.

We may now state our main existence result.

Theorem 4.4. *Let u_0 satisfy* (4.50). *Then, there exists $u \in C([0,\infty); W^{1,p}_{\text{loc}}(\mathbb{R}^2))^2$, a weak solution of* (4.1)–(4.2), *satisfying* (4.51)–(4.52).

Remarks 4.10. 1) If $p < 2$, the proof below shows that one can build $u \in C([0,\infty); L^q(\mathbb{R}^2))^2$ with $q = \dfrac{2p}{2-p}$, choosing u_0 (up to a constant) in $L^q(\mathbb{R}^2)$.

2) It is possible to extend the above result to the case when $\nabla u_0 \in L^{p_1} + L^{p_2}$ where $1 < p_1, p_2 < \infty$.

3) If $\operatorname{curl} u_0$ belongs, in addition to the assumption (4.50), to $L^\infty(\mathbb{R}^2)$, then one can check (by standard arguments) the uniqueness of solutions (normalized by requesting for instance $\int_{B_1} u \, dx = \int_{B_1} u_0 \, dx$, see the proof of Theorem 4.4 below for more details).

4) If $\operatorname{curl} u_0$ only belongs to $L^\infty(\mathbb{R}^2)$, we do not know if the above result (or an appropriate modification of it) holds.

5) Let us mention that when $p > 2$, u grows at infinity at most like $|x|^{1-2/p}$. More precisely, we have for some $C \geq 0$

$$\sup_{\mathbb{R}^2} \frac{|u(x)|}{(1+|x|^2)^{\frac{1}{2}-\frac{1}{p}}} \leq C\Big\{\|\nabla u\|_{L^p(\mathbb{R}^2)} + \|u\|_{L^p(B_1)}\Big\}. \tag{4.53}$$

Indeed, if $u \in W^{1,p}_{\text{loc}}(\mathbb{R}^2)$, $\nabla u \in L^p(\mathbb{R}^2)$, then we have, setting $\alpha = 1 - \frac{2}{p}$,

$$\sup_{x \neq x'} \frac{|u(x)-u(x')|}{|x-x'|^\alpha} \leq C_0 \|\nabla u\|_{L^p(\mathbb{R}^2)} \tag{4.54}$$

for some $C_0 > 0$ independent of u. Therefore, we have

$$\begin{aligned}
|u(x)| &\leq C_0 \|\nabla u\|_{L^p(\mathbb{R}^2)} |x|^\alpha + |u(0)| \\
&\leq C_0 \|\nabla u\|_{L^p(\mathbb{R}^2)} |x|^\alpha + \Big|u(0) - \pi^{-1/p}\|u\|_{L^p(B_1)}\Big| \\
&\quad + \pi^{-1/p}\|u\|_{L^p(B_1)} \\
&\leq C_0 \|\nabla u\|_{L^p(\mathbb{R}^2)} (1+|x|^\alpha) + \pi^{-1/p}\|u\|_{L^p(B_1)}
\end{aligned}$$

using (4.54). The growth $|x|^\alpha$ is essentially optimal since $u = (1+|x|^2)^{\beta/2}$ satisfies $\nabla u \in L^p(\mathbb{R}^2)$ for all $0 < \beta < \alpha$. $\quad\square$

Before we present the proof of Theorem 4.4, let us first make some preliminary remarks. We introduce the Banach space $\tilde{D}^{1,p}(\mathbb{R}^2) = \{u \in W^{1,p}_{\text{loc}}(\mathbb{R}^2), \nabla u \in L^p(\mathbb{R}^2)^2\}$ equipped with one of the equivalent norms $\|\nabla u\|_{L^p(\mathbb{R}^2)} + \|u\|_{L^p(B_1)}$ or $\|\nabla u\|_{L^p(\mathbb{R}^2)} + |\int_{B_1} u|$ (we could as well replace B_1 by any ball B_R for $0 < R < \infty$) and we restrict our attention, as in Theorem 4.4, to the case when $1 < p < \infty$. When $1 < p < 2$, by Sobolev embeddings, we see that $\tilde{D}^{1,p}(\mathbb{R}^2) = \mathbb{R} + \{u \in L^q(\mathbb{R}^2), \nabla u \in L^p(\mathbb{R}^2)^2\}$ where $q = \frac{2p}{2-p}$. We shall need, in the course of proving Theorem 4.4, some technical results on this space given by

Lemma 4.2. Let $u \in \tilde{D}^{1,p}(\mathbb{R}^2)^2$ satisfy: $\operatorname{div} u = 0$ a.e. in $\mathbb{R}^2$.

(i) We have

$$\frac{\partial u}{\partial x_j} = -(-\Delta)^{-1}\frac{\partial}{\partial x_j}\nabla^\perp \operatorname{curl} u \qquad \text{where} \quad \nabla^\perp = \begin{pmatrix} \frac{\partial}{\partial x_2} \\ -\frac{\partial}{\partial x_1} \end{pmatrix}, \tag{4.55}$$

or in other words $\frac{\partial u}{\partial x_j}$ *is given in terms of* $\operatorname{curl} u$ *by a singular integral convolution operator whose kernel is* $\frac{1}{2\pi} \frac{\partial}{\partial x_j} \left(\frac{x^\perp}{|x|^2} \right)$ *and* $x^\perp = \left(\begin{smallmatrix} x_2 \\ -x_1 \end{smallmatrix} \right)$.

(ii) *There exists* $u^n \in C_0^\infty(\mathbb{R}^2)^2$ *such that* $\operatorname{div} u^n = 0$ *on* $\mathbb{R}^2$, ∇u^n *converges to* ∇u *in* $L^p(\mathbb{R}^2)$ *and* u^n *converges to* u *in* $L^p(B_R)$ *(for instance for all* $R \in (0, \infty)$ *if* $p > 2$, *while there exists* $c \in \mathbb{R}^2$ *such that* u^n *converges to* $u - c$ *in* $L^q(\mathbb{R}^2)^2$ *when* $1 < p < 2$ *(and* $q = 2p/(2-p)$*).*

Proof of Lemma 4.2. (i) The proof of (4.55) is straightforward since one checks easily that $f_j = \frac{\partial u}{\partial x_j} + (-\Delta)^{-1} \frac{\partial}{\partial x_j} \nabla^\perp \operatorname{curl} u \in L^p(\mathbb{R}^2)$ (using the fact that the singular integral which is a composition of appropriate Riesz transforms is bounded on $L^p(\mathbb{R}^2)$) and $\operatorname{curl} f_j = 0$, $\operatorname{div} f_j = 0$ in $\mathcal{D}'(\mathbb{R}^2)$. Hence, f_j is harmonic on $\mathbb{R}^2$ (harmonic gradient in fact) and thus $f_j \equiv 0$ a.e. in $\mathbb{R}^2$.

(ii) We have already shown in Appendix A the case when $p = 2$. When $1 < p < 2$, the proof is rather easy since $u = c + v$ where $c \in \mathbb{R}^2$, $v \in L^q(\mathbb{R}^2)^2$, $\nabla v \in L^p(\mathbb{R}^2)$, $\operatorname{div} v = 0$ a.e. in $\mathbb{R}^2$ $(q = \frac{2p}{2-p})$. Then, by "standard" density results (or proofs), we can build $v_n \in C_0^\infty(\mathbb{R}^2)$ such that $\operatorname{div} v_n = 0$ in $\mathbb{R}^2$, v_n converges in $L^q(\mathbb{R}^2)$ to v, ∇v_n converges in $L^p(\mathbb{R}^2)$ to ∇v and the proof is complete in this case.

One possible proof in the case when $p > 2$ consists in several layers of approximations. First of all, we truncate and regularize $\omega = \operatorname{curl} u$ and obtain $\omega^k \in C_0^\infty(\mathbb{R}^2)$ which converges, as k goes to $+\infty$, to ω in $L^p(\mathbb{R}^2)$. Next, we consider u^k defined by $u^k \in L^r(\mathbb{R}^2)^2$ for all $r > 2$, $u^k \in C_b^\infty(\mathbb{R}^2)^2$ and decays at infinity like $\frac{1}{|x|}$, $\operatorname{div} u^k = 0$ in $\mathbb{R}^2$, $\operatorname{curl} u^k = \omega^k$ in $\mathbb{R}^2$ $\left(u^k = \frac{1}{2\pi} \frac{x^\perp}{|x|^2} * \omega^k \right)$. Using, as in (i) above, the boundedness in $L^p(\mathbb{R}^2)$ of Riesz transforms, we deduce that $\nabla u^k \in L^r(\mathbb{R}^2)$ for $1 < r \le \infty$ and that ∇u^k converges in $L^p(\mathbb{R}^2)$ to $-(-\Delta)^{-1} \nabla \nabla^\perp \omega = \nabla u$ in view of (4.55). Therefore, if we set $c^k = f_{B_1} u - u^k \, dx$, $u^k + c^k$ converges in $\tilde{\mathcal{D}}^{1,p}(\mathbb{R}^2)$ to u as k goes to $+\infty$.

Next (using for instance the case $p = 2$ already treated in Appendix A), there exists $u^{k,m} \in C_0^\infty(\mathbb{R}^2)^2$ such that $\operatorname{div} u^{k,m} = 0$ in $\mathbb{R}^2$, $u^{k,m}$ converges in $\tilde{\mathcal{D}}^{1,p}(\mathbb{R}^2)^2$ to u^k as m goes to $+\infty$, for each fixed $k \ge 1$.

Finally, we choose for each $k \ge 1$ some $\varphi^k \in C_0^\infty(\mathbb{R}^2)^2$ such that $\operatorname{div} \varphi^k = 0$ in $\mathbb{R}^2$, $\varphi^k(0) = c^k$ and we set $\varphi^{k,m} = \varphi^k(\frac{x}{m})$ for $m \ge 1$. Observe that for $k \ge 1$ fixed, $\varphi^{k,m}$ converges to c^k in $\tilde{\mathcal{D}}^{1,p}(\mathbb{R}^2)^2$ since we have

$$\int_{\mathbb{R}^2} |\nabla \varphi^{k,m}|^p \, dx = m^{2-p} \int_{\mathbb{R}^2} |\nabla \varphi^k|^p \, dx \to 0 \qquad \text{as} \quad m \to +\infty.$$

In conclusion, we have shown that $u^{k,m} + \varphi^{k,m}$ converges in $\tilde{\mathcal{D}}^{1,p}(\mathbb{R}^2)$ as m goes to $+\infty$ to $u^k + c^k$ which, in turn, converges to u in $\tilde{\mathcal{D}}^{1,p}(\mathbb{R}^2)$ as k goes to $+\infty$, and this completes the proof of Lemma 4.2. $\square$

Remark 4.11. It is possible to give a different proof of part (i) of Lemma 4.2 (when $p \geq 2$) using the classical approach to density results, the fact that the range of the "divergence-map" from $\tilde{D}^{1,p}(\mathbb{R}^2)$ into $L^p(\mathbb{R}^2)$ is $L^p(\mathbb{R}^2)$ and the density of $C_0^\infty(\mathbb{R}^2)$ in $\tilde{D}^{1,p}(\mathbb{R}^2)$. This last fact, however, requires some justification: we can either adapt the argument given above or argue in a slightly more direct way as follows. First, approximating if necessary u by $T_R(u)$—observe that $T_R(u)$ converges to u in $\tilde{D}^{1,p}(\mathbb{R}^2)$ as R goes to $+\infty$—we can assume without loss of generality that $u \in L^\infty \cap \tilde{D}^{1,p}(\mathbb{R}^2)$. Then, we consider $u_n = \varphi(\frac{x}{n})u$ where $\varphi \in C_0^\infty(\mathbb{R}^2)$, $0 \leq \varphi \leq 1$, $\varphi \equiv 1$ on B_1, $\mathrm{Supp}\,\varphi \subset B_2$.

If $p > 2$, u_n converges to u in $\tilde{D}^{1,p}(\mathbb{R}^2)$ since we have

$$\nabla u_n - \nabla u = \left[1 - \varphi\left(\frac{x}{n}\right)\right]\nabla u + \frac{1}{n}\nabla\varphi\left(\frac{x}{n}\right)u$$

and

$$\int_{\mathbb{R}^2}\left|\frac{1}{n}\nabla\varphi\left(\frac{x}{n}\right)u\right|^p dx \leq Cn^{2-p}\|u\|_{L^\infty}^p .$$

Then, smoothing u_n by convolution allows to conclude.

If $p = 2$, we build in the above way $u_n \in C_0^\infty(\mathbb{R}^2)$ such that u_n converges to u weakly in $\tilde{D}^{1,2}(\mathbb{R}^2)$ and this is enough to conclude. □

Proof of Theorem 4.4

Step 1. The case when $1 < p < 2$. This is in fact the easy case since there exists $c \in \mathbb{R}^2$ such that $u_0 - c \in L^{\frac{2p}{2-p}}(\mathbb{R}^2)^2$, and we observe that because of the galilean invariance of the Euler equation we can always take c to be 0: indeed, u is a solution of (4.1)–(4.2) if and only if v defined by $v(x,t) = u(x+ct,t) - c$ on $\mathbb{R}^2 \times [0,\infty)$ is a solution of (4.1)–(4.2) corresponding to $u_0 - c$. Therefore, without loss of generality, we may assume that $u_0 \in L^{\frac{2p}{2-p}}(\mathbb{R}^2)^2$, $\nabla u_0 \in L^p(\mathbb{R}^2)^2$. Then, as in Lemma 4.2 (and its proof), we introduce $u_0^n \in C_0^\infty(\mathbb{R}^2)^2$ such that $\mathrm{div}\,u_0^n = 0$ in $\mathbb{R}^2$, u_0^n converges to u_0 in $L^{\frac{2p}{2-p}}(\mathbb{R}^2)$ and ∇u_0^n converges to ∇u_0 in $L^p(\mathbb{R}^2)$ as n goes to $+\infty$. In view of Theorem 4.1, we can solve (4.1)–(4.2) with u_0 replaced by u_0^n and we find a smooth solution u^n on $\mathbb{R}^2 \times [0,\infty)$ such that $\omega^n = \mathrm{curl}\,u^n$ solves (uniquely)

$$\left.\begin{array}{ll}\dfrac{\partial\omega^n}{\partial t} + u^n\cdot\nabla\omega^n = 0 & \text{in}\quad \mathbb{R}^2\times(0,\infty),\\[2mm] \omega^n|_{t=0} = \omega_0^n\;(=\mathrm{curl}\,u_0^n) & \text{in}\quad \mathbb{R}^2.\end{array}\right\} \qquad (4.56)$$

Since, for each $n \geq 1$, u^n is bounded on $\mathbb{R}^2 \times [0,T]$, we deduce that $\omega^n \in C_0^\infty(\mathbb{R}^2 \times [0,T])$ for all $T \in (0,\infty)$. In addition, because of (4.12), ω^n

is bounded in $C([0,\infty); L^p(\mathbb{R}^2))$ and thus ∇u^n is bounded in $C([0,\infty);$ $L^p(\mathbb{R}^2))$ by Sobolev inequalities. This implies that u^n is bounded in $C([0,\infty); L^{\frac{2p}{2-p}}(\mathbb{R}^2))$. Extracting a subsequence if necessary, we may assume that u^n converges weakly to some u in $L^{\frac{2p}{2-p}}(\mathbb{R}^2 \times (0,T))$ ($\forall\, T \in (0,\infty)$). Since ω_0^n converges in $L^p(\mathbb{R}^2)$ to $\omega_0 = \operatorname{curl} u_0$, we can now use the general convergence results of R.J. DiPerna and P.L. Lions [128]—see section 2.2 in chapter 2 for results of a similar nature—to deduce that ω^n converges in $C([0,T]; L^p(\mathbb{R}^2))$ ($\forall\, T \in (0,\infty)$) to the unique renormalized solution ω of (4.51) and of course $\omega \equiv \operatorname{curl} u$. Then exactly as above we deduce that $u^n, \nabla u^n$ converge respectively to $u, \nabla u$ in $C([0,T]; L^{\frac{2p}{2-p}}(\mathbb{R}^2))$, $C([0,T]; L^p(\mathbb{R}^2))$ for all $T \in (0,\infty)$. This is enough to conclude.

Step 2. The case when $p \geq 2$. We use Lemma 4.2 to introduce $u_0^n \in C_0^\infty(\mathbb{R}^2)^2$ such that u_0^n converges to u_0 in $\tilde{D}^{1,p}(\mathbb{R}^2)^2$ as n goes to $+\infty$. We then follow the argument given in step 1 and obtain u^n, ω^n, as in step 1, which satisfy: $\nabla u^n, \omega^n$ are bounded in $C([0,\infty; L^p(\mathbb{R}^2))$. We then set

$$c^n(t) = \fint_{B_1} (u^n(t) - u_0^n)\, dx$$

and define $\tilde{u}^n$ by $\tilde{u}^n(x,t) = u^n\big(x + \int_0^t c^n(s)\, ds\ ,\ t\big) - c^n(t)$. We next set $\tilde{p}^n = p^n + \frac{dc^n}{dt} \cdot x$, $\tilde{\omega}^n = \omega^n\big(x + \int_0^t c^n(s)\, ds,\ t\big)$. Then, one checks that $(\tilde{u}^n, \tilde{p}^n)$ solves (4.1)–(4.2) with u_0 replaced by u_0^n and that (4.56) holds for $\tilde{\omega}^n$ with u^n replaced by $\tilde{u}^n$. Let us also remark that $\nabla \tilde{u}^n$ and $\tilde{\omega}^n$ are still bounded in $C([0,\infty); L^p(\mathbb{R}^2))$, and in addition we have now, because of the choice of $c^n(t)$: for all $t \geq 0$, $\fint_{B_1} \tilde{u}^n(t)\, dx = \fint_{B_1} u_0^n\, dx \underset{n}{\to} \fint_{B_1} u_0\, dx$.

Therefore, $\tilde{u}^n$ is bounded in $C([0,\infty); \tilde{D}^{1,p}(\mathbb{R}^2))$, and extracting a subsequence if necessary, we may always assume that $\tilde{u}^n$ converges to some $u \in L^\infty(0,\infty; \tilde{D}^{1,p}(\mathbb{R}^2))$ in $L^q(0,R; W^{1,p}(B_R))$ weakly for all $R \in (0,\infty)$, $1 < q < \infty$ (and $D\tilde{u}^n$ converges weakly to Du in $L^q(0,T; L^p(\mathbb{R}^2))$) for all $T \in (0,\infty)$, $1 < q < \infty$, and $\fint_{B_1} u(x,t)\, dx = \fint_{B_1} u_0(x)\, dx$ for all $t > 0$).

Next, as in step 1, we wish to deduce the convergence of $\tilde{\omega}^n$ to ω ($= \operatorname{curl} u$) in $C([0,T]; L^p(\mathbb{R}^2))$ ($\forall\, T \in (0,\infty)$) from (4.56) where ω is the unique solution of (4.51) (in $C([0,T]; L^p(\mathbb{R}^2))$). In order to apply the convergence results due to R.J. DiPerna and P.L. Lions [128] (see also section 2.2 in chapter 2), we simply need to check that we have for all $T \in (0,\infty)$

$$|u|(1+|x|)^{-1} \in L^1(0,T; L^1(\mathbb{R}^2)) + L^\infty(\mathbb{R}^2 \times (0,T)). \qquad (4.57)$$

Once this is checked, we conclude as in step 1 that ∇u^n converges to ∇u in $C([0,T]; L^p(\mathbb{R}^2))$ for all $T \in (0,\infty)$ and thus u^n converges to u in

$C([0,T]; \tilde{\mathcal{D}}^{1,p}(\mathbb{R}^2))$ for all $T \in (0, \infty)$, as n goes to $+\infty$. These convergences then yield the conclusions of Theorem 4.4.

When $p > 2$, (4.57) follows immediately from the fact that $u \in L^\infty(0, \infty; \tilde{\mathcal{D}}^{1,p}(\mathbb{R}^2))$ and (4.53): indeed, $|u|(1+|x|)^{-\alpha} \in L^\infty(\mathbb{R}^2 \times (0, \infty))$ for some $0 < \alpha < 1$. When $p = 2$, we write $u = u_1 + u_2$ where u_1, u_2 are defined by

$$
\left.
\begin{aligned}
&\operatorname{div} u_1 = \operatorname{div} u_2 = 0, \quad \operatorname{curl} u_1 = \omega \, 1_{(|\omega| \le 1)}, \\
&\operatorname{curl} u_2 = \omega \, 1_{(|\omega| > 1)} \quad \text{a.e. in} \quad \mathbb{R}^2 \times (0, \infty), \\
&\fint_{B_1} u_1 \, dx = \fint_{B_1} u_0 \, dx, \quad \fint_{B_1} u_2 \, dx = 0 \quad \text{for} \quad t \ge 0; \\
&u_1, u_2 \in L^\infty(0, \infty; \tilde{\mathcal{D}}^{1,2}(\mathbb{R}^2)).
\end{aligned}
\right\}
$$

Clearly, $\omega \, 1_{(|\omega| \le 1)} \in L^\infty(0, \infty; L^2 \cap L^\infty(\mathbb{R}^2))$, hence, as we have done several times before, $u_1 \in L^\infty(0, \infty; \tilde{\mathcal{D}}^{1,p}(\mathbb{R}^2))$ for $2 \le p < \infty$ and, as we just saw above, $\frac{u_1}{1+|x|} \in L^\infty(\mathbb{R}^2 \times (0, \infty))$.

On the other hand, since $\omega \in L^\infty(0, \infty; L^2(\mathbb{R}^2))$, $\omega \, 1_{(|\omega| > 1)} \in L^\infty(0, \infty; L^1 \cap L^2(\mathbb{R}^2))$ and thus $u_2 \in L^\infty(0, \infty; \tilde{\mathcal{D}}^{1,p}(\mathbb{R}^2))^2$ for $1 < p < 2$. In particular, as seen above, $u_2 = c(t) + w$ where $w \in L^\infty(0, \infty; L^q(\mathbb{R}^2))^2$ for $2 < q < \infty$ and thus $c(t) = \fint_{B_1} (u_2 - w)(x, t) \, dx = \fint_{B_1} w(x; t) \, dx \in L^\infty(0, \infty)$. Therefore, u_2 satisfies (4.57) and we conclude. $\square$

4.3 Estimates in three dimensions?

The incompressible (homogeneous) Euler equations (4.1)–(4.2) in three dimensions ($N = 3$) are far from being understood. Indeed, the only information that is available concerns short time existence. More precisely, in the case of Dirichlet boundary conditions or in the periodic case (or when $\Omega = \mathbb{R}^N$), it is known that if u_0 is smooth enough ($u_0 \in X$ where $X = H^s$ with $s > \frac{5}{2}$, $C^{1,\alpha}$ with $0 < \alpha < 1$), then there exists a maximal time interval $[0, T^*)$ ($T^* \le +\infty$) such that there exists a unique solution $u \in C([0,T]; X)$ ($\forall \, T \in (0, T^*)$) of (4.1)–(4.2) (and with the appropriate boundary conditions of course) and if $T^* < \infty$, $\|u(t)\|_X$ goes to $+\infty$ as t goes to T_*-.

Of course, the crucial question which is still completely open is to decide whether $T_* < \infty$ or not. J.T. Beale, T. Kato and A. Majda [28] (see also G. Ponce [391] for a variant involving the deformation tensor instead of the vorticity) have established a fine criterion for the finite time blow-up of smooth solutions: let $T \in (0, \infty)$; if there exists a (unique) solution $u \in C([0,T]; X)$ such that $\int_0^T \|\operatorname{curl} u(t)\|_{L^\infty} \, dt < \infty$, then $T^* > T$. Or in other words, if $T_* < \infty$, then $\int_0^{T_*} \|\mathrm{curl}\, u(t)\|_{L^\infty}\, dt = +\infty$. We do not wish to re-prove this statement here. Let us simply mention that the main idea behind this criterion is the following: if $\int_0^T \|\mathrm{curl}\, u(t)\|_{L^\infty}\, dt < \infty$ then one can bound $\|u(t)\|_X$ uniformly on $[0, T)$ and, using the short time existence result, one can continue the solution on a larger interval than $[0, T]$.

The appearance of singularities (i.e. the breakdown of smooth solutions) in finite time is an outstanding open mathematical problem, whose solution would have a serious impact on three-dimensional incompressible fluid mechanics. After many years of intensive numerical simulations which were inconclusive, two recent independent numerical experiments by R. Grauer and T. Sideris [193] and later by R. Kerr [262] indicate possible breakdowns of smooth solutions in finite time (and certainly violent growths of $\|\mathrm{curl}\, u(t)\|_{L^\infty}$ that, for all practical mathematical purposes, make it difficult to believe in a priori estimates on $\mathrm{curl}\, u$).

The striking difference between the cases $N = 2$ and $N = 3$ is also illustrated by the so-called "$2 + 1/2$" dimensional flows, used in R.J. DiPerna and A. Majda [129] to provide examples of weakly convergent sequences of solutions of 3D Euler equations whose weak limits do not satisfy (4.1), namely solutions of (4.1) say in the periodic case (for instance) such that u is independent of x_3. In that case, (u_1, u_2) solve the Euler equation (4.1) in two dimensions (in the periodic case) while $w = u_3$ simply solves the following transport equation

$$\frac{\partial w}{\partial t} + \mathrm{div}\,(uw) = 0 \quad \text{in} \quad \mathbb{R}^2 \times (0, \infty), \quad w|_{t=0} = w_0 \quad \text{in} \quad \mathbb{R}^2 \quad (4.58)$$

and w, w_0 are periodic in x_1, x_2 (of periods $T_1, T_2 > 0$).

The decoupling between (u_1, u_2) and w allows us to solve first for (u_1, u_2), using the results of sections 4.1–4.2, and then to solve (4.58) in a classical way when u is smooth (or Lipschitz, or almost Lipschitz). When $w_0 = \mathrm{curl}\, u_0 \left(= \frac{\partial}{\partial x_2} u_1(0) - \frac{\partial}{\partial x_1} u_2(0) \right) \in L^p(\Omega)$ ($\Omega = (0, T_1) \times (0, T_2)$) with $p > 1$, resp. $w_0|\log w_0| \in L^1(\Omega)$, then we have seen that there exists at least one solution of the Euler equation $u \in C([0, \infty); W^{1,p}(\Omega))$, resp. $W^{1,1}(\Omega)$. In this case, there exists, as soon as $w_0 \in L^q(\Omega)$ ($1 \le q \le \infty$), a unique solution of (4.58) which belongs to $C([0, \infty); L^q(\Omega))$ if $q < \infty$ (and to $C([0, \infty); L^r(\Omega))$ for all $1 \le r < \infty$ and to $L^\infty(\Omega \times (0, \infty))$ if $q = +\infty$)—see R.J. DiPerna and P.L. Lions [128].

These flows provide examples of global weak solutions of the 3D Euler equations which are smooth (and thus unique) if the initial conditions are smooth. But these flows, as was observed by R.J. DiPerna and the author [124], also show that solutions (even smooth ones) of the 3D Euler equations cannot be estimated in $W^{1,p}$, for $1 < p < \infty$, on any time interval $(0, h)$ if the initial data is only assumed to be bounded in $W^{1,p}$. In other words, intermediate a priori estimates or, more precisely, $W^{1,p}$ a priori estimates do not hold in three dimensions—let us emphasize the fact that they do hold in two dimensions, see sections 4.1–4.2 above. Indeed, we choose for (u_1, u_2) stationary flows of the 2D Euler equations namely $(u_1(x_2), 0)$ where u_1 is smooth, periodic in x_2 (of period T_2). Then, the solution of (4.58) is simply given by $w_0(x_1 - tu_1(x_2), x_2)$. The lack of a priori estimates in $W^{1,p}$ is then clear choosing for instance $u_1^\varepsilon(x_2)$ to behave like $(\varepsilon^2 + x_2^2)^{\theta/2}$ near $x_2 = 0$ (and smooth elsewhere uniformly in ε) and $w_0^\varepsilon(x_1, x_2)$ to behave like $(\varepsilon^2 + |x|^2)^{\theta - 1/2}$ near 0 (and smooth elsewhere uniformly in ε) where $\varepsilon \in (0,1]$, $\theta \in (0,1)$, and we take $\theta \neq \frac{1}{2}$, otherwise log modifications have to be made. Then, obviously, $(u_1^\varepsilon, w_0^\varepsilon)$ are bounded in $W^{1,p}$, uniformly in ε, for $1 \leq p < \frac{1}{1-\theta}$ while for $q \geq 1$ we have

$$\int_\Omega \left| \frac{\partial w^\varepsilon(t)}{\partial x_2} \right|^q dx_1 dx_2$$

$$= \int_\Omega t^q \left| \frac{\partial w_0^\varepsilon}{\partial x_1}(x_1 - tu_1^\varepsilon(x_2), x_2) \right|^q |(u_1^\varepsilon)'(x_2)|^q dx_1 dx_2$$

$$= t^q \int_\Omega \left| \frac{\partial w_0^\varepsilon}{\partial x_1}(y_1, x_2) \right|^q |(u_1^\varepsilon)'(x_2)|^q dy_1 dx_2$$

$$\geq \nu t^q \int_{B_\delta} \frac{|x_1|^q}{(\varepsilon^2 + |x|^2)^{(\frac{3}{2}-\theta)q}} \cdot \frac{|x_2|^q}{(\varepsilon^2 + x_2^2)^{(1-\frac{\theta}{2})q}} dx_1 dx_2$$

for some $\delta > 0$ sufficiently small, independent of ε, where $\nu = |2\theta - 1|^q \theta^q > 0$. Since we have for some constant $C(\theta, q) > 0$ (if $1 \leq q < \frac{1}{1-\theta}$)

$$\int_{B_\delta} \frac{|x_1|^q}{|x|^{(3-2\theta)q}} \cdot \frac{1}{|x_2|^{(1-\theta)q}} dx_1 dx_2 = C(\theta, q) \int_0^\delta \frac{1}{r^{3(1-\theta)q}} r \, dr,$$

we deduce that $\frac{\partial w^\varepsilon}{\partial x_2}$ is bounded in L^q uniformly in ε (for any fixed $t > 0$) if and only if $q < \frac{2}{3(1-\theta)}$, and in particular it is not bounded if $\frac{2}{3(1-\theta)} \leq q < \frac{1}{1-\theta}$.

This construction shows that, for each $1 < p < \infty$, $t \in (0, \infty)$, $\varepsilon \in (0,1)$, $\delta \in (0,1)$, there exists a smooth (periodic) solution u of Euler equations in three dimensions such that $\|u(0)\|_{W^{1,p}} \leq \varepsilon$ and $\|u(t)\|_{W^{1,p}} \geq \frac{1}{\delta}$. This also gives examples of smooth flows such that $\|\text{curl}\, u(t)\|_{L^\infty} \geq t\|\text{curl}\, u(0)\|_{L^\infty}^2$.

Let us conclude by mentioning that there are other known smooth regimes for three-dimensional Euler equations like the axisymmetric case without swirl (see A. Majda [316] and Ph. Serfati [420] for more details). But, even in that case, J.M. Delort [119] observed that the "vortex sheet" problem (see section 4.2 for a slightly more detailed presentation of this problem) is mathematically quite different from the pure two-dimensional case.

Finally, let us observe that the lack of intermediate a priori estimates and more specifically of bounds that yield the compactness of appropriate sequences of solutions (or approximated solutions) has made impossible until now the construction of weak solutions in $C([0, \infty); L^2)$ or even $L^\infty(0, \infty; L^2)$ satisfying (4.1) in the sense of distributions. Weaker notions of solutions are considered in the next section.

4.4 Dissipative solutions

As we have seen in the preceding section, even the global existence of weak solutions of the Euler equations is not known in three dimensions. On the other hand, an obvious bound in $C([0, \infty); L^2)$ follows trivially from the (formal) conservation of energy. It is then natural to attempt building up solutions in a weaker sense than in the distributions sense. A very weak notion (relying on relaxed Young measures or relaxed measure-valued solutions) is proposed by R.J. DiPerna and A. Majda [129] but the relevance of this notion is not entirely clear since it is not known that "solutions" in the sense of [129] coincide with smooth solutions as long as the latter do exist.

We shall propose here a different notion of "very weak" solution that we call dissipative solutions, for reasons we shall explain later on. This notion seems to be new and the idea behind the notion appears in P.L. Lions [306] in the context of Boltzmann's equation. We wish to emphasize immediately that we are not convinced that such a notion is neither relevant nor useful. Its only merits are: 1) such solutions exist, 2) as long as a "smooth" solution exists with the same initial condition, any such dissipative solution coincides with it, 3) we shall use it in some small Mach number limits in chapter 9 to pass to the limit from some compressible models to the Euler equations.

From now on, we take $N \geq 3$ and we only consider the periodic case in order to simplify the presentation and keep the ideas clear, although everything we do below can be extended or adapted to the case when $\Omega = \mathbb{R}^N$ or to the case of Dirichlet boundary conditions. This is why all the functions considered below are always assumed to be periodic. The initial condition u_0 (see 4.2) is always assumed to belong to $L^2(\Omega)^N$ and to satisfy (4.3) in $\mathbb{R}^N$.

Let us first explain the formal idea of this new notion. Let us consider a smooth test function v on $\mathbb{R}^N \times [0, \infty)$ such that $\operatorname{div} v = 0$ on $\mathbb{R}^N \times [0, \infty)$. We define

$$E(v) = -\frac{\partial v}{\partial t} - P((v \cdot \nabla)v) \qquad (4.59)$$

(recall that P is the projection onto periodic divergence-free vector fields), and we write $v_0 = v|_{t=0}$.

If u is a solution of (4.1)–(4.2), then we can write

$$\left(\frac{\partial}{\partial t} + u \cdot \nabla\right)(u-v) + (u-v) \cdot \nabla v + \nabla \pi \; = \; E(v) \qquad \text{in} \quad \mathbb{R}^N \times (0, \infty)$$

for some scalar function π. Then, multiplying this equality by $u - v$ and integrating (over the period), we find

$$\frac{d}{dt} \int_\Omega |u-v|^2 \, dx \; = \; -2\int_\Omega (d(u-v), u-v) \, dx + \int_\Omega 2E(v) \cdot (u-v) \, dx \quad (4.60)$$

where $d(= d(v)) = \left(\frac{1}{2}(\partial_i v_j + \partial_j v_i)\right)_{ij}$. We then set (for each $t \geq 0$)

$$\|d^-\|_\infty \; = \; \left\|\left(\sup_{|\xi|=1} -(d\xi, \xi)\right)^+\right\|_{L^\infty(\Omega)},$$

and we deduce from (4.60)

$$\frac{d}{dt} \int_\Omega |u-v|^2 \, dx \; \leq \; 2\|d^-\|_\infty \int_\Omega |u-v|^2 \, dx + 2\int_\Omega E(v) \cdot (u-v) \, dx. \quad (4.61)$$

Hence, in particular, we have for all $t \geq 0$

$$\left.\begin{aligned}
\int_\Omega |u-v|^2 \, (x,t) \, dx \; &\leq \; \exp\left(2\int_0^t \|d^-\|_\infty \, ds\right) \int_\Omega |u_0 - v_0|^2 \, dx \\
&+ 2\int_0^t ds \int_\Omega dx \, \exp\left(2\int_s^t \|d^-\|_\infty \, dv\right) E(v) \cdot (u-v).
\end{aligned}\right\} \quad (4.62)$$

The definition of dissipative solutions of (4.1)–(4.2) given below consists precisely in requesting (4.62) to hold for an appropriate class of v.

The reason why we call these solutions dissipative solutions is the following: if we take $v \equiv 0$ then obviously $E(v) \equiv 0$, $d \leq 0$ and (4.60) is nothing but the (formal) conservation of energy while (4.61) and (4.62) are standard relaxed energy inequalities which allow for a possible loss (dissipation) of energy though various losses of L^2 compactness (via oscillations, concentrations, etc.).

We may now give the precise

Definition 4.1. *Let* $u \in L^\infty(0, \infty; L^2)^N \cap C([0, \infty); L^2 - w)^N$. *Then* u *is a dissipative solution of* (4.1)–(4.2) *if* $u(0) = u_0$, div $u = 0$ *in* $\mathcal{D}'(\mathbb{R}^N \times (0, \infty))$ *and* (4.62) *holds for all* $v \in C([0, \infty); L^2)^N$ *such that* $d \in L^1(0, T; L^\infty)$, $E(v) \in L^1(0, T; L^2)$ $(\forall \, T \in (0, \infty))$ *and* div $v = 0$ *in* $\mathcal{D}'(\mathbb{R}^N \times (0, \infty))$.

Remark 4.12. 1) It is worth pointing out that the main regularity requirement for v in the above formulation, namely $d \in L^1(0,T;L^\infty)$, is the same as in the blow-up criterion obtained by G. Ponce [391]—and, in fact, for similar reasons.

2) If $d \in L^p(0,T;L^\infty)$ (or $\operatorname{curl} v \in L^p(0,T;L^\infty)$) for some $p \in [1,\infty]$ then $Dv \in L^p(0,T;L^q)$ for all $1 < q < \infty$ and $Dv \in L^p(0,T;BMO)$: in particular, there exists $A(t) \geq 0 \in L^p(0,T)$ such that if $|x-y| \leq 1/2$

$$|v(x,t) - v(y,t)| \leq A(t)|x-y|\,|\log|x-y||. \tag{4.63}$$

3) When $E(v) = 0$ (for instance) and $Dv_0 \in L^q$ for some $1 < q < \infty$ then $d \in L^1(0,T;L^\infty)$ implies formally that $Dv \in L^\infty(0,T;L^q)$. Indeed, we observe that for $i \neq j$

$$\left(\frac{\partial}{\partial t} + v \cdot \nabla\right)(\partial_j v_i - \partial_i v_j) = \partial_j v_k\,\partial_k v_i - \partial_i v_k\,\partial_k v_j$$
$$= 2\,\partial_j v_k\,d_{ik} - 2\,\partial_i v_k\,d_{jk}.$$

Therefore, we deduce

$$\frac{d}{dt}\,\|\operatorname{curl} v\|_{L^q}^q \leq C_0 q\,\|d\|_{L^\infty}\,\|Dv\|_{L^q}\,\|\operatorname{curl} v\|_{L^q}^{q-1}$$
$$\leq C(q)\,\|d\|_{L^\infty}\,\|\operatorname{curl} v\|_{L^q}^q$$

and our claim follows easily.

4) Let us observe that if $d \in L^1(0,T;L^\infty)$ then $E(v) \in L^1(0,T;L^2)$ if and only if $\frac{\partial v}{\partial t} \in L^1(0,T;L^2)$. Indeed, $\frac{\partial v}{\partial t} + E(v) = -P((v \cdot \nabla)v)$ and $Dv \in L^1(0,T;L^q)$ for all $1 \leq q < \infty$. In particular, $(v \cdot \nabla)v$ and thus $P((v \cdot \nabla)v) \in L^1(0,T;L^r)$ for all $1 \leq r < 2$ since $v \in L^\infty(0,T;L^2)$. Furthermore, for each $1 \leq i \leq N$, we have

$$(v \cdot \nabla)v_i = v_j\,\partial_j v_i = v_j(\partial_j v_i + \partial_i v_j) - \partial_i\left(\frac{1}{2}|v|^2\right)$$
$$= 2d \cdot v - \partial_i\left(\frac{1}{2}|v|^2\right).$$

Therefore, $P((v \cdot \nabla)v) = 2P(d \cdot v) \in L^1(0,T;L^2)$ since $d \in L^1(0,T;L^\infty)$ and $v \in L^\infty(0,T;L^2)$. $\square$

An obvious consequence of the definition is the following

Proposition 4.1. *If there exists a solution $v \in C([0,T];L^2)$ of (4.1)–(4.2) on $\mathbb{R}^N \times (0,T)$ satisfying $d \in L^1(0,T;L^\infty)$ for some $T \in (0,\infty)$, then any dissipative solution u of (4.1)–(4.2) is equal to v on $\mathbb{R}^N \times [0,T]$.* $\square$

Indeed, $E(v) \equiv 0$ and we conclude!

Before we give and prove our existence result for dissipative solutions, let us mention that, in Definition 4.1, we can replace the regularity requirements on v by much stronger ones. In other words, it suffices to check (4.62) for smooth test functions v. Indeed, if (4.62) holds for smooth v, then we can check it also holds for the class of v described in Definition 4.1 and in Remark 4.12 (4) by a straightforward regularization procedure. More precisely, let $\rho_\varepsilon = \frac{1}{\varepsilon^N} \rho(\frac{\cdot}{\varepsilon})$, $\rho \in C_0^\infty(\mathbb{R}^N)$, $\int_{\mathbb{R}^N} \rho \, dx = 1$, $\operatorname{Supp} \rho \subset B_1$. We set $v_\varepsilon = v * \rho_\varepsilon$. Obviously v_ε converges to v in $C([0,T]; L^2)$, $\frac{\partial v_\varepsilon}{\partial t} \in L^1(0, T; C_b^k)$ for all $k \geq 0$, $\operatorname{div} v_\varepsilon = 0$ in $\mathbb{R}^N \times (0, \infty)$, $v_\varepsilon \in C([0,T]; C_b^k)$ for all $k \geq 0$, $T \in (0, \infty)$ and

$$
\begin{aligned}
E(v_\varepsilon) &= -\frac{\partial v_\varepsilon}{\partial t} - P((v_\varepsilon \cdot \nabla) v_\varepsilon) \\
&= \left[-\frac{\partial v}{\partial t} - P[(v \cdot \nabla)v] \right] * \rho_\varepsilon + \left[P[(v \cdot \nabla)v] * \rho_\varepsilon - P((v_\varepsilon \cdot \nabla)v_\varepsilon) \right] \\
&= E(v) * \rho_\varepsilon + 2\left[[P(d \cdot v)] * \rho_\varepsilon - P(d_\varepsilon \cdot v_\varepsilon) \right] \\
&= E(v) * \rho_\varepsilon + 2P\{ (d \cdot v) * \rho_\varepsilon - d_\varepsilon \cdot v_\varepsilon \}
\end{aligned}
$$

in view of Remark 4.12 (4). Obviously, $E(v) * \rho_\varepsilon$ converges to $E(v)$, as ε goes to 0_+, in $L^1(0, T; L^2)$ ($\forall\, T \in (0, \infty)$) and so does $E(v_\varepsilon)$ provided we show that $(d \cdot v) * \rho_\varepsilon - d_\varepsilon \cdot v_\varepsilon$ converges to 0, as ε goes to 0_+, in $L^1(0, T; L^2)$ ($\forall\, T \in (0, \infty)$). Since $d \in L^1(0, T; L^\infty)$, $v \in C([0,T]; L^2)$ ($\forall\, T \in (0, \infty)$), this is straightforward. Observe in addition that $\|d_\varepsilon^-\|_\infty \leq \|d^-\|_\infty$ for all $\varepsilon > 0$, a.e. $t > 0$ and that $\|d_\varepsilon^-\|_\infty$ converges to $\|d^-\|_\infty$ in $L^1(0, T)$ ($\forall\, T \in (0, \infty)$). We then apply (4.62) with v replaced by v_ε and we conclude letting ε go to 0 in view of the convergences collected above. Let us observe that by a second layer of approximation (regularizing in t), we can take v smooth on $\mathbb{R}^N \times [0, \infty)$ in (4.62).

We can prove now the existence of dissipative solutions.

Proposition 4.2. *There exists at least one dissipative solution of (4.1)–(4.2).*

Proof of Proposition 4.2. We consider u_ν, a weak solution of the Navier–Stokes equation (see section 3.1),

$$
\frac{\partial u_\nu}{\partial t} + (u_\nu \cdot \nabla) u_\nu - \nu \Delta u_\nu + \nabla p_\nu = 0, \quad \operatorname{div} u_\nu = 0 \quad \text{in } \mathbb{R}^N \times (0, \infty) \quad (4.64)
$$

satisfying (4.2) and the energy inequality

$$
\frac{d}{dt} \int_\Omega \frac{1}{2} |u_\nu(t)|^2 \, dx + \nu \int_\Omega |\nabla u_\nu|^2 \, dx \leq 0 \quad \text{in } \mathcal{D}'(0, \infty). \quad (4.65)
$$

Recall that $u_\nu \in L^2(0,T;H^1)$ ($\forall\, T \in (0,\infty)$) $\cap\, L^\infty(0,\infty;L^2) \cap C([0,\infty);$ $L^2 - w)$ (in addition, $u_\nu(t)$ goes to u_0 in $L^2(\Omega)$ as t goes to 0_+).

We next consider v as in Definition 4.1, and, as shown above just before the statement of Proposition 4.2, we can take v arbitrarily smooth on $\mathbb{R}^N \times [0,\infty)$. Then, multiplying (4.64) by v we find

$$\left. \begin{aligned} \frac{d}{dt}\int_\Omega u_\nu \cdot v\, dx - \int_\Omega u_\nu \cdot \frac{\partial v}{\partial t} + (u_\nu \cdot \nabla)v \cdot u_\nu\, dx \\ + \nu \int_\Omega \nabla u_\nu \cdot \nabla v\, dx = 0 \quad \text{in}\quad \mathcal{D}'(0,\infty), \end{aligned} \right\}$$

and thus, by definition of $E(v)$ and since $\operatorname{div} u_\nu = \operatorname{div} v = 0$, we deduce

$$\left. \begin{aligned} \frac{d}{dt}\int_\Omega u_\nu \cdot v\, dx + \int_\Omega u_\nu \cdot E(v)\, dx - \int_\Omega [(u_\nu - v)\cdot\nabla]v \cdot (u_\nu - v)\, dx \\ + \nu \int_\Omega \nabla u_\nu \cdot \nabla v\, dx = 0 \quad \text{in}\quad \mathcal{D}'(0,\infty). \end{aligned} \right\} \quad (4.66)$$

In addition, we have for all $t \geq 0$

$$\left. \begin{aligned} \frac{d}{dt}\int_\Omega \frac{|v|^2}{2}\, dx = \int_\Omega v \cdot \frac{\partial v}{\partial t}\, dx = \int_\Omega v[-E(v) - (v\cdot\nabla)v]\, dx \\ = -\int_\Omega v \cdot E(v)\, dx. \end{aligned} \right\} \quad (4.67)$$

Combining (4.65), (4.66) and (4.67), we deduce

$$\left. \begin{aligned} \frac{d}{dt}\int_\Omega \frac{1}{2}|u_\nu - v|^2\, dx = -\int_\Omega [(u_\nu - v)\cdot\nabla]v \cdot (u_\nu - v)\, dx \\ + \nu\int_\Omega \nabla u_\nu \cdot \nabla v\, dx + \int_\Omega E(v)\cdot(u_\nu - v)\, dx \end{aligned} \right\}$$

or

$$\frac{d}{dt}\int_\Omega |u_\nu - v|^2\, dx = -2\int_\Omega (d(u_\nu - v), u_\nu - v)\, dx$$

$$+ 2\nu\int_\Omega \nabla u_\nu \cdot \nabla v + 2\int_\Omega E(v)\cdot(u_\nu - v)\, dx$$

$$\leq 2\|d^-\|_\infty \int_\Omega |u_\nu - v|^2\, dx + C\nu \left(\int_\Omega |\nabla u|^2\, dx\right)^{1/2}$$

$$+ 2\int_\Omega E(v)\cdot(u_\nu - v)\, dx$$

and finally for all $T \in (0,\infty)$ and for all $t \in [0,T]$

$$
\left.
\begin{aligned}
\int_\Omega |u_\nu - v|^2(x,t)\, dx &\le e^{2\int_0^t \|d^-\|_\infty ds} \int_\Omega |u_0 - v_0|^2\, dx \\
&+ 2\int_0^t ds \int_\Omega dx\, e^{2\int_s^t \|d^-\|_\infty d\sigma}\, E(v)\cdot(u_\nu - v) + C_T \nu^{1/2},
\end{aligned}
\right\} \quad (4.68)
$$

for some positive constant C_T which is independent of ν. Here, we used the fact that (4.65) yields a bound, uniform in ν, on $\nu \int_0^\infty dt \cdot \int_\Omega dx\, |\nabla u_\nu|^2$.

Finally, we observe that $\frac{\partial u_\nu}{\partial t} = -\partial_j P(u_{\nu,j}u) - \nu^{1/2}\partial_j P(\nu^{1/2}\partial_j u)$ and thus $\frac{\partial u_\nu}{\partial t}$ is bounded in $L^2(0,\infty; H^{-1}) + L^\infty(0,\infty; W^{-(1+s),1})$ for all $s > 0$. Extracting a subsequence if necessary (see Appendix C for more details) we may assume that u_ν converges to some u weakly in $L^\infty(0,\infty; L^2) - *$ and weakly in L^2 uniformly in $t \in [0,T]$ for all $T \in (0,\infty)$. And $u \in L^\infty(0,\infty; L^2)\cap C([0,\infty); L^2 - w)$, div $u = 0$ in $\mathcal{D}'(\mathbb{R}^N \times (0,\infty))$, $u|_{t=0} = u_0$ in $\mathbb{R}^N$ and passing to the limit in (4.68), as ν goes to 0_+, we recover (4.62) and Proposition 4.2 is shown. $\square$

4.5 Density-dependent Euler equations

In this section we briefly review the "state of the art" concerning the density-dependent Euler equations or, in other words, the inhomogeneous incompressible Euler equations.

We look for a non-negative scalar function $\rho(x,t)$ (the density of the fluid) and for a divergence-free vector field $u(x,t)$ (the velocity of the fluid) which are solutions of

$$
\left.
\begin{aligned}
\frac{\partial \rho}{\partial t} + \text{div}\,(\rho u) &= 0, \quad \text{div}\, u = 0 \\
\frac{\partial \rho u_i}{\partial t} + \text{div}\,(\rho u u_i) + \nabla p &= 0, \quad \text{for}\ \ 1 \le i \le N
\end{aligned}
\right\} \quad (4.69)
$$

for some scalar function p, the so-called hydrostatic pressure. We consider as usual the periodic case (the unknowns are periodic and the equations hold in $\mathbb{R}^N \times (0,\infty)$), the case when $\Omega = \mathbb{R}^N$ or the case of Dirichlet boundary conditions where the equations are set in a smooth, bounded, open connected set in $\mathbb{R}^N$ and $u \cdot n = 0$ on $\partial\Omega$ where n denotes the unit outward normal. Of course, we complement (4.69) with initial conditions

$$
\rho|_{t=0} = \rho_0, \quad \rho u|_{t=0} = m_0 \quad (4.70)
$$

where $\rho_0 \ge 0$. In fact, exactly as in chapter 2 for the inhomogeneous incompressible Navier–Stokes equations, the precise meaning of the initial condition for ρu has to be interpreted correctly but in the simple case when we assume that ρ_0 satisfies for some $0 < \alpha \leq \beta < \infty$

$$\alpha \leq \rho_0 \leq \beta \qquad (4.71)$$

then, we may set $u_0 = \frac{m_0}{\rho_0}$ and we require u_0 to satisfy (4.3) (in $\mathbb{R}^N$ or in Ω in the case of Dirichlet boundary conditions).

Obviously, (4.69) contains the usual incompressible Euler equations as a particular case: take $\rho \equiv 1$ in (4.69). Therefore, we cannot expect to know more about this system of equations than for the incompressible Euler equations and in particular the case when $N \geq 3$ seems rather hopeless as far as the understanding of global solutions is concerned. But even if we concentrate on $N = 2$ in the rest of this short section, this will not help much since very little is known even in this case. Of course, (4.69) is well posed locally in t provided (4.71) (and (4.3)) holds and we choose ρ_0, u_0 to be smooth enough but, even when $N = 2$, it is not known whether smooth solutions persist for all $t \geq 0$ or break down in finite time. Again, there seems to be some numerical evidence of finite time breakdown but this has yet to be confirmed systematically.

There are very few known a priori estimates. Of course, (4.69) implies, at least formally, that we have for all $t \geq 0$ and for all $0 < a < b < \infty$:

$$\text{meas} \{x \ / \ a < \rho(x,t) < b\} = \text{meas} \{x \ / \ a < \rho_0(x) < b\}. \qquad (4.72)$$

In particular, if ρ_0 satisfies (4.71) then $\rho(t)$ also satisfies (4.71) for all $t \geq 0$.

The conservation of energy is obtained by multiplying the second equation of (4.69) by u and integrating by parts using the first equation, and reads

$$\frac{d}{dt} \int \rho |u|^2 \, dx = 0 \qquad \text{for} \quad t \geq 0. \qquad (4.73)$$

When (4.71) holds, this yields an estimate on u in $C([0,\infty); L^2)$.

To the best of our knowledge, these are the only known a priori estimates even when $N = 2$. It is also worth remarking that the failure of intermediate a priori estimates that we showed in section 4.3 on the incompressible Euler equations when $N \geq 3$ can be established for (4.69) when $N = 2$ using in fact the same examples: $u_1 \equiv u_1(x_2)$, $u_2 \equiv 0$, $p \equiv 0$ and $\rho(x,t) = \rho_0(x_1 - tu_1(x_2), x_2)$.

Let us conclude by mentioning a remarkable identity which however does not help—or at least does not seem to help—to analyse (4.65) mathematically. Still for $N = 2$, we consider the vorticity $\omega = \partial_2 u_1 - \partial_1 u_2$ and we write at least formally

$$\frac{\partial u}{\partial t} + (u \cdot \nabla)u + \frac{1}{\rho}\nabla p = 0$$

hence, taking the curl (and using the fact that div $u = 0$)

$$\frac{\partial \omega}{\partial t} + (u \cdot \nabla)\omega + \partial_2\left(\frac{1}{\rho}\right)\partial_1 p - \partial_1\left(\frac{1}{\rho}\right)\partial_2 p = 0. \qquad (4.74)$$

In particular, we deduce for all smooth function β from $\mathbb{R}$ into $\mathbb{R}$

$$\frac{\partial}{\partial t}\{\omega\beta(\rho)\} + (u \cdot \nabla)\{\omega\beta(\rho)\} = \beta'(\rho)\left\{\partial_1\left(\frac{1}{\rho}\right)\partial_2(p) - \partial_2\left(\frac{1}{\rho}\right)\partial_1 p\right\}$$

$$= \frac{\beta'(\rho)}{\rho^2}\left\{\partial_2\rho\partial_1 p - \partial_1\rho\partial_2 p\right\} = \partial_2[\gamma(\rho)]\partial_1 p - \partial_1[\gamma(\rho)]\partial_2 p$$

$$= \partial_2\{\gamma(\rho)\partial_1 p\} - \partial_1\{\gamma(\rho)\partial_2 p\}$$

where $\gamma' = \frac{\beta(t)}{t^2}$. Then, we integrate the resulting equation and, at least when $\Omega = \mathbb{R}^N$ or in the periodic case, we deduce immediately

$$\frac{d}{dt}\int \omega\beta(\rho)\,dx = 0 \qquad \text{for all} \quad t \geq 0 \qquad (4.75)$$

for any function β from $\mathbb{R}$ into $\mathbb{R}$.

4.6 Hydrostatic approximations

There exists a huge literature on the so-called hydrostatic approximations of incompressible models and a considerable number of models have been proposed (see for instance J. Pedlosky [386] on geophysical flows, and P. Constantin, A. Majda and E.G. Tabak [103], J.L. Lions, R. Temam and S. Wang [296],[297],[298]) with applications to oceanography, meteorology, geophysical flows and the huge variety of quasi-geostrophic models. Some models have been analysed mathematically or implemented numerically but, to the best of our knowledge, very little seems to be known on the model we discuss in this section which is the inviscid version of a very classical model. Our motivation for restricting our attention to this model is mostly mathematical since we hope it could help to understand some of the singular features of the classical three-dimensional incompressible Euler equations.

Let us first present this "hydrostatic" inviscid model. We consider a three dimensional strip $S = \{(x_1, x_2, z) \in \mathbb{R}^3 \,/\, 0 < z < 1\}$ and we look, to simplify the presentation and the notation, at a situation where all unknown functions are required to be periodic in x_1 and in x_2 (of periods respectively T_1 and $T_2 > 0$). We could consider as well the case when

$(x_1, x_2) \in \omega$ where ω is a smooth bounded connected open set in $\mathbb{R}^2$ and then we impose "Dirichlet boundary conditions" on $\partial\omega \times [0, 1]$ but we shall not do so here.

From now on, all differential operators ∇, div, Δ, curl refer to the two-dimensional operators (acting on x_1, x_2) and we also define $\partial_1 = \frac{\partial}{\partial x_1}$, $\partial_2 = \frac{\partial}{\partial x_2}$, $\partial_3 = \partial_z = \frac{\partial}{\partial z}$.

We look for a velocity field $(u, v) = (u_1(x_1, x_2, z), u_2(x_1, x_2, z), v(x_1, x_2, z))$ ($\in \mathbb{R}^3$ for all $(x_1, x_2, z) = (x, z) \in S$) and for a scalar function p (the pressure as usual) satisfying

$$\left.\begin{array}{l} \dfrac{\partial u}{\partial t} + (u \cdot \nabla)u + v\dfrac{\partial}{\partial z}u + \nabla p = 0, \quad \dfrac{\partial p}{\partial z} = 0, \\[3mm] \text{div}\, u + \dfrac{\partial v}{\partial z} = 0 \quad \text{in} \quad S \times (0, \infty) \end{array}\right\} \tag{4.76}$$

$$v|_{z=0} = v|_{z=1} = 0 \quad \text{on} \quad \mathbb{R}^2 \tag{4.77}$$

(recall that all functions are required to be periodic in x_1 and in x_2).

Observe that (4.72) is nothing but the system of incompressible Euler equations in S "with $v = u_3$" satisfying Dirichlet conditions on $\{z = 0, 1\}$, where the third equation on v is replaced by $\frac{\partial p}{\partial z} = 0$. In other words, we neglect in the third equation for v the term $\frac{\partial v}{\partial t} + u \cdot \nabla v + v\frac{\partial v}{\partial z}$ and simply write $\frac{\partial p}{\partial z} = 0$ (the so-called hydrostatic approximation; in the presence of gravity terms we can simply replace $\frac{\partial p}{\partial z} = 0$ by $\frac{\partial p}{\partial z} = a$ for some fixed constant a but this does not modify (4.72) since we can then consider $p - az$).

Let us immediately mention that (4.76)–(4.77) contains as a particular case the usual two-dimensional incompressible Euler equations: indeed, take $v = 0$; then $u = u(x_1, x_2)$ solves the 2D Euler equation. Let us also observe that the energy $\int_{\Omega \times (0,1)} |u|^2 \, dx \, dz$ is conserved (at least formally): indeed, multiply (4.76) by u and integrate over $\Omega \times (0, 1)$ to find

$$\frac{d}{dt}\int_{\Omega \times (0,1)} \frac{|u|^2}{2}\, dx\, dz + \int_{\Omega}\left(\frac{|u|^2}{2} + p\right)v(x, 1) - \left(\frac{|u|^2}{2} + p\right)v(x, 0)\, dx = 0$$

and our claim is proved in view of (4.77).

There are various equivalent formulations of (4.77). Let us mention at least one. We define $\overline{\varphi} = \overline{\varphi}(x) = \int_0^1 \varphi(x, z)\, dz$ for an arbitrary function φ (periodic in x) on S. Then, if we integrate (4.76) in z from 0 to 1, we find using (4.77)

$$\frac{\partial \overline{u}}{\partial t} + \text{div}\, \overline{[u \otimes u]} + \nabla p = 0, \quad \text{div}\, \overline{u} = 0 \quad \text{in} \quad \mathbb{R}^2 \tag{4.78}$$

(recall that $p = p(x, y)$ in view of (4.76)). In particular, we deduce taking the divergence of (4.78)

$$-\Delta p = \partial^2_{ij}(\overline{u_i u_j}) \quad \text{in} \quad \mathbb{R}^2 \tag{4.79}$$

and p (which is periodic) can be normalized to satisfy $\int_\Omega p\, dx = 0$ for all $t \geq 0$. Then, it is possible to rewrite (4.76)–(4.77) as follows: $u = u(x, z)$, $p = p(x)$ solve

$$\left. \begin{array}{l} \dfrac{\partial u}{\partial t} + (u \cdot \nabla)u - \left(\displaystyle\int_0^z \operatorname{div} u(x, \xi)d\xi \right)u_z + \nabla p = 0 \quad \text{in } S \times (0, \infty) \\[2mm] \operatorname{div} \overline{u} = 0 \quad \text{in } \mathbb{R}^2, \quad -\Delta p = \partial^2_{ij}(\overline{u_i u_j}) \quad \text{in } \mathbb{R}^2, \\[2mm] \displaystyle\int_\Omega p\, dx = 0 \quad \text{for} \quad t \geq 0. \end{array} \right\} \tag{4.80}$$

Indeed, we have checked above that a solution of (4.76)–(4.77) satisfies (4.80). Conversely, we set $v = -\int_0^z \operatorname{div} u(x, \xi)\, d\xi$, and obviously (4.76) holds, $v|_{z=0} = 0$, and it just remains to check that $v|_{z=1} = 0$ in order to prove our claim. But, $v|_{z=1} = -\int_0^1 \operatorname{div} u(x, \xi)\, d\xi = -\operatorname{div} \overline{u} = 0$ in $\mathbb{R}^2$, and we conclude.

In fact, it is possible to simplify (4.80) slightly. First of all, we prescribe initial conditions

$$u|_{t=0} = u_0 \quad \text{in} \quad S \tag{4.81}$$

where u_0 (periodic in x) satisfies

$$\operatorname{div} \overline{u}_0 = 0 \quad \text{in} \quad \mathbb{R}^2. \tag{4.82}$$

Then, we claim that (4.80) is equivalent to

$$\left. \begin{array}{l} \dfrac{\partial u}{\partial t} + (u \cdot \nabla)u - \left(\displaystyle\int_0^z \operatorname{div} u(x, \xi)\, d\xi \right)u_z + \nabla p = 0 \quad \text{in } S \times (0, \infty) \\[2mm] -\Delta p = \partial^2_{ij}(\overline{u_i u_j}) \quad \text{in } \mathbb{R}^2, \quad \displaystyle\int_\Omega p\, dx = 0 \quad \text{for } t \geq 0. \end{array} \right\} \tag{4.83}$$

Indeed, we just have to check that $v = -\int_0^z \operatorname{div} u(x, \xi)\, d\xi$ satisfies: $v|_{z=1} = 0$. In order to do so, we write the first equation of (4.83) in conserved form

$$\frac{\partial u}{\partial t} + \operatorname{div}(u \otimes u) + \partial_z(vu) + \nabla p = 0 \quad \text{in} \quad S \times (0, \infty).$$

Then, we integrate this equation in z from $z = 0$ to $z = 1$, we take its divergence and we obtain in view of (4.79)

$$\frac{\partial}{\partial t}(\operatorname{div} \overline{u}) + \operatorname{div}\{v(x, 1)\, u(x, 1)\} = 0 \quad \text{in} \quad \mathbb{R}^2 \times (0, \infty)$$

or, in other words, setting $w(x,t) = v(x,1,t) = -(\operatorname{div}\overline{u})(x,t)$

$$\frac{\partial w}{\partial t} - \operatorname{div}\{u(x,1,t)w\} = 0 \quad \text{in} \quad \mathbb{R}^2 \times (0,\infty). \qquad (4.84)$$

Then, (4.82) shows that $w|_{t=0} \equiv 0$ on $\mathbb{R}^2$. This fact combined with the transport equation (4.84) allows us to conclude: $v(x,1,t) = w(x,t) = 0$ on $\mathbb{R}^2 \times (0,\infty)$.

We continue our formal discussion of this model to propose a heuristic derivation of it from the three-dimensional Euler equations. We feel that any rigorous justification of this derivation would be a useful contribution. This derivation is similar to the one proposed by O. Besson and M.R. Laydi [61] in some viscous situations.

We consider the usual three-dimensional incompressible Euler equations in a shallow strip $S^\varepsilon = \{X \in \mathbb{R}^3 \ / \ X = (x,z) \in \mathbb{R}^2 \times (0,\varepsilon)\}$ where $\varepsilon > 0$:

$$\left.\begin{aligned}
\frac{\partial U}{\partial t} + U \cdot \nabla_X U + \nabla_X P &= 0 \quad \text{in} \quad S^\varepsilon \times (0,\infty), \\
\operatorname{div}_X(U) &= 0 \quad \text{in} \quad S^\varepsilon \times (0,\infty)
\end{aligned}\right\} \qquad (4.85)$$

with the following boundary conditions

$$U|_{z=0} = U|_{z=\varepsilon} = 0 \qquad (4.86)$$

and, for instance, periodic boundary conditions in x.

We then assume that the initial conditions take the following form

$$U|_{t=0} = \left(u_0\left(x,\frac{z}{\varepsilon}\right), \varepsilon v_0\left(x,\frac{z}{\varepsilon}\right)\right) = U_0^\varepsilon \quad \text{in} \quad S^\varepsilon \qquad (4.87)$$

where u_0, v_0 are given functions on S with values respectively in $\mathbb{R}^2, \mathbb{R}$. Then, requesting that U_0^ε satisfies the boundary conditions (4.86) and $\operatorname{div}_X U_0^\varepsilon = 0$ in S^ε amounts to requiring

$$\operatorname{div} u_0 + \partial_z v_0 = 0 \quad \text{in} \ S, \quad v_0|_{z=0} = v_0|_{z=1} = 0 \quad \text{in} \ \mathbb{R}^2. \qquad (4.88)$$

Next, if U^ε is a solution of (4.85)–(4.86) corresponding to the initial condition U_0^ε, we may try to write: $U^\varepsilon(x,z,t) = \left(u^\varepsilon\left(x,\frac{z}{\varepsilon},t\right), \varepsilon v^\varepsilon\left(x,\frac{z}{\varepsilon},t\right)\right)$, $P^\varepsilon(x,z,t) = p^\varepsilon\left(x,\frac{z}{\varepsilon},t\right)$ in $S^\varepsilon \times (0,\infty)$ where $u^\varepsilon, v^\varepsilon$ are now defined on $S \times (0,\infty)$ with values respectively in $\mathbb{R}^2, \mathbb{R}$. In that case, (4.85) and (4.86) become

$$\left.\begin{aligned}
\frac{\partial u^\varepsilon}{\partial t} + (u^\varepsilon \cdot \nabla)u^\varepsilon + v^\varepsilon \partial_z u^\varepsilon + \nabla p^\varepsilon &= 0 \qquad \text{in} \quad S \times (0,\infty), \\
\operatorname{div} u^\varepsilon + \frac{\partial v^\varepsilon}{\partial z} &= 0 \\
\varepsilon^2\left(\frac{\partial v^\varepsilon}{\partial t} + (u^\varepsilon \cdot \nabla)v^\varepsilon + v^\varepsilon \partial_z v^\varepsilon\right) + \partial_z p^\varepsilon &= 0 \qquad \text{in} \quad S \times (0,\infty), \\
\operatorname{div} u^\varepsilon + \frac{\partial v^\varepsilon}{\partial z} &= 0 \qquad \text{in} \ S \times (0,\infty),
\end{aligned}\right\} \qquad (4.89)$$

$$v^{\varepsilon}|_{z=0} \;=\; v^{\varepsilon}|_{z=1} \;=\; 0 \quad \text{in} \quad \mathbb{R}^2 \times (0, \infty). \tag{4.90}$$

Then, at least formally, we expect that, as ε goes to 0_+, $(u^{\varepsilon}, v^{\varepsilon}, p^{\varepsilon})$ "converges" to (u, v, p) solving (4.76)–(4.77). This is why we think that the study of the model (4.76)–(4.77) might shed some light on the three-dimensional incompressible Euler equations.

APPENDIX A

Truncation of Divergence-free Vector Fields

in Sobolev Spaces

We sketch here a general procedure to approximate in $W^{1,p}$ $(1 < p < \infty)$ divergence-free vector fields, vanishing on the boundary in the case of a bounded region, by compactly supported divergence-free vector fields.

More precisely, we consider $u \in W_0^{1,p}(\Omega)$ (resp. $W^{1,p}(\mathbb{R}^N)$) where Ω is a bounded, connected, Lipschitz domain in $\mathbb{R}^N$ $(N \geq 2)$. We assume

$$\operatorname{div} u = 0 \quad \text{a.e. in} \quad \Omega \quad (\text{resp. } \mathbb{R}^N). \tag{A.1}$$

We then set $\Omega_\delta = \{x \in \Omega, \, \operatorname{dist}(x, \partial\Omega) > \delta\}$ (resp. $B_{1/\delta} = \{x \in \mathbb{R}^N \, / \, |x| < 1/\delta\}$) for $\delta > 0$ small enough in case Ω is smooth, otherwise we choose Ω_δ to be a smooth, connected domain satisfying $\{x \in \Omega, \, \operatorname{dist}(x, \partial\Omega) > \delta\} \subset \Omega_\delta \subset \overline{\Omega}_\delta \subset \Omega$.

Next, we solve the following (linear) Stokes problem in Ω_δ

$$\left. \begin{array}{l} -\Delta u_\delta + u_\delta + \nabla p_\delta = -\Delta u + u \quad \text{in} \quad \Omega_\delta, \\ u_\delta \in W_0^{1,p}(\Omega_\delta), \quad \operatorname{div} u_\delta = 0 \quad \text{a.e. in} \quad \Omega_\delta. \end{array} \right\} \tag{A.2}$$

If Ω is bounded, we can of course skip the zero-order terms u and u_δ in equation (A.2). In view of classical results on Stokes problems (see [472] for instance), there exists a unique solution (u_δ, p_δ) of (A.2) in $W_0^{1,p}(\Omega_\delta) \times (L^p(\Omega_\delta)/\mathbb{R})$. If we request Ω_δ to be "Lipschitz uniformly in δ", which is the case if we simply take $\Omega_\delta = \{x \in \Omega \, , \, \operatorname{dist}(x, \partial\Omega) > \delta\}$ or $B_{1/\delta}$, we can in fact normalize p_δ in such a way that

$$\|p_\delta\|_{L^p(\Omega_\delta)} \leq C\|u\|_{W^{1,p}(\Omega_\delta)}. \tag{A.3}$$

On the other hand, we always have

$$\|u_\delta\|_{W^{1,p}(\Omega_\delta)} \leq C\|u\|_{W^{1,p}(\Omega_\delta)}, \tag{A.4}$$

where C denotes various positive constants independent of $\delta > 0$, u.

In fact, if $p = 2$, (A.4) takes a simpler form since, multiplying (A.1) by u_δ, we obtain

$$\int_{\Omega_\delta} |\nabla u_\delta|^2 + |u_\delta|^2 \, dx \ \leq \ \int_{\Omega_\delta} \nabla u \cdot \nabla u_\delta + u u_\delta \, dx \qquad (A.5)$$

and thus

$$\int_{\Omega_\delta} |\nabla u_\delta|^2 + |u_\delta|^2 \, dx \ \leq \ \int_{\Omega_\delta} |\nabla u|^2 + |u|^2 \, dx. \qquad (A.6)$$

We next claim that, as δ goes to 0_+, u_δ converges to u in $W_0^{1,p}(\Omega)$ (resp. $W^{1,p}(\mathbb{R}^N)$). In order to make this claim meaningful, we have to extend u_δ to Ω (resp. $\mathbb{R}^N$) by 0: in doing so, we preserve the nullity of div u_δ now in Ω (resp. in $\mathbb{R}^N$). We next prove the (strong) convergence in $W_0^{1,p}(\Omega)$ by two slightly different arguments, the first one in the case $p = 2$ where we use the simple relation (A.6) while the second one will be valid for all $1 < p < \infty$. First of all, we observe that in all cases, u_δ converges to u weakly in $W_0^{1,p}(\Omega)$ (resp. $W^{1,p}(\mathbb{R}^N)$): indeed, extracting a subsequence if necessary, we may assume that u_δ converges to $\tilde{u}$ weakly in $W_0^{1,p}(\Omega)$ (resp. $W^{1,p}(\mathbb{R}^N)$) while p_δ converges to some π weakly in $L^p(\Omega)$ (resp. $L^p(\mathbb{R}^N)$). Then we have

$$-\Delta(u-\tilde{u}) + (u-\tilde{u}) - \nabla\pi \ = \ 0 \qquad \text{in} \quad \Omega \quad (\text{resp. in } \mathbb{R}^N), \qquad (A.7)$$

$$\left.\begin{array}{l} u-\tilde{u} \in W_0^{1,p}(\Omega) \quad (\text{resp. } W^{1,p}(\mathbb{R}^N)), \\[4pt] \text{div}\,(u-\tilde{u}) = 0 \quad \text{a.e. in } \mathbb{R}^N, \end{array}\right\} \qquad (A.8)$$

and this implies, by the uniqueness for the Stokes problem, $u = \tilde{u}$, $\pi = 0$ (recall that $\pi \in L^p$) if $\Omega = \mathbb{R}^N$, π is a constant otherwise.

Then, if $p = 2$, (A.6) implies the strong convergence ! In the general case, using the linearity of the construction and the bound (A.4), we deduce that it is enough to prove the strong convergence whenever $u \in W_0^{1,q}(\Omega)$ (resp. $u \in W^{1,q}(\mathbb{R}^N) \cap W^{1,r}(\mathbb{R}^N)$) for some $q \in (p, +\infty)$ (resp. $q \in (p, \infty)$, $r \in (1, p)$, $r < 2 < q$). We use here the density of "smoother" divergence-free vector fields in $W_0^{1,p}$, although it is possible to give slightly more complicated proofs of the strong convergence which do not use this fact. We first claim that $u_\delta - u$ converges to 0 in C^1, say, on compact subsets of Ω. Indeed, taking the divergence of the equation (A.2), we obtain

$$-\Delta p_\delta \ = \ 0 \qquad \text{in} \quad \Omega_\delta \qquad (A.9)$$

and we have already shown that p_δ converges weakly to a constant (resp. 0) in L^p. Hence, ∇p_δ converges to 0 in C^1, say, on compact subsets of Ω. Since $-\Delta(u_\delta - u) + u_\delta - u = -\nabla p_\delta$ and $u_\delta - u$ converges weakly to 0 in

$W^{1,p}$, we deduce easily from the regularity results on Laplace's equation the convergence of $u_\delta - u$ to 0 in C^1, say, on compact subsets of Ω.

On the other hand, u_δ is bounded in $W_0^{1,q}(\Omega)$ (resp. $W^{1,q}(\mathbb{R}^N) \cap W^{1,r}(\mathbb{R}^N)$). Then, if Ω is bounded, we write, for $\delta < \delta_0$, using Hölder's inequality and the $W^{1,q}$ bound

$$\int_\Omega |\nabla(u_\delta - u)|^p + |u_\delta - u|^p \, dx$$

$$\leq \operatorname{meas}(\Omega_{\delta_0}) \sup_{\Omega_{\delta_0}}\left(|\nabla(u_\delta - u)|^p + |u_\delta - u|^p\right) + C \operatorname{meas}(\Omega - \Omega_{\delta_0})^{1/(q/p)'}$$

and we conclude letting first δ go to 0_+ then δ_0 go to 0_+.

If $\Omega = \mathbb{R}^2$, we first observe that there exists μ, nondecreasing, Lipschitz on $[0,\infty)$, such that $\mu \geq 1$ on $[0,\infty)$, $\mu(t) \to +\infty$ as $t \to +\infty$ and $\int_{\mathbb{R}^N} \mu(|x|)(|\nabla u|^2 + |u|^2) \, dx < \infty$. Then, multiplying (A.2) by $\mu(|x|) \, u_\delta$, we easily deduce that we have

$$\int_{\Omega_\delta} \mu(|x|)\left(|\nabla u_\delta|^2 + |u_\delta|^2\right) \leq C. \tag{A.10}$$

Then, we write for all $R \in (0,\infty)$, $\varepsilon \in (0,\infty)$, $\delta \in (0, 1/R)$, defining $f_\delta = |\nabla(u_\delta - u)|^p + |u_\delta - u|^p$,

$$\int_{\mathbb{R}^N} f_\delta \, dx \leq \operatorname{meas}(B_R) \sup_{B_R} f_\delta + C\varepsilon^{p-r} + C\varepsilon^{q-p} + \int_{(\varepsilon \leq f_\delta \leq 1/\varepsilon)} f_\delta \, 1_{B_R^c} \, dx$$

where we used the $W^{1,r} \cap W^{1,q}$ bound. Hence, we have, denoting by C_ε various positive constants depending only on ε and p

$$\int_{\mathbb{R}^N} f_\delta \, dx \leq C\varepsilon^{p-r} + C\varepsilon^{q-p} + \operatorname{meas}(B_R) \sup_{B_R} f_\delta$$

$$+ C_\varepsilon \left(\int_{B_R^c} |\nabla(u_\delta - u)|^2 + |u_\delta - u|^2 \, dx\right)$$

$$\leq C\varepsilon^{p-r} + C\varepsilon^{q-p} + \operatorname{meas}(B_R) \sup_{B_R} f_\delta + C_\varepsilon \, \mu(R)^{-1}$$

and we conclude upon letting first δ go to 0_+, then ε go to 0_+ and finally R go to $+\infty$.

Remark A.1. The above procedure yields a constructive approximation of u by $C_0^\infty(\Omega)$ (resp. $C_0^\infty(\mathbb{R}^N)$): indeed, one just needs to smooth u_δ by convolution.

In chapter 2, we use a variant of the preceding truncation in $\mathbb{R}^N$ that we describe now. First of all, let $\mathcal{D}^{1,2}(\mathbb{R}^N) = \{u \in L^{\frac{2N}{N-2}}(\mathbb{R}^N), \nabla u \in$

$L^2(\mathbb{R}^N)\}$ if $N \geq 3$. $(\mathcal{D}^{1,2}(\mathbb{R}^N)$ is a Hilbert space for the scalar product $\int_{\mathbb{R}^N} \nabla u \cdot \nabla v \, dx.)$ Then let $\tilde{\mathcal{D}}^{1,2}(\mathbb{R}^2) = \{u \in H^1_{loc}(\mathbb{R}^2), \nabla u \in L^2(\mathbb{R}^2)\}$. $\tilde{\mathcal{D}}^{1,2}(\mathbb{R}^2)$ is a Hilbert space for the scalar product $\int_{\mathbb{R}^2} \nabla u \cdot \nabla v + 1_{B_1} uv \, dx$ and an equivalent norm is given by $\|\nabla u\|_{L^2(\mathbb{R}^2)} + |\fint_{B_1} u \, dx|$. These facts on $\tilde{\mathcal{D}}^{1,2}(\mathbb{R}^2)$ are easy consequences of standard inequalities like

$$\left(\int_{B_R} |u|^2 \, dx\right)^{1/2} \leq C_R \left(\int_{B_R} |\nabla u|^2 + 1_{B_1} |u|^2 \, dx\right)^{1/2}$$

$$\leq C_R \left\{\|\nabla u\|_{L^2(B_R)} + \left|\fint_{B_1} u\right|\right\}$$

where $C_R > 0$ only depends on R which is arbitrary in $[1, +\infty)$. We denote by $|||\cdot|||$ the norm in $\mathcal{D}^{1,2}(\mathbb{R}^N)$ or $\tilde{\mathcal{D}}^{1,2}(\mathbb{R}^2)$. Next, we introduce essentially as before for $u \in \mathcal{D}^{1,2}(\mathbb{R}^N)$ if $N \geq 3$, $u \in \tilde{\mathcal{D}}^{1,2}(\mathbb{R}^2)$ if $N = 2$ the solution u_R of

$$\left. \begin{array}{l} -\Delta u_R + \nabla p_R = -\Delta u \quad \text{in} \quad B_R, \\ u_R \in H^1_0(B_R), \quad \operatorname{div} u_R = 0 \quad \text{a.e. in } B_R \end{array} \right\} \tag{A.11}$$

where $R \in [1, +\infty)$. We shall also use, when $N = 2$, the following variant

$$\left. \begin{array}{l} -\Delta \tilde{u}_R + 1_{B_1} \tilde{u}_R + \nabla p_R = -\Delta u + 1_{B_1} u \quad \text{in } B_R, \\ \tilde{u}_R \in H^1_0(B_R), \quad \operatorname{div} \tilde{u}_R = 0 \quad \text{a.e. in } B_R. \end{array} \right\} \tag{A.12}$$

If $N \geq 3$, we introduce the linear map $T_R(u) = u_R$ and, if $N = 2$, we consider two linear maps

$$T_R(u) = u_R + \fint_{B_1} (u - u_R), \quad \tilde{T}_R(u) = \tilde{u}_R. \tag{A.13}$$

Theorem A.1. *Assume that (A.1) holds.*
1) $T_R(u)$ *converges to* u, *as* R *goes to* $+\infty$, *in* $\mathcal{D}^{1,2}(\mathbb{R}^N)$ *if* $N \geq 3$ *while* $T_R(u)$ *and* $\tilde{T}_R(u)$ *converge to* u, *as* R *goes to* $+\infty$, *in* $\tilde{\mathcal{D}}^{1,2}(\mathbb{R}^2)$.
2) *We have for all* $R_0 \in (0, \infty)$

$$\sup_{|||u||| \leq 1} \|u - T_R(u)\|_{L^2(B_{R_0})} \to 0 \quad \text{as} \quad R \to +\infty \tag{A.14}$$

$$\sup_{|||u||| \leq 1} \|u - \tilde{T}_R(u)\|_{L^2(B_{R_0})} \to 0 \quad \text{as} \quad R \to +\infty. \tag{A.15}$$

Proof of Theorem A.1. First of all, we observe that, exactly as for (A.5), we obtain easily

$$\int_{\mathbb{R}^N} |\nabla T_R(u)|^2 \, dx \leq \int_{\mathbb{R}^N} |\nabla u|^2 \, dx \tag{A.16}$$

and

$$\int_{\mathbb{R}^2} |\nabla \tilde{T}_R(u)|^2 + 1_{B_1}|\tilde{T}_R(u)|^2 \, dx \leq \int_{\mathbb{R}^2} |\nabla u|^2 + 1_{B_1}|u|^2 \, dx \qquad \text{(A.17)}$$

where, of course, we extend u_R and $\tilde{u}_R$ by 0 to $\mathbb{R}^N$. Therefore, $T_R(u)$, $\tilde{T}_R(u)$ are bounded, uniformly in $R \geq 1$, in $\mathcal{D}^{1,2}(\mathbb{R}^N)$, $\tilde{\mathcal{D}}^{1,2}(\mathbb{R}^2)$ respectively. Since, if $N = 2$, $\int_{B_1} T_R(u) \, dx = \int_{B_1} u \, dx$, we have only to show the weak convergence to u in view of (A.16) and (A.17) or, in other words, extracting subsequences if necessary, that $\overline{u} = u$ if $\overline{u}$ is the weak limit in $\mathcal{D}^{1,2}(\mathbb{R}^N)$ (or $\tilde{\mathcal{D}}^{1,2}(\mathbb{R}^2)$) of $T_R(u)$, $\tilde{T}_R(u)$ as R goes to $+\infty$.

We begin with $T_R(u)$ and we observe that, if we normalize p_R by $\int_{B_R} p_R \, dx = 0$, we have

$$\left. \begin{array}{l} \partial_i p_R = \partial_j f_R^{i,j} \quad \text{in} \quad \mathcal{D}'(B_R), \\[2mm] \displaystyle\int_{B_R} p_R \, dx = 0, \quad \|f_R^{i,j}\|_{L^2(B_R)} \leq C \end{array} \right\} \qquad \text{(A.18)}$$

for some $C \geq 0$ independent of R (depending only on $\|\|u\|\|$). This yields

$$\|p_R\|_{L^2(B_R)} \leq C \qquad \text{(A.19)}$$

and we may thus assume, extracting a subsequence if necessary, that p_R converges weakly in $L^2(\mathbb{R}^N)$ to some p. We then obtain, setting $w = \overline{u} - u$,

$$\left. \begin{array}{l} -\Delta w + \nabla p = 0 \quad \text{in } \mathcal{D}'(\mathbb{R}^N), \quad p \in L^2(\mathbb{R}^N), \\[2mm] \text{div } w = 0 \quad \text{a.e. in } \mathbb{R}^N \end{array} \right\} \qquad \text{(A.20)}$$

and $w \in \mathcal{D}^{1,2}(\mathbb{R}^N)$ if $N \geq 3$, $w \in \tilde{\mathcal{D}}^{1,2}(\mathbb{R}^2)$ if $N = 2$ and then $\int_{B_1} w \, dx = 0$. Taking the divergence of (A.20), we immediately deduce that $p \equiv 0$ and thus $w \equiv 0$ if $N \geq 3$, $w \equiv$ constant if $N = 2$. Finally, $w \equiv 0$ if $N = 2$ since $\int_{B_1} w \, dx = 0$.

The proof for $\tilde{T}_R(u)$ is a bit more delicate. Multiplying (A.12) by w_R, where $1 \leq R' \leq R$, $w = \overline{u} - u$ and w_R is the solution of (A.11) corresponding to w (instead of u), we obtain

$$\int_{B_R} \nabla(\tilde{T}_R(u) - u) \cdot \nabla w_{R'} \, dx + \int_{B_1} (\tilde{T}_R(u) - u) \cdot w_{R'} \, dx = 0$$

and, letting R go to $+\infty$,

$$\int_{\mathbb{R}^2} \nabla w \cdot \nabla w_{R'} \, dx + \int_{B_1} w \cdot w_{R'} \, dx = 0. \qquad \text{(A.21)}$$

If we show that $\int_{B_1} w \, dx = 0$, we see that (A.21) holds with $w_{R'}$ replaced by $T_{R'}(w)$ and thus, letting R' go to $+\infty$, we deduce

$$\int_{\mathbb{R}^2} |\nabla w|^2 \, dx + \int_{B_1} |w|^2 \, dx = 0$$

hence $w \equiv 0$. In order to show our claim on $\int_{B_1} w \, dx$, we take the curl of (A.12) and we find

$$-\Delta \{\mathrm{curl}\,(\tilde{T}_R(u) - u)\} + \mathrm{curl}\,(1_{B_1}(\tilde{T}_R(u) - u)) = 0 \quad \text{in} \quad \mathcal{D}'(B_R)$$

and, letting R go to $+\infty$, we obtain

$$-\Delta \,\mathrm{curl}\, w + \mathrm{curl}\,(1_{B_1} w) = 0 \quad \text{in } \mathcal{D}'(\mathbb{R}^2), \ \mathrm{curl}\, w \in L^2(\mathbb{R}^2).$$

Hence

$$
\begin{aligned}
\mathrm{curl}\, w &= \frac{1}{2\pi} \int_{B_1} \frac{y_2 - x_2}{|y - x|^2} w_1(y) \, dy - \frac{1}{2\pi} \int_{B_1} \frac{y_1 - x_1}{|y - x|^2} w_2(y) \, dy \\
&= \frac{1}{2\pi}\left\{ -\frac{x_2}{|x|^2}\left(\int_{B_1} w_1 \, dy \right) + \frac{x_1}{|x|^2}\left(\int_{B_1} w_2 \, dy \right) + o\!\left(\frac{1}{|x|}\right) \right\},
\end{aligned}
$$

$$\text{as } |x| \to +\infty.$$

Since $\mathrm{curl}\, w \in L^2(\mathbb{R}^2)$, we deduce easily that $\int_{B_1} w \, dy = 0$, and this completes the proof of part 1.

Since the embedding of $\mathcal{D}^{1,2}(\mathbb{R}^N)$, $\tilde{\mathcal{D}}^{1,2}(\mathbb{R}^2)$ into $L^2(B_{R_0})$ is compact, we have only to show, in order to prove part 2, that $T_{R_n}(u_n)$, $T_{R_n}(u_n)$ converge weakly in $\mathcal{D}^{1,2}(\mathbb{R}^N)$, $\tilde{\mathcal{D}}^{1,2}(\mathbb{R}^2)$ respectively to u whenever u_n converges weakly to u in these spaces, and this fact is shown exactly as in part 1. $\square$

We conclude this appendix with the introduction and the study of some related truncations. To this end, we consider $f \in L^\infty(\mathbb{R}^N) \cap L^{\frac{N}{2},\infty}(\mathbb{R}^N)$, $f \geq 0$, $f \not\equiv 0$ and we assume in all that follows that $N \geq 3$. If $u \in \mathcal{D}^{1,2}(\mathbb{R}^N)$, $R \in (0,\infty)$, we define $\Theta_R(u) = u_R$ to be the solution of

$$\left.\begin{aligned}
- \Delta u_R + f u_R + \nabla p_R &= -\Delta u + f u \quad \text{in} \quad \mathcal{D}'(B_R), \\
u_R \in H_0^1(B_R), \quad \mathrm{div}\, u_R &= 0 \quad \text{a.e. in } B_R,
\end{aligned}\right\} \quad (A.22)$$

and we have

Theorem A.2. *Assume that (A.1) holds.*
1) $\Theta_R(u)$ converges to u in $\mathcal{D}^{1,2}(\mathbb{R}^N)$ as R goes to $+\infty$.

2) *Let u^n converge weakly in $\mathcal{D}^{1,2}(\mathbb{R}^N)$ to some u. Let $f^n \geq 0$, $f^n \in L^{\frac{N}{2},\infty}(\mathbb{R}^N)$, $f^n \not\equiv 0$ be bounded in $L^\infty(\mathbb{R}^N)$ and assume that $\sqrt{f^n}u^n$ is bounded in $L^2(\mathbb{R}^N)$ and that f^n converges to f in $L^1(B_M)$ for all $M \in (0,\infty)$. We denote by $\Theta_R^n(u^n)$ the solution of (A.22) with u, f replaced by u^n, f^n respectively. Then, $\Theta_R^n(u^n)$ converges weakly to u^n in $\mathcal{D}^{1,2}(\mathbb{R}^N)$ uniformly in n as R goes to $+\infty$ and, in particular, we have*

$$\sup_n \|u^n - \Theta_R^n(u^n)\|_{L^2(B_{R_0})} \to 0 \quad \text{as} \quad R \to +\infty. \tag{A.23}$$

Proof of Theorem A.2. 1) We have (extending as usual $u_R = \Theta_R(u)$ by 0)

$$\left.\begin{aligned}
\int_{\mathbb{R}^N} |\nabla u_R|^2 + f|u_R|^2 \, dx &= \int_{\mathbb{R}^N} \nabla u \cdot \nabla u_R + fu \cdot u_R \, dx \\
&\leq \int_{\mathbb{R}^N} |\nabla u|^2 + f|u|^2 \, dx.
\end{aligned}\right\} \tag{A.24}$$

Notice that $f|u|^2 \in L^1(\mathbb{R}^N)$ since $f \in L^{\frac{N}{2},\infty}(\mathbb{R}^N)$, $|u|^2 \in L^{\frac{N}{N-2},1}(\mathbb{R}^N)$. It is thus enough to show that if u_R converges weakly to some $\tilde{u}$ in $\mathcal{D}^{1,2}(\mathbb{R}^N)$ then $\tilde{u} \equiv u$. Writing $w = -\Delta(\tilde{u}-u) + f(\tilde{u}-u) \in H^{-1}(\mathbb{R}^N)$, we deduce from (A.22) that we have

$$<w,\phi>= 0, \quad \text{for all} \quad \phi \in C_0^\infty(\mathbb{R}^N), \text{ div } \phi = 0 \quad \text{on } \mathbb{R}^N. \tag{A.25}$$

Therefore, by classical results, there exists $p \in L^{\frac{2N}{N-2},2}(\mathbb{R}^N)+L^2(\mathbb{R}^N)$ such that $w = -\nabla p$: once we know that $w = -\nabla p$, the integrability of p is easily seen by Fourier transforms. Indeed $\hat{P} = \frac{|\xi|}{(1+|\xi|^2)^{1/2}}\hat{p} \in L^2(\mathbb{R}^N)$ and thus $\hat{p}1_{(|\xi|\geq 1)} \in L^2(\mathbb{R}^N)$, while $\hat{p}1_{(|\xi|<1)} = 1_{(|\xi|<1)}\frac{(1+|\xi|^2)^{1/2}}{|\xi|}\hat{P} \in L^{\frac{2N}{N+2},2}(\mathbb{R}^N)$ since $\frac{1}{|\xi|} \in L^{N,\infty}(\mathbb{R}^N)$.

Furthermore, we have clearly

$$-\Delta p = \text{div } w = \text{div}\,(f(\tilde{u}-u)) \quad \text{in} \quad \mathcal{D}'(\mathbb{R}^N)$$

and $f(\tilde{u}-u) \in L^{\frac{2N}{N+2},2}(\mathbb{R}^N)$ since $f \in L^{N,\infty}(\mathbb{R}^N)$, $(\tilde{u}-u) \in L^{\frac{2N}{N-2},2}(\mathbb{R}^N)$. Hence, we obtain

$$p \in L^2(\mathbb{R}^N), \quad \nabla p \in L^{\frac{2N}{N+2},2}(\mathbb{R}^N). \tag{A.26}$$

We then set $z = (u-\tilde{u})$ and multiply w by $z\varphi_n$ where $\varphi_n = \varphi(\frac{\cdot}{n})$, $n \geq 1$ and $\varphi \in C_0^\infty(\mathbb{R}^N)$, $\text{Supp}\,\varphi \subset B_2$, $0 \leq \varphi \leq 1$, $\varphi \equiv 1$ on B_1. We obtain easily

$$\int_{\mathbb{R}^N} \left(|\nabla z|^2 + f|z|^2 \right) \varphi_n \, dx$$

$$= \int_{\mathbb{R}^N} |z|^2 \left(\frac{1}{2} \Delta \varphi_n \right) + p \, z \cdot \nabla \varphi_n \, dx$$

$$\leq \frac{C}{n^2} \int_{(n \leq |x| \leq 2n)} |z|^2 \, dx + \frac{C}{n} \int_{(n \leq |x| \leq 2n)} |p| \, |z| \, dx$$

$$\leq \frac{C}{n^2} \int_{(n \leq |x| \leq 2n)} |z|^2 \, dx + C \left(\int_{\mathbb{R}^N} |p|^2 \, dx \right)^{1/2} \left\{ \frac{1}{n^2} \int_{(n \leq |x| \leq 2n)} |z|^2 \, dx \right\}.$$

In view of (A.26), we conclude easily that $z \equiv 0$ once we observe

$$\frac{1}{n^2} \int_{(n \leq |x| \leq 2n)} |z|^2 \, dx \leq \frac{C}{n^2} \left(\int_{(n \leq |x| \leq 2n)} |z|^{\frac{2N}{N-2}} \, dx \right)^{\frac{N-2}{N}} (n^N)^{\frac{2}{N}}$$

$$= C \int_{(n \leq |x| \leq 2n)} |z|^{\frac{2N}{N-2}} \, dx \underset{n}{\to} 0.$$

2) Since, for each $n \geq 1$, $\Theta_R^n(u^n)$ converges in $\mathcal{D}^{1,2}(\mathbb{R}^N)$ to u^n in view of the proof of part 1, we have only to show that $\tilde{u}^n = \Theta_{R^n}^n(u^n)$ converges weakly in $\mathcal{D}^{1,2}(\mathbb{R}^N)$ to u as n and R_n go to $+\infty$, and (A.23) then follows in view of the Rellich–Kondrakov compactness theorems.

Observe first that because of (A.23) and the assumptions made, $\tilde{u}^n$ is bounded in $\mathcal{D}^{1,2}(\mathbb{R}^N)$ and thus, without loss of generality, we may assume that $\tilde{u}^n$ converges weakly in $\mathcal{D}^{1,2}(\mathbb{R}^N)$ to some $\tilde{u}$. Next, we simply observe that $f^n \tilde{u}^n$, $f^n u^n$ converge respectively in L^1_{loc} (and even L^p_{loc} for $1 \leq p < \frac{2N}{N-2}$) to $f\tilde{u}$, fu since f^n, $\tilde{u}^n$ and u^n converge in L^2_{loc}. At this stage, we conclude that $\tilde{u} \equiv u$ exactly as in part 1. □

Let us remark that part 2 of Theorem A.2 yields the following variant of (A.23)

$$\sup_n \left\{ \|u^n - \Theta_R^n(u^n)\|_{L^2(B_{R_0})} \left(1 + \|u^n\|^2_{\mathcal{D}^{1,2}(\mathbb{R}^N)} + \|\sqrt{f^n}\, u^n\|^2_{L^2(\mathbb{R}^N)} \right)^{-1/2} \right\}$$

$$\to 0 \qquad \text{as} \quad R \to +\infty$$

$$\tag{A.27}$$

for all $u^n \in \mathcal{D}^{1,2}(\mathbb{R}^N)$. Indeed, we just apply part 2 with $u^n \left(1 + \|u^n\|^2_{\mathcal{D}^{1,2}(\mathbb{R}^N)} + \|\sqrt{f^n}\, u^n\|^2_{L^2(\mathbb{R}^N)} \right)^{-1/2}$ instead of u^n.

APPENDIX B

Two Facts on $\tilde{\mathcal{D}}^{1,2}(\mathbb{R}^2)$

Recall (see Appendix A) that $\tilde{\mathcal{D}}^{1,2}(\mathbb{R}^2) = \{u \in H^1_{\text{loc}}(\mathbb{R}^2) ,\ \nabla u \in L^2(\mathbb{R}^2)\}$. We begin with a remark.

Lemma B.1. *We have $\tilde{\mathcal{D}}^{1,2}(\mathbb{R}^2) \cap (L^1(\mathbb{R}^2) + L^2(\mathbb{R}^2)) \hookrightarrow H^1(\mathbb{R}^2)$, and, more precisely, there exists a constant $C > 0$ such that for all $u \in \tilde{\mathcal{D}}^{1,2}(\mathbb{R}^2)$ satisfying $u = u_1 + u_2$ with $u_1 \in L^1(\mathbb{R}^2)$, $u_2 \in L^2(\mathbb{R}^2)$*

$$\|u\|_{L^2(\mathbb{R}^2)} \leq C\Big\{\|u_1\|_{L^1(\mathbb{R}^2)}^{1/2}\|\nabla u\|_{L^2(\mathbb{R}^2)}^{1/2} + \|u_2\|_{L^2(\mathbb{R}^2)}\Big\}. \qquad (\text{B.1})$$

Proof of Lemma B.1. We first remark that $|u| = \tilde{u}_1 + \tilde{u}_2$ with $\tilde{u}_1 = \min(|u_1|, |u|) \in L^1(\mathbb{R}^2)$, $0 \leq \tilde{u}_2 \leq |u_2|$ and $\tilde{u}_2 \in L^2(\mathbb{R}^2)$. Next, we also have: $|u| \leq 2w + |u|\, 1_{(|u|\leq 2)}$ where $w = (|u| - 1)^+ \in \tilde{\mathcal{D}}^{1,2}(\mathbb{R}^2)$, and

$$\int_{\mathbb{R}^2} |u|^2\, 1_{(|u|\leq 2)}\, dx \leq 2\int_{\mathbb{R}^2} \tilde{u}_1^2\, 1_{(\tilde{u}_1 \leq 2)}\, dx + 2\int_{\mathbb{R}^2} \tilde{u}_2^2\, dx$$

$$\leq 4\int_{\mathbb{R}^2} \tilde{u}_1\, dx + 2\int_{\mathbb{R}^2} \tilde{u}_2^2\, dx$$

hence

$$\big\|u\, 1_{(|u|\leq 2)}\big\|_{L^2(\mathbb{R}^2)} \leq 2\Big(\|u_1\|_{L^1(\mathbb{R}^2)}^{1/2} + \|u_2\|_{L^2(\mathbb{R}^2)}\Big). \qquad (\text{B.2})$$

On the other hand, we also have

$$\int_{\mathbb{R}^2} w\, dx \leq \int_{\mathbb{R}^2} |u|\, 1_{(|u|\geq 1)}\, dx$$

$$\leq \int_{\mathbb{R}^2} \tilde{u}_1\, dx + \int_{\mathbb{R}^2} \tilde{u}_2\Big(1_{(\tilde{u}_1 \geq 1/2)} + 1_{(\tilde{u}_2 \geq 1/2)}\Big)\, dx$$

$$\leq \|u_1\|_{L^1(\mathbb{R}^2)} + \|u_2\|_{L^2(\mathbb{R}^2)}\left\{\text{meas}\Big\{\tilde{u}_1 \geq \frac{1}{2}\Big\}^{1/2} + \text{meas}\Big\{\tilde{u}_2 \geq \frac{1}{2}\Big\}^{1/2}\right\}$$

$$\leq \|u_1\|_{L^1(\mathbb{R}^2)} + \|u_2\|_{L^2(\mathbb{R}^2)}\left\{\sqrt{2}\|u_1\|_{L^1(\mathbb{R}^2)}^{1/2} + 2\|u_2\|_{L^2(\mathbb{R}^2)}\right\},$$

and we deduce from Gagliardo–Nirenberg type inequalities

$$\|w\|_{L^2(\mathbb{R}^2)} \leq C\|w\|_{L^1(\mathbb{R}^2)}^{1/2}\|\nabla w\|_{L^2(\mathbb{R}^2)}^{1/2} \leq C\|w\|_{L^1(\mathbb{R}^2)}^{1/2}\|\nabla u\|_{L^2(\mathbb{R}^2)}^{1/2}$$

therefore we find

$$\begin{aligned}
\|w\|_{L^2(\mathbb{R}^2)} \leq C\|\nabla u\|_{L^2(\mathbb{R}^2)}^{1/2}\Big\{&\|u_1\|_{L^1(\mathbb{R}^2)}^{1/2} \\
&+ \|u_2\|_{L^2(\mathbb{R}^2)}^{1/2}\|u_1\|_{L^1(\mathbb{R}^2)}^{1/4} + \|u_2\|_{L^2(\mathbb{R}^2)}\Big\}
\end{aligned}$$

or

$$\|w\|_{L^2(\mathbb{R}^2)} \leq C\|\nabla u\|_{L^2(\mathbb{R}^2)}^{1/2}\Big\{\|u_1\|_{L^1(\mathbb{R}^2)}^{1/2} + \|u_2\|_{L^2(\mathbb{R}^2)}\Big\}. \tag{B.3}$$

Combining (B.2) and (B.3) we obtain

$$\left.\begin{aligned}
\|u\|_{L^2(\mathbb{R}^2)} \leq C\Big\{&\|u_2\|_{L^2(\mathbb{R}^2)} + \|\nabla u\|_{L^2(\mathbb{R}^2)}^{1/2}\|u_1\|_{L^1(\mathbb{R}^2)}^{1/2}\Big\} \\
&+ C\|u_1\|_{L^1(\mathbb{R}^2)}^{1/2} + C\|\nabla u\|_{L^2(\mathbb{R}^2)}^{1/2}\|u_2\|_{L^2(\mathbb{R}^2)}.
\end{aligned}\right\} \tag{B.4}$$

Then, replacing u by λu and thus u_1, u_2 by $\lambda u_1, \lambda u_2$ where $\lambda > 0$ is arbitrary, we also deduce

$$\begin{aligned}
\|u\|_{L^2(\mathbb{R}^2)} \leq C\Big\{&\|u_2\|_{L^2(\mathbb{R}^2)} + \|\nabla u\|_{L^2(\mathbb{R}^2)}^{1/2}\|u_1\|_{L^1(\mathbb{R}^2)}^{1/2}\Big\} \\
&+ \frac{C}{\sqrt{\lambda}}\|u_1\|_{L^1(\mathbb{R}^2)}^{1/2} + C\sqrt{\lambda}\|\nabla u\|_{L^2(\mathbb{R}^2)}^{1/2}\|u_2\|_{L^2(\mathbb{R}^2)}.
\end{aligned}$$

Since this inequality holds for all $\lambda > 0$, we choose $\lambda = \|u_1\|_{L^1(\mathbb{R}^2)}^{1/2} \cdot \|\nabla u\|_{L^2(\mathbb{R}^2)}^{-1/2}\|u_2\|_{L^2(\mathbb{R}^2)}^{-1}$ and we find

$$\begin{aligned}
\|u\|_{L^2(\mathbb{R}^2)} \leq C\Big\{&\|u_2\|_{L^2(\mathbb{R}^2)} + \|\nabla u\|_{L^2(\mathbb{R}^2)}^{1/2}\|u_1\|_{L^1(\mathbb{R}^2)}^{1/2}\Big\} \\
&+ C\|u_1\|_{L^1(\mathbb{R}^2)}^{1/4}\|\nabla u\|_{L^2(\mathbb{R}^2)}^{1/4}\|u_2\|_{L^2(\mathbb{R}^2)}^{1/2},
\end{aligned}$$

and this inequality yields (B.1). □

Remark B.1. A similar (and simpler) proof shows a similar result in $\mathcal{D}^{1,2}(\mathbb{R}^N)$. More precisely, if $N \geq 3$, $u \in \mathcal{D}^{1,2}(\mathbb{R}^N)$ with $u = u_1 + u_2$, $u_1 \in L^1(\mathbb{R}^N)$, $u_2 \in L^2(\mathbb{R}^N)$ then $u \in L^2(\mathbb{R}^N)$ (and thus $H^1(\mathbb{R}^N)$) and we have for some $C > 0$ depending only on N

$$\|u\|_{L^2(\mathbb{R}^N)} \leq C\Big\{\|u_1\|_{L^1(\mathbb{R}^N)}^{\theta}\|\nabla u\|_{L^2(\mathbb{R}^N)}^{1-\theta} + \|u_2\|_{L^2(\mathbb{R}^N)}\Big\} \tag{B.5}$$

where $\theta = \frac{2}{N+2}$.

We next turn to weighted L^p bounds for elements of $\tilde{\mathcal{D}}^{1,2}(\mathbb{R}^2)$. Recall (see Appendix A) that we choose the scalar product $\left(\int_{\mathbb{R}^2} \nabla u \cdot \nabla v + 1_{B_1} u \cdot v \, dx\right)$ on $\tilde{\mathcal{D}}^{1,2}(\mathbb{R}^2)$ for which this space is a Hilbert space, and we denote by $|||u|||$ the corresponding norm.

Theorem B.1. *For all* $m \in [2, +\infty)$, *there exists a positive constant* C *such that we have for all* $u \in \tilde{\mathcal{D}}^{1,2}(\mathbb{R}^2)$

$$\left(\int_{\mathbb{R}^2} \frac{|u|^m}{<x>^2} (\log <x>)^{-\theta} \, dx\right)^{1/m} \leq C|||u||| \qquad (B.6)$$

if $\theta \in (\frac{m}{2} + 1, \infty)$.

Proof of Theorem B.1. We begin with the case $m = 2$ and prove in fact

$$\left(\int_{S^1} |u(r\omega)|^2 \, d\omega\right)^{1/2} \leq 2\left(\int_{B_1} |u|^2 \, dx\right)^{1/2} + \|\nabla u\|_{L^2(\mathbb{R}^2)}^{1/2} (\log 2r)^{1/2}, \quad (B.7)$$

for all $r \geq 1$, where we write $x = r\omega$, $\omega \in S^1$, $r = |x|$, and the case $m = 2$ of (B.5) follows easily upon integrating in Ω.

In order to show (B.7), we write for all $r > 0$

$$\left.\begin{aligned}\frac{d}{dr}\left(\int_{S^1} |u(r\omega)|^2 \, d\omega\right)^{1/2} &\leq \left(\int_{S^1} \left|\frac{\partial u}{\partial r}(r\omega)\right|^2 d\omega\right)^{1/2} \\ &\leq \left(\int_{S^1} |\nabla u(r\omega)|^2 \, d\omega\right)^{1/2}\end{aligned}\right\} \qquad (B.8)$$

and observe that there exists $r_0 \in [\frac{1}{2}, 1]$ such that

$$\left(\int_{S^1} |u(r_0\omega)|^2 \, d\omega\right)^{1/2} \leq 2\left(\int_{B_1} |u|^2 \, dx\right)^{1/2}.$$

Hence, integrating (B.8) from r_0 to $r \geq 1$, we deduce

$$\left(\int_{S^1} |u(r\omega)|^2 \, d\omega\right)^{1/2}$$

$$\leq 2\left(\int_{B_1} |u|^2 \, dx\right)^{1/2} + \int_{r_0}^r \left(\int_{S^1} |\nabla u(t\omega)|^2 \, d\omega\right)^{1/2} dt$$

$$\leq 2\left(\int_{B_1} |u|^2 \, dx\right)^{1/2} + \left(\int_{r_0}^r \int_{S^1} |\nabla u(t\omega)|^2 \, t \, d\omega \, dt\right)^{1/2} \left(\log \frac{r}{r_0}\right)^{1/2}$$

and (B.7) follows. The above calculations are obvious if u is smooth and (B.7) is thus shown for all $u \in \tilde{\mathcal{D}}^{1,2}(\mathbb{R}^2)$ by density.

Let us also remark that (B.7) implies for all $R \geq 1$

$$\int_{B_R} |u|^2 \, dx \leq C R^2 \log 2R \, |||u|||^2 \tag{B.9}$$

$$\int_{B_R} |u| \, dx \leq C R^2 (\log 2R)^{1/2} \, |||u|||. \tag{B.10}$$

We next want to show (B.6) for $m \in (2, \infty)$. We observe that we have for all $R \geq 1$

$$\left. \begin{array}{l} \left(\int_{B_R} \left| u - \fint_{B_R} u \right|^m dx \right)^{1/m} \\ \leq C \left(\int_{B_R} \left| u - \fint_{B_R} u \right|^2 dx \right)^{1/m} \left(\int_{B_R} |\nabla u|^2 \, dx \right)^{\frac{m-2}{2m}}. \end{array} \right\} \tag{B.11}$$

This is one form of the Gagliardo–Nirenberg inequalities (the fact that C does not depend on R can be easily deduced from a simple scaling argument). Combining (B.9), (B.10) and (B.11), we deduce for all $R \geq 2$

$$\int_{B_R} |u|^m \, dx \leq C R^2 (\log 2R)^{m/2} \, |||u|||^m. \tag{B.12}$$

Next, by a simple integration by parts, we obtain for all $R \geq 2$

$$\int_{B_R} \frac{|u|^m}{<x>^2} (\log <x>)^{-\theta} \, dx$$

$$= \left(\int_{B_R} |u|^m \, dx \right) (1+R^2)^{-1} (\log \sqrt{1+R^2})^{-\theta}$$

$$+ \int_0^R \left(\int_{B_r} |u|^m \, dx \right) \left\{ \frac{2r}{(1+r^2)^2} (\log \sqrt{1+r^2})^{-\theta} \right.$$

$$\left. + \frac{\theta}{1+r^2} (\log \sqrt{1+r^2})^{-(\theta+1)} \frac{r}{1+r^2} \right\} dr$$

$$\leq C|||u|||^m \left(1 + \int_0^R \frac{1}{\sqrt{1+r^2}} (\log \sqrt{1+r^2})^{\frac{m}{2}-\theta} \, dr \right)$$

$$\leq C|||u|||^m$$

if $\theta > \frac{m}{2} + 1$. $\square$

APPENDIX C

Compactness in Time with Values
in Weak Topologies

Let X be a separable reflexive Banach space and let f^n be bounded in $L^\infty(0, T; X)$ for some $T \in (0, \infty)$. We assume that $f^n \in C([0, T]; Y)$ where Y is a Banach space such that $X \hookrightarrow Y$, Y' is separable and dense in X'. Furthermore, we assume that for all $\varphi \in Y'$, $<\varphi, f^n(t)>_{Y' \times Y}$ is uniformly continuous in $t \in [0, T]$ uniformly in $n \geq 1$.

We then choose a closed ball B_{R_0} of X containing all the values of $f^n(t)$ for $t \in [0, T]$, $n \geq 1$. The weak topology of X, since X' is separable, makes B_{R_0} a compact metric space, and we denote by $C([0, T], X - w)$ the space of continuous functions on $[0, T]$ with values in $B_{R_0} \subset X$ equipped with the weak topology.

Lemma C.1. *Under the above conditions, f^n is relatively compact in $C([0, T]; X - w)$.*

Proof of Lemma C.1. Let $(\varphi_k)_{k \geq 1}$ be a dense sequence in Y'. We consider the "weak topology" distance on B_{R_0} given by

$$d(f, g) = \sum_{k \geq 1} \frac{1}{2^k} \frac{|<\varphi_k, f - g>_{Y' \times Y}|}{1 + |<\varphi_k, f - g>_{Y' \times Y}|}.$$

Using the Ascoli–Arzéla theorem, we have only to show

$$\sup_{n \geq 1} d(f^n(t), f^n(s)) \to 0 \qquad \text{as} \quad t, s \in [0, T], \ |t - s| \to 0,$$

and this is obvious since, by assumption, we have for each $k \geq 1$

$$\sup_{n \geq 1} |<\varphi_k, f^n(t) - f^n(s)>_{Y' \times Y}| \to 0 \qquad \text{as} \quad |t - s| \to 0.$$

Hence, for all $k \geq 1$

$$d(f^n(t), f^n(s)) \leq \sup_{1 \leq j \leq k} \sup_n |<\varphi_j, f^n(t) - f^n(s)>_{Y' \times Y}| + \frac{1}{2^k}$$

and we conclude upon letting first $|t - s|$ go to 0 and then k go to $+\infty$. $\square$

APPENDIX D

Weak L^1 Estimates for Solutions of the Heat Equation

We wish to prove in this appendix a result concerning solutions of the heat equation in $\mathbb{R}^N$ that we used in section 3.3. The main result we want to present in this direction is given by

Theorem D.1. Let $f \in L^1(0, T; L^{q,r}(\mathbb{R}^N))$ where $1 < q < \infty$, $1 \le r < \infty$. Let $u \in C([0, T]; L^{q,r}(\mathbb{R}^N))$ be the solution of

$$\frac{\partial u}{\partial t} - \frac{1}{2} \Delta u = f \quad in \quad \mathbb{R}^N \times (0, T), \quad u|_{t=0} = 0 \quad in \quad \mathbb{R}^N. \tag{D.1}$$

Then, we have

$$\frac{\partial u}{\partial t}, D_x^2 u \in L^{q,r}(\mathbb{R}^N) \quad a.e. \quad t \in (0, T) \tag{D.2}$$

$$\left. \begin{aligned} \operatorname{meas} \left\{ t \in (0, T) \; / \; \|\varphi\|_{L^{q,r}(\mathbb{R}^N)} \ge \lambda \right\} &\le C \|f\|_{L^1(0, T; L^{q,r}(\mathbb{R}^N))} \lambda^{-1} \\ &\text{for all} \quad \lambda > 0 \end{aligned} \right\} \tag{D.3}$$

for some $C > 0$ independent of f, where $\varphi = \frac{\partial \varphi}{\partial t}$, $D_x^2 \varphi$.

Remarks D.1. 1) Here and below, $L^{q,r}(\mathbb{R}^N)$ denotes the usual Lorentz spaces $(L^{q,q}(\mathbb{R}^N) = L^q(\mathbb{R}^N))$.

2) The existence of the solution u of (D.1) and its continuity in t with values in $L^{q,r}(\mathbb{R}^N)$ follows easily from the representation formula

$$\left. \begin{aligned} u(x, t) = \int_0^t ds \int_{\mathbb{R}^N} dy \, f(y, s) \exp\left(-\frac{|x-y|^2}{2(t-s)}\right) (2\pi(t-s))^{-N/2} \\ \text{a.e. in} \; \mathbb{R}^N, \quad \forall t \in [0, T]. \end{aligned} \right\} \tag{D.4}$$

3) The normalization constant $\frac{1}{2}$ in front of Δ is irrelevant and the result is valid if we replace $-\frac{1}{2}\Delta$ by $-\nu\Delta$ with $\nu > 0$ or, more generally, by an arbitrary uniformly elliptic second-order operator with bounded uniformly continuous coefficients.

4) If we take $r = q$, the estimate (D.3) together with the classical $L^q(\mathbb{R}^N \times (0, T))$ result for $D^2 u$ and $\frac{\partial u}{\partial t}$ shows, by interpolation, that

$\frac{\partial u}{\partial t}$, $D^2 u \in L^p(0,T; L^q(\mathbb{R}^N))$ whenever $1 < q < \infty, 1 < p \le q$. Then, by an easy duality argument, we see that this result is also valid if $1 < q, p < \infty$ and we recover the "classical" $L^p(L^q)$ regularity result for solutions of the heat equation, a result that we used in section 4.3. □

Proof of Theorem D.1. We first use and recall the following variant of the fundamental "covering lemma" due to A.P. Calderón and A. Zygmund [**78**] (see also A. Zygmund [**497**] and E. Stein [**454**]) contained in L. Hörmander [**222**] (Lemma 3.1, observe that the proof in [**222**] immediately extends to $L^{q,r}(\mathbb{R}^N)$ spaces): for each $M > 0$, we can find g, h_k, disjoint intervals I_k in $[0,T]$ such that

$$f = g + \sum_{k=1}^{\infty} h_k,$$

$$\|g\|_{L^1(0,T;L^{q,r}(\mathbb{R}^N))} + \sum_{k=1}^{\infty} \|h_k\|_{L^1(0,T;L^{q,r}(\mathbb{R}^N))} \le 3\|f\|_{L^1(0,T;L^{q,r}(\mathbb{R}^N))}$$

$$\left.\vphantom{\sum_{k=1}^{\infty}}\right\} \quad (D.5)$$

$$\|g\|_{L^{q,r}(\mathbb{R}^N)} \le 2M \qquad \text{a.e.} \quad t \in (0,T) \qquad (D.6)$$

$$\int_0^T h_k(t,x)\, dt = 0 \qquad \text{a.e. in} \quad \mathbb{R}^N,$$
$$h_k(t,x) = 0 \qquad \text{if } t \notin I_k, \, \forall\, k \ge 1, \left.\vphantom{\int_0^T}\right\} \qquad (D.7)$$

$$\sum_{k=1}^{\infty} \text{meas}\,(I_k) \le \frac{1}{M}\|f\|_{L^1(0,T;L^{q,r}(\mathbb{R}^N))}.$$

Then, we write, defining $p = (2\pi)^{-N/2} \exp\left(-\frac{|x|^2}{2t}\right)$,

$$\frac{\partial u}{\partial t} - f = \int_0^t ds \int_{\mathbb{R}^N} dy \frac{\partial p}{\partial t}(t-s,y)g(s,y) + \sum_{k=1}^{\infty} \int_0^t ds \int_{\mathbb{R}^N} dy \frac{\partial p}{\partial t}(t-s,y)h_k(s,y)$$
$$(D.8)$$

and denote by v and w_k the first term and the generic term in the series of the right-hand side. We also write $V(t) = \|v(t,\cdot)\|_{L^{q,r}(\mathbb{R}^N)}$, $W_k(t) = \|w_k(t,\cdot)\|_{L^{q,r}(\mathbb{R}^N)}$ a.e. $t \in (0,T)$. We are going to show the estimate (D.3) for $\frac{\partial u}{\partial t}$ or equivalently for $\varphi = \frac{\partial u}{\partial t} - f$, the proof being exactly the same for $D_x^2 u$ (or using elliptic regularity and the equation (D.1)). To this end, we estimate v and w_k in various ways.

From the classical $L^p(\mathbb{R}^N \times (0,T))$ estimates for solutions of the heat equation (and by interpolation) we obtain, denoting by C various constants independent of f and M,

$$v \in L^{q,r}(0,T; L^{q,r}(\mathbb{R}^N)) \text{ and } \|V\|_{L^{q,r}(0,T)} \le C\|G\|_{L^{q,r}(0,T)} \qquad (D.9)$$

where we define $G(t) = \|g\|_{L^{q,r}(\mathbb{R}^N)}$ a.e. on $(0,T)$. Notice that (D.6) means

$$0 \leq G \leq 2M \qquad \text{a.e.} \quad t \in (0,T) \tag{D.10}$$

while (D.5) yields, normalizing without loss of generality $\|f\|_{L^1(0,T;L^{q,r}(\mathbb{R}^N))}$ to be 1,

$$\|G\|_{L^1(0,T)} \leq 3. \tag{D.11}$$

Then, $\|G\|_{L^{q,r}(0,T)} \leq C \left(\int_0^\infty \mu(t)^{r/q} t^{r-1} \, dt \right)^{1/r}$ where $\mu(t) = \text{meas}\,\{s \in (0,T) \,/\, G(s) > t\}$ $(0 \leq \mu \leq T)$. Notice that $\mu(t) = 0$ if $t \geq 2M$ (by (D.10)) and that $\int_0^\infty \mu(t) \, dt = \|G\|_{L^1(0,T)} \leq 3$ by (D.11). Hence, if $1 \leq r \leq q$, we deduce, using Hölder's inequality,

$$\|G\|_{L^{q,r}(0,T)} \leq C \left(\int_0^\infty \mu \, dt \right)^{1/q} \left(\int_0^{2M} t^{q(r-1)/(q-r)} \, dt \right)^{\frac{q-r}{qr}}$$

$$\leq C M^{1-1/q}.$$

Now, if $r \geq q$, we write

$$\|g\|_{L^{q,r}(0,T)} \leq C \left(\int_0^{2M} \mu(t) \{\mu(t)t\}^{\frac{r}{q}-1} t^{r\left(1-\frac{1}{q}\right)} \, dt \right)^{1/r}$$

$$\leq C \left(\int_0^\infty \mu \, dt \right)^{1/r} M^{1-1/q} \left[\sup_{t>0} \{\mu(t)t\} \right]^{\frac{1}{q}-\frac{1}{r}}$$

$$\leq C \left(\int_0^\infty \mu \, dt \right)^{1/q} M^{1-1/q} \leq C M^{1-1/q}.$$

In both cases we have shown

$$\|G\|_{L^{q,r}(0,T)} \leq C M^{1-1/q} \tag{D.12}$$

and thus, in particular, we deduce from (D.9) that we have for all $\lambda > 0$

$$\text{meas}\,\{t \in [0,T] \,/\, V(t) > \lambda\} \leq C M^{q-1} \lambda^{-q}. \tag{D.13}$$

Next, denoting by I_k^2 the interval with the same centre as I_k and with a double length (if $I_k = (t_k, t_{k+1})$, $I_k^2 = \left(\frac{3t_k - t_{k+1}}{2}, \frac{3t_{k+1} - t_k}{2} \right)$), we set $I = \bigcup_{k \geq 1} I_k^2$. We remark that (D.8) implies

$$\text{meas}\,\{I\} = 2 \sum_{k=1}^\infty \text{meas}\,(I_k) \leq 2/M. \tag{D.14}$$

Then, we claim that we have

$$\int_{(I_k^2)^C} dt\, W_k(t) = \|w_k\|_{L^1(I_k^2;L^{q,r}(\mathbb{R}^N))} \leq C\|h_k\|_{L^1(0,T;L^{q,r}(\mathbb{R}^N))} . \quad (D.15)$$

Admitting temporarily this claim, we complete the proof of the theorem. Indeed, if $t \notin I$

$$\|\varphi\|_{L^{q,r}(\mathbb{R}^N)} \leq V(t) + W(t) \qquad \text{where} \quad W(t) = \sum_{k=1}^{\infty} W_k(t)$$

hence, for all $\lambda > 0$, we deduce using (D.14), (D.13), (D.15) and (D.5)

$$\text{meas}\left\{ t \in (0,T) \ / \ \|\varphi\|_{L^{q,r}(\mathbb{R}^N)} > \lambda \right\}$$
$$\leq \text{meas}(I) + \text{meas}\left\{ t \notin I \ / \ \|\varphi\|_{L^{q,r}(\mathbb{R}^N)} > \lambda \right\}$$
$$\leq \frac{2}{M} + \text{meas}\left\{ t \notin I \ / \ W(t) > \frac{\lambda}{2} \right\} + \text{meas}\left\{ t \in (0,T) \ / \ V(t) > \frac{\lambda}{2} \right\}$$
$$\leq \frac{2}{M} + \frac{2}{\lambda} \int_{I^C} W(t)\, dt + C\, M^{q-1} \lambda^{-q}$$
$$\leq \frac{2}{M} + \frac{2}{\lambda} \sum_{k=1}^{\infty} \int_{(I_k^2)^C} W_k(t)\, dt + C\, M^{q-1} \lambda^{-q}$$
$$\leq \frac{2}{M} + \frac{C}{\lambda} \sum_{k=1}^{\infty} \|h_k\|_{L^1(0,T;L^{q,r}(\mathbb{R}^N))} + C\, M^{q-1} \lambda^{-q}$$
$$\leq \frac{2}{M} + \frac{C}{\lambda} + C\, M^{q-1} \lambda^{-q}$$

and we conclude choosing $M = \lambda$.

The only point remaining to check is the claim (D.15). We recall that we have on $\mathbb{R}^N$

$$\left. \begin{array}{ll} w_k(t) \equiv 0 & \text{if} \quad t \notin I,\, t \leq t_k; \\[2mm] w_k(t) = \displaystyle\int_{I_k} \frac{\partial p}{\partial t}(t-s) * h_k(s)\, ds & \text{if} \quad t \notin I,\, t > t_{k+1} \end{array} \right\} \quad (D.16)$$

and because of (D.7) if $t > t_{k+1}$

$$w_k(t) = \int_{I_k} \left[\frac{\partial p}{\partial t}(t-s) - \frac{\partial p}{\partial t}(t-t_k) \right] * h_k(s)\, ds. \quad (D.17)$$

We then wish to estimate the norm of $\frac{\partial p}{\partial t}(t-s) - \frac{\partial p}{\partial t}(t-t_k)$ as a multiplier in $L^{q,r}(\mathbb{R}^N)$ (denoting by $\|\cdot\|_{M^{q,r}}$ this norm) and we shall prove below that

$$\left\| \frac{\partial p}{\partial t}(t-s) - \frac{\partial p}{\partial t}(t-t_k) \right\|_{M^{q,r}} \leq C\, \frac{s-t_k}{(t-s)^2}. \quad (D.18)$$

If this is the case, we complete the proof of (D.15) easily: indeed, we have

$$\int_{(I_k^2)^C} W_k(t)\, dt$$

$$\leq \int_{\frac{3t_{k+1}-t_k}{2}}^{T} dt \left\| \int_{I_k} \left[\frac{\partial p}{\partial t}(t-s) - \frac{\partial p}{\partial t}(t-t_k) \right] * h_k(s)\, ds \right\|_{L^{q,r}(\mathbb{R}^N)}$$

$$\leq C \int_{\frac{3t_{k+1}-t_k}{2}}^{T} dt \int_{t_k}^{t_{k+1}} ds \, \frac{s-t_k}{(t-s)^2} \|h_k(s)\|_{L^{q,r}(\mathbb{R}^N)}$$

$$\leq C \int_{t_k}^{t_{k+1}} ds \, \|h_k(s)\|_{L^{q,r}(\mathbb{R}^N)} \left\{ \int_{\frac{3t_{k+1}-t_k}{2}}^{+\infty} \frac{s-t_k}{(t-s)^2}\, dt \right\}$$

$$\leq C \int_{t_k}^{t_{k+1}} ds \, \|h_k(s)\|_{L^{q,r}(\mathbb{R}^N)} \left\{ \frac{s-t_k}{3t_{k+1}-t_k-2s} \right\}$$

$$\leq C \int_{t_k}^{t_{k+1}} ds \, \|h_k(s)\|_{L^{q,r}(\mathbb{R}^N)} = C\|h_k\|_{L^1(0,T;L^{q,r}(\mathbb{R}^N))}$$

since

$$\frac{s-t_k}{3t_{k+1}-t_k-2s} \leq \frac{t_{k+1}-t_k}{3(t_{k+1}-t_k)} = \frac{1}{3} \quad \text{if} \quad s \in I_k\,.$$

Next, in order to prove the estimate (D.18), we compute the Fourier transform of $\frac{\partial p}{\partial t}(t-s) - \frac{\partial p}{\partial t}(t-t_k)$, namely $\left\{ e^{-(t-s)\frac{|\xi|^2}{2}} - e^{(t-t_k)\frac{|\xi|^2}{2}} \right\}|\xi|^2$.

We observe that $\frac{2}{t-s}\left\{ e^{-|\xi|^2} - e^{-\left(1+\frac{s-t_k}{t-s}\right)|\xi|^2} \right\}|\xi|^2$ defines a multiplier which has the same $M^{q,r}$ norm by the dilations invariance of this norm. Therefore, (D.18) is proven if we check that $\left(e^{-|\xi|^2} - e^{-(1+\lambda)|\xi|^2} \right)|\xi|^2 = |\xi|^2 e^{-|\xi|^2}(1 - e^{-\lambda|\xi|^2})$ defines a multiplier whose M^q norm (and thus any $M^{q,r}$ norm by interpolation) is bounded by $C\lambda$ for all λ and for all $q \in (1,\infty)$. This is straightforward since we have for all $|\alpha| = m \geq 0$

$$\left| \partial^\alpha \left[(1-e^{-\lambda|\xi|^2})\, e^{-|\xi|^2}\, |\xi|^2 \right] \right| (1+|\xi|^2)^{m/2}$$

$$\leq C \sum_{j=0}^{m} (1+|\xi|^2)^{\frac{m+j+2}{2}}\, e^{-|\xi|^2}\, \lambda (1+|\xi|^2)^{\frac{m-j+2}{2}}$$

$$\leq C\lambda(1+|\xi|^2)^{m+2}\, e^{-|\xi|^2} \leq C(m)\lambda$$

and we conclude using classical results on Fourier multipliers. $\square$

APPENDIX E

A Short Proof of the Existence and

Uniqueness of Renormalized Solutions

for Parabolic Equations

The goal of this appendix is a short and direct proof of Theorem 3.10. We thus consider solutions of

$$\frac{\partial f}{\partial t} + u \cdot \nabla f - \Delta f = F \quad \text{in} \quad \Omega \times (0,T) \tag{E.1}$$

$$f|_{t=0} = f_0 \quad \text{in} \quad \Omega, \quad \frac{\partial f}{\partial n} = 0 \quad \text{on} \quad \partial\Omega \times (0,T) \tag{E.2}$$

where $f_0 \in L^1(\Omega)$, $F \in L^1(\Omega \times (0,T))$ and Ω is, say, a bounded, smooth open set in $\mathbb{R}^N$ ($N \geq 2$); we could as well treat the case when $\Omega = \mathbb{R}^N$, or the periodic case or the case when the Neumann boundary condition is replaced by a Dirichlet boundary condition. Finally, we assume that u satisfies

$$\left. \begin{array}{l} u \in L^2(\Omega \times (0,T))^N, \quad \text{div}\, u = 0 \quad \text{in} \quad \mathcal{D}'(\Omega \times (0,T)), \\ u \cdot n = 0 \quad \text{on} \quad \partial\Omega \times (0,T) \end{array} \right\} \tag{E.3}$$

(recall that $u \cdot n \in L^2(0,T; H^{-1/2}(\Omega))$).

More general conditions on u (and $\text{div}\, u$) are possible that yield the same result as below but these conditions are more than enough for the application we have in mind, namely Theorem 3.10. For the same reason, we shall assume

$$F \geq 0 \quad \text{a.e. in} \quad \Omega \times (0,T) \tag{E.4}$$

even if it is not necessary for the result to follow; however, this condition allows us to give rather simple proofs.

As explained in chapter 3, section 3.4, we cannot simply use distributions theory to solve (E.1)–(E.2) since we would have to define the product uf (writing $u \cdot \nabla f$ as $\text{div}\,(uf)$ since $\text{div}\, u = 0$) ; and since we only assume u to be in $L^2(\Omega \times (0,T))^N$, we would need to know that $f \in L^2(\Omega \times (0,T))$. However, we cannot expect f to be in $L^2(\Omega \times (0,T))$ since F and f only belong to L^1.

As we saw in section 3.4, we expect f (and we can obtain corresponding formal a priori estimates) to satisfy:

$$f \in C([0,T]; L^1(\Omega)) \cap L^1(0,T; L^q(\Omega)) \quad \text{for all} \quad q \in \left[1, \frac{N}{N-2}\right) \quad \text{(E.5)}$$

$$\left.\begin{array}{l} T_R(f) \in L^2(0,T; H^1(\Omega)) \quad \text{for all} \quad R \in (0,\infty), \\[2mm] \displaystyle\lim_{R\to+\infty} \frac{1}{R} \int_\Omega dx \int_0^T dt\, |\nabla T_R(f)|^2 = 0, \end{array}\right\} \quad \text{(E.6)}$$

where $T_R(t) = \max\left(\min\left(t, R\right), -R\right)$ for $t \in \mathbb{R}$, $R \in (0,\infty)$, ∇ denotes the spatial gradient (in x) and

$$\nabla_x f \in L^r(\Omega \times (0,T)) \quad \text{for all} \quad r \in \left[1, \frac{N+1}{N}\right). \quad \text{(E.7)}$$

In particular, we know that $\nabla T_R(f) = \nabla f\, 1_{(|f|<R)}$ a.e. in $\Omega \times (0,T)$.

In order to solve (E.1)–(E.2), we shall use the notion of *renormalized solutions* introduced by R.J. DiPerna and P.L. Lions [**125**] in the context of the Fokker–Planck–Boltzmann equations, see also L. Boccardo, I. Diaz, D. Giachetti and F. Murat [**67**], P.L. Lions and F. Murat [**308**] for nonlinear elliptic equations, D. Blanchard [**65**], D. Blanchard and F. Murat [**66**] (and the references therein) for parabolic equations. The idea is simple: we write down the equation satisfied by $\beta(f)$ where $\beta \in C_0^\infty(\mathbb{R})$ using (E.1); let us just mention that many equivalent formulations are possible. First of all, we notice that $\beta(f) \in C([0,T]; L^1(\Omega)) \cap L^\infty(\Omega \times (0,T))$ since β is bounded, $\beta(f) \in L^2(0,T; H^1(\Omega))$ since $\nabla\beta(f) = \beta'(f)\nabla f$ a.e. and $|\beta'(f)\nabla f| \leq \sup_{\mathbb{R}} |\beta'(t)|\,|\nabla T_R(f)|$ where $R > 0$ is chosen in a such a way that $\text{Supp}\,(\beta') \subset [-R, R]$. Similarly, $\beta''(f)|\nabla f|^2 \in L^1(\Omega \times (0,T))$. Then, formally, we obtain

$$\frac{\partial\beta(f)}{\partial t} + \text{div}\,\{u\beta(f)\} - \Delta\beta(f) + \beta''(f)|\nabla f|^2 = \beta'(f)F \quad \text{in } \Omega \times (0,T). \quad \text{(E.8)}$$

Obviously, $\beta(f)$ should satisfy (E.2) with f_0 replaced by $\beta(f_0)$, and this combined with (E.8) yields the following weak formulation: we have for all $\beta \in C_0^\infty(\mathbb{R})$ and for all $\varphi \in C^1(\overline{\Omega} \times [0,T])$ (for instance)

$$\left.\begin{array}{l} \displaystyle\frac{d}{dt} \int_\Omega \beta(f(x,t))\varphi(x,t)\, dx + \int_\Omega \beta(f)\left\{-\frac{\partial\varphi}{\partial t} - u\cdot\nabla\varphi\right\} \\[3mm] + \nabla\beta(f)\cdot\nabla\varphi + \beta''(f)|\nabla f|^2\varphi - \beta'(f)F\varphi\, dx = 0 \quad \text{in } \mathcal{D}'(0,T), \\[3mm] \beta(f) \in C([0,T]; L^1(\Omega)), \quad \beta(f)|_{t=0} = \beta(f_0) \quad \text{a.e. in} \quad \Omega. \end{array}\right\} \quad \text{(E.9)}$$

In conclusion, we say that f is a renormalized solution of (E.1)–(E.2) if it satisfies (E.5)–(E.7) and (E.9).

Theorem E.1. *There exists a unique renormalized solution of* (E.1)–(E.2).

Remarks E.1. 1) If we know that $u \in L^p(\Omega \times (0,T))$ for some $p \geq 2$ and that $f \in L^q(\Omega \times (0,T))$ for some $q \in [1,+\infty]$ such that $\frac{1}{p} + \frac{1}{q} \leq 1$ (thus $|u|f \in L^1(\Omega \times (0,T))$), then (E.9) combined with (E.6) implies that f is a "standard weak" solution of (E.1)–(E.2); we could as well treat cases where u, f have different integrabilities in x and in t. Before proving this claim, let us observe that the integrability of f can be estimated in terms of integrability requirements upon F and f_0. In order to prove the above fact, we use (E.8) with $\beta = \beta_n$ as given in the proof of Theorem E.1 below and we let n go to $+\infty$ using (E.6) to show that $\beta_n''(f)|\nabla f|^2$ converges to 0 in $L^1(\Omega \times (0,T))$.

2) The proof of Theorem E.1 below also shows that the unique renormalized solution of (E.1)–(E.2) is the limit (in $C([0,T]; L^1(\Omega))$ for instance) of the solutions of regularized problems (regularize u, F or f_0).

3) The uniqueness proof given below also yields the expected fact that if f_1 and f_2 are renormalized solutions of (E.1)–(E.2) with F, f_0 replaced respectively by F_1, F_2, $f_{0,1}$, $f_{0,2}$ and if $\lambda, \mu \in \mathbb{R}$ then $\lambda f_1 + \mu f_2$ is the renormalized solution of (E.1)–(E.2) with F, f_0 replaced by $\lambda F_1 + \mu F_2$, $\lambda f_{0,1} + \mu f_{0,2}$ respectively. This is of course a very natural fact since (E.1)–(E.2) is a linear problem; however, the notion of renormalized solutions is a nonlinear one! □

Proof of Theorem E.1

Step 1. Uniqueness. Let f_1, f_2 be two renormalized solutions of (E.1)–(E.2). We shall write equations like (E.8) for various quantities involving f_1 and f_2 that satisfy the Neumann boundary condition contained in (E.2); whenever we do so, we really mean that the weak formulation like (E.9) holds. With this convention, we have by assumption

$$\left.\begin{array}{l} \dfrac{\partial}{\partial t}\{\beta(f_1) - \beta(f_2)\} + \mathrm{div}\,[u\{\beta(f_1) - \beta(f_2)\}] - \Delta[\beta(f_1) - \beta(f_2)] \\[2mm] \quad + \beta''(f_1)|\nabla f_1|^2 - \beta''(f_2)|\nabla f_2|^2 \; = \; F[\beta'(f_1) - \beta'(f_2)]. \end{array}\right\}$$

Notice also that since $\mathrm{div}\,u = 0$, $u \in L^2(\Omega \times (0,T))$, $\beta(f_1) - \beta(f_2) \in L^2(0,T; H^1(\Omega))$ the second-term $\mathrm{div}\,[u\{\beta(f_1) - \beta(f_2)\}]$ may be rewritten as $u \cdot \nabla(\beta(f_1) - \beta(f_2))$ (one simply needs to argue by density on u).

Next, let $\gamma \in C_0^\infty(\mathbb{R})$; $g = \gamma(\beta(f_1) - \beta(f_2)) \in C([0,T]; L^1(\Omega)) \cap L^2(0,T; H^1(\Omega))$ satisfies $g|_{t=0} = 0$ in Ω and

$$\left. \begin{array}{l} \dfrac{\partial g}{\partial t} + \text{div}\,(ug) - \Delta g + \{\beta''(f_1)|\nabla f_1|^2 - \beta''(f_2)|\nabla f_2|^2\}\gamma'(\beta(f_1) \\[2mm] - \beta(f_2)) + \gamma''(\beta(f_1) - \beta(f_2))|\nabla\beta(f_1) - \nabla\beta(f_2)|^2 \\[2mm] \qquad = F[\beta'(f_1) - \beta'(f_2)]\gamma'(\beta(f_1) - \beta(f_2)). \end{array} \right\} \quad \text{(E.10)}$$

This computation is obvious formally: to justify it, we simply observe that we have as above

$$\text{div}\,(ug) \;=\; u\cdot\nabla g \;=\; u\cdot\nabla[\beta(f_1) - \beta(f_2)]\,\gamma'(\beta(f_1) - \beta(f_2))$$

while if $\frac{\partial h}{\partial t} - \Delta h \in L^1(\Omega \times (0,T))$, $h \in L^2(0,T;H^1(\Omega))$, $\gamma(h)$ satisfies

$$\frac{\partial\gamma(h)}{\partial t} - \Delta\gamma(h) + \gamma''(h)|\nabla h|^2 \;=\; \left\{\frac{\partial h}{\partial t} - \Delta h\right\}\gamma'(h).$$

This last fact requires some explanations: first of all, regularizing h in t by convolution, we see it is enough to prove the above claim when h is smooth in t ($h \in C^1([0,T];H^1(\Omega))$) since we then easily pass to the limit and prove the claim. When h is smooth in t, we just have to show that if $-\Delta h \in L^1(\Omega)$, $h \in H^1(\Omega)$ then $-\Delta\gamma(h) = \gamma''(h)|\nabla h|^2 + \gamma'(h)(-\Delta h)$. In order to check this fact, we use the weak formulation (incorporating the Neumann boundary condition), namely

$$\int_\Omega \nabla h \cdot \nabla\varphi\,dx \;=\; \int_\Omega (-\Delta h)\varphi\,dx \qquad \text{for all}\quad \varphi \in C^1(\overline{\Omega}).$$

This equality holds by density for $\varphi \in H^1(\Omega) \cap L^\infty(\Omega)$ (approximate such a φ by $\varphi_n \in C^1(\overline{\Omega})$ such that $\varphi_n \underset{n}{\to} \varphi$ in $H^1(\Omega) \cap L^p(\Omega)$ for all $1 \le p < \infty$, φ_n bounded in $L^\infty(\Omega)$). Then, we take $\varphi = \gamma'(h)\psi$ where $\psi \in C^1(\overline{\Omega})$ and we find

$$\int_\Omega (-\Delta h)\gamma'(h)\psi\,dx \;=\; \int_\Omega \nabla h \cdot \{\gamma'(h)\nabla\psi + \gamma''(h)\nabla h\psi\}\,dx$$

$$= \int_\Omega \nabla\gamma(h)\cdot\nabla\psi + \gamma''(h)|\nabla h|^2\psi\,dx,$$

and this completes the proof of (E.10).

We next use (E.10) with β replaced by $\beta_n(t) = n\beta_1\left(\frac{t}{n}\right)$ $(n \ge 1)$ where $\beta_1 \in C_0^\infty(\mathbb{R})$, $\beta_1(t) = t$ if $|t| \le 1$, $\beta_1(t) = 0$ if $|t| \ge 2$ and γ replaced by $\gamma_n(t) = \gamma_0(t)\,\zeta(t/n)$ where $\zeta \in C_0^\infty(\mathbb{R}) \ge 0$, $\zeta(t) = 1$ if $|t| \le 1$, $\zeta(t) = 0$ if $|t| \ge 2$, $\gamma_0'' \in C_0^\infty(\mathbb{R})$, $\gamma_0'' \ge 0$ and $\gamma_0'' > 0$ near 0, $\gamma_0'(0) = 0$, $\gamma_0(0) = 0$. Writing $g_n = \gamma_n(\beta_n(f_1) - \beta_n(f_2))$, we deduce from (E.10)

$$\left. \begin{array}{l} \dfrac{\partial g_n}{\partial t} + \text{div}\,(ug_n) - \Delta g_n \;\le\; \dfrac{C}{n}\left\{|\nabla f_1|^2\,\mathbf{1}_{(|f_1|<2n)} + |\nabla f_2|^2\,\mathbf{1}_{(|f_2|<2n)}\right\} \\[3mm] \qquad + F[\beta_n'(f_1) - \beta_n'(f_2)]\gamma_n''(\beta_n(f_1) - \beta_n(f_2)). \end{array} \right\}$$

where we used the fact that $|\beta_n''| \le \frac{C}{n}$, $\gamma_n'' = \gamma_0'' \zeta + \frac{2}{n} \gamma_0' \zeta'\left(\frac{t}{n}\right) + \frac{1}{n^2} \gamma_0 \zeta''\left(\frac{t}{n}\right) \ge -\frac{C}{n}$ and where C denotes various positive constants independent of $n \ge 1$.

We integrate this inequality over Ω (using the weak formulation of it with $\varphi \equiv 1!$) and we deduce for all $t \in [0, T]$ and for all $n \ge 1$

$$\int_\Omega \gamma_n(\beta_n(f_1) - \beta_n(f_2))(x, t)\, dx$$

$$\le \frac{C}{n} \int_0^T dt \int_\Omega dx \left\{ |\nabla f_1|^2\, 1_{(|f_1| < 2n)} + |\nabla f_2|^2\, 1_{(|f_2| < 2n)} \right\}$$

$$+ \int_0^T dt \int_\Omega dx\, |F[\beta_n'(f_1) - \beta_n'(f_2)]\, \gamma_n'(\beta_n(f_1) - \beta_n(f_2))|$$

$$\le \frac{C}{n} \int_0^T dt \int_\Omega dx \left\{ |\nabla f_1|^2\, 1_{(|f_1| < 2n)} + |\nabla f_2|^2\, 1_{(|f_2| < 2n)} \right\}$$

$$+ C \int_0^T dt \int_\Omega dx\, |F| \left| \beta_1'\left(\frac{f_1}{n}\right) - \beta_1'\left(\frac{f_2}{n}\right) \right|.$$

Because of (E.6) the first term of the right-hand side goes to 0 as n goes to $+\infty$, and the second term also goes to 0 since $\beta_1'\left(\frac{f_1}{n}\right)$, $\beta_1'\left(\frac{f_2}{n}\right)$ converge a.e. in $\Omega \times (0, T)$ to 1 and are bounded uniformly in n. From the definition of γ_n and β_n (and the fact that $f_1, f_2 \in C([0, T]; L^1(\Omega))$) one checks easily that $\gamma_n(\beta_n(f_1) - \beta_n(f_2))$ converges to $\gamma_0(f_1 - f_2)$ in $C([0, T]; L^1(\Omega))$ as n goes to $+\infty$. Hence, $\gamma_0(f_1 - f_2) \equiv 0$ and $f_1 - f_2 \equiv 0$ a.e. in $\Omega \times (0, T)$ since, by construction, $\gamma_0(t) > 0$ if $|t| > 0$.

Step 2. Approximation, bounds and convergence in $C([0,T];L^1(\Omega))$.
We next want to show the existence of a renormalized solution of (E.1)–(E.2). To this end, we consider the following regularized problem

$$\left. \begin{aligned} \frac{\partial f^\varepsilon}{\partial t} + u^\varepsilon \cdot \nabla f^\varepsilon - \Delta f^\varepsilon &= F^\varepsilon \quad \text{in} \quad \Omega \times (0, T), \\ \frac{\partial f^\varepsilon}{\partial n} &= 0 \quad \text{on} \quad \partial\Omega \times (0, T), \quad f^\varepsilon|_{t=0} = f_0^\varepsilon \quad \text{in} \quad \Omega \end{aligned} \right\} \quad \text{(E.11)}$$

where $\varepsilon \in (0, 1]$, $F^\varepsilon \in C_0^\infty(\Omega \times (0, T))$, $F^\varepsilon \underset{\varepsilon}{\to} F$ in $L^1(\Omega \times (0, T))$, $F^\varepsilon \ge 0$ a.e., $f_0^\varepsilon \in C_0^\infty(\Omega)$, $f_0^\varepsilon \underset{\varepsilon}{\to} f_0$ in $L^1(\Omega)$, $u^\varepsilon \in C_0^\infty(\Omega \times (0, T))$, $\operatorname{div} u^\varepsilon = 0$ in $\Omega \times (0, T)$, $u^\varepsilon \underset{\varepsilon}{\to} u$ in $L^2(\Omega \times (0, T))$. The equation (E.11) is now a standard parabolic problem that admits a unique smooth solution f^ε (say $C^2(\overline{\Omega} \times [0, T])$).

One easily checks that f^ε is bounded, uniformly in $\varepsilon \in (0, 1]$, in $C([0, T]; L^1(\Omega))$. We next claim that (E.6) holds for f^ε uniformly in ε, from which bound we deduce a bound in $L^1(0, T; L^q(\Omega))$ $\left(\forall\, q < \frac{N}{N-2} \right)$ by Theorem

E.2 below and a bound upon ∇f in $L^r(\Omega \times (0,T))$ for $r \in \left[1, \frac{N+1}{N}\right)$ by the arguments developed in chapter 4. Next, in order to prove our claim, we multiply (E.11) by $T_R(f^\varepsilon)$, integrate by parts over Ω and find

$$
\left.
\begin{aligned}
\sup_{t \in [0,T]} \|S_R(f^\varepsilon(t))\|_{L^1(\Omega)} + \int_0^T dt \int_\Omega dx \, |\nabla T_R(f^\varepsilon)|^2 \\
\leq \int_\Omega dx \, S_R(f_0^\varepsilon) + \int_0^T dt \int_\Omega dx \, F^\varepsilon \, T_R(f^\varepsilon)
\end{aligned}
\right\}
\tag{E.12}
$$

where $S_R(t) = \frac{1}{2}t^2$ if $|t| \leq R$, $= R|t| - \frac{R^2}{2}$ if $|t| \geq R$. Since $S_R(t) \leq R|t|$ on $\mathbb{R}$ and $|T_R(f^\varepsilon)| \leq R$ we deduce that $T_R(f^\varepsilon)$ is bounded uniformly in ε in $L^2(0,T; H^1(\Omega))$ for each $R \in (0,\infty)$. In addition, (E.12) yields for all $M \in (0,\infty)$

$$
\frac{1}{R} \int_0^T dt \int_\Omega dx \, |\nabla T_R(f^\varepsilon)|^2
$$
$$
\leq \int_\Omega \frac{1}{R} S_R(f_0^\varepsilon) \, dx + \int_0^T dt \int_\Omega dx \, |F^\varepsilon| \frac{1}{R} |T_R(f^\varepsilon)|
$$
$$
\leq \frac{M^2}{R} \operatorname{meas}(\Omega) + \int_{(|f_0^\varepsilon|>M)} |f_0^\varepsilon| \, dx + \frac{M}{R} \|F^\varepsilon\|_{L^1(\Omega \times (0,T))}
$$
$$
+ \int_0^T dt \int_\Omega dx \, |F^\varepsilon| 1_{(|f^\varepsilon|>M)} \, .
$$

Our claim then follows since on the one hand f_0^ε converges to f_0 in $L^1(\Omega)$ and thus $\int_{(|f_0^\varepsilon|>M)} |f_0^\varepsilon| \, dx$ goes to 0 as M goes to $+\infty$ uniformly in $\varepsilon \in (0,1]$, and on the other hand F^ε converges to F in $L^1(\Omega)$, meas $\{(x,t) \in \Omega \times (0,T) \, / \, |f^\varepsilon(x,t)| > M\} \leq \frac{T}{M} \sup_{0 \leq t \leq T} \|f^\varepsilon(t)\|_{L^1(\Omega)}$ hence $\int_0^T dt \int_\Omega dx$ $|F^\varepsilon| 1_{(|f^\varepsilon|>M)}$ goes to 0 as M goes to $+\infty$ uniformly in $\varepsilon \in (0,1]$. In conclusion, we have shown since $\frac{1}{R} S_R(t) \geq \frac{1}{2} |t| 1_{|t| \geq R}$

$$
\left.
\begin{aligned}
\frac{1}{R} \int_0^T dt \int_\Omega dx |\nabla f^\varepsilon|^2 \, 1_{(|f^\varepsilon|<R)} \;\to\; 0 \\
\text{as } R \to +\infty, \text{ uniformly in } \varepsilon \in (0,1].
\end{aligned}
\right\}
\tag{E.13}
$$

$$
\left.
\begin{aligned}
\sup_{t \in [0,T]} \left\| f^\varepsilon(t) \, 1_{|f^\varepsilon(t)| \geq R} \right\| \\
\text{as } R \to +\infty, \text{ uniformly in } \varepsilon \in (0,T].
\end{aligned}
\right\}
\tag{E.14}
$$

We finally show that f^ε is a Cauchy sequence in $C([0,T]; L^1(\Omega))$. To this end, we follow the uniqueness proof with f_1, f_2 replaced by f^ε, f^η for

$\varepsilon, \eta \in (0, 1]$, and we obtain, setting $g_n = \gamma_n(\beta_n(f^\varepsilon) - \beta_n(f^\eta))$,

$$\frac{\partial g_n}{\partial t} + \operatorname{div}(u^\varepsilon g_n) - \Delta g_n$$

$$\leq \frac{C}{n}\Big\{ |\nabla f^\varepsilon|^2 \, 1_{(|f^\varepsilon| < 2n)} + |\nabla f^\eta|^2 \, 1_{(|f^\eta| < 2n)} \Big\}$$

$$+ C|u^\varepsilon - u^\eta| \, |\nabla f^\eta| \, 1_{(|f^\eta| < 2n)}$$

$$+ [F^\varepsilon \beta_n'(f^\varepsilon) - F^\eta \beta_n'(f^\eta)] \, \gamma_n'(\beta_n(f^\varepsilon) - \beta_n(f^\eta))$$

$$\leq \frac{C}{n}\Big\{ |\nabla f^\varepsilon|^2 \, 1_{(|f^\varepsilon| < 2n)} + |\nabla f^\eta|^2 \, 1_{(|f^\eta| < 2n)} \Big\} + Cn|u^\varepsilon - u^\eta|^2$$

$$+ C|F^\varepsilon - F| + C|F^\eta - F| + C|F| \, 1_{(|f^\varepsilon| > n)} + C|F| \, 1_{(|f^\eta| > n)}.$$

Hence, integrating in x and t, we deduce using (E.13) and the definitions of $u^\varepsilon, F^\varepsilon, f_0^\varepsilon$

$$\sup_{t \in [0,T]} \big\| \gamma_n\{\beta_n(f^\varepsilon) - \beta_n(f^\eta)\} \big\|_{L^1(\Omega)} \leq \delta_n + \omega_n(\varepsilon, \eta) \qquad \text{(E.15)}$$

where $\delta_n \underset{n}{\to} 0$, $\omega_n(\varepsilon, \eta) \to 0$ as $\varepsilon, \eta \to 0_+$ for each $n \geq 1$ fixed. We also used to derive (E.15) the following observations

$$\sup_{\varepsilon \in (0,1]} \int_0^T dt \int_\Omega dx \, |F| \, 1_{(|f^\varepsilon| > n)} \to 0 \qquad \text{as} \quad n \to +\infty$$

since $\sup_{\varepsilon \in (0,1]} \operatorname{meas}_{x,t}(|f^\varepsilon| > n) \to 0$ as $n \to +\infty$, and

$$0 \leq \gamma_n(\beta_n(f_0^\varepsilon) - \beta_n(f_0^\eta)) \leq C|f_0^\varepsilon - f_0^\eta|.$$

Next, we observe that we have

$$|f^\varepsilon - f^\eta| \leq |\beta_n(f^\varepsilon) - \beta_n(f^\eta)| + (|f^\varepsilon| + |f^\eta|)\Big(1_{(|f^\varepsilon| > n)} + 1_{(|f^\eta| > n)}\Big)$$

and that for each $\delta \in (0, 1)$, there exists $C_\delta > 0$ such that $|t| \leq \delta + C_\delta \, \gamma_n(t) + |t| \, 1_{|t| > n}$ for all $t \in \mathbb{R}$. Hence,

$$|f^\varepsilon - f^\eta| \leq \delta + C_\delta \gamma_n(\beta_n(f^\varepsilon) - \beta_n(f^\eta))$$

$$+ |\beta_n(f^\varepsilon) - \beta_n(f^\eta)| \, 1_{(|\beta_n(f^\varepsilon) - \beta_n(f^\eta)| > n)}$$

$$+ (|f^\varepsilon| + |f^\eta|)\Big(1_{(|f^\varepsilon| > n)} + 1_{(|f^\eta| > n)}\Big)$$

$$\leq \delta + C_\delta \gamma_n(\beta_n(f^\varepsilon) - \beta_n(f^\eta))$$

$$+ C(|f^\varepsilon| + |f^\eta|)\Big(1_{(|\beta_n(f^\varepsilon)| > \frac{n}{2})} + 1_{(|\beta_n(f^\eta)| > \frac{n}{2})}\Big)$$

$$+ (|f^\varepsilon| + |f^\eta|)\Big(1_{(|f^\varepsilon| > n)} + 1_{|f^\eta| > n)}\Big).$$

Since $|\beta_n(t)| > \frac{n}{2}$ implies $\beta_1\left(\frac{t}{n}\right) > \frac{1}{2}$ and $t > \frac{n}{2}$, we obtain finally

$$\left.\begin{aligned}
|f^\varepsilon - f^\eta| \;\leq\; &\delta + C_\delta\,\gamma_n(\beta_n(f^\varepsilon) - \beta_n(f^\eta)) \\
&+ C(|f^\varepsilon| + |f^\eta|)\left(1_{\left(|f^\varepsilon| > \frac{n}{2}\right)} + 1_{\left(|f^\eta| > \frac{n}{2}\right)}\right).
\end{aligned}\right\} \tag{E.16}$$

Combining (E.15) and (E.16), we deduce

$$\sup_{t\in[0,T]} \|f^\varepsilon(t) - f^\eta(t)\|_{L^1(\Omega)} \;\leq\; \delta + C_\delta\delta_n + C_\delta\omega(\varepsilon,\eta) + \gamma_n \tag{E.17}$$

where $\gamma_n \underset{n}{\to} 0$, and we used the following facts

$$\left.\begin{aligned}
\sup_{t\in[0,T]} \operatorname{meas}_x\left(|f^\varepsilon(t)| > \frac{n}{2}\right) \;\to\; 0 \\
\text{as } n \text{ goes to } +\infty, \text{ uniformly in } \varepsilon \in (0,1].
\end{aligned}\right\}$$

$$\sup_{t\in[0,T]} \int_A |f^\varepsilon(t)|\,dx \;\leq\; M\operatorname{meas}(A) + \sup_{[0,T]}\left\|f^\varepsilon(t)\,1_{(|f^\varepsilon(t)|>M)}\right\|_{L^1(\Omega)}$$

and the last quantity goes to 0 as R goes to $+\infty$ uniformly in $\varepsilon \in (0,1]$ in view of (E.14).

Letting ε, η go to $+\infty$ and finally δ to 0_+, we deduce from (E.17) that f^ε converges uniformly in $t \in [0,T]$ in $L^1(\Omega)$ and in $L^1(0,T;L^q(\Omega))$ to some $f \in C([0,T];L^1(\Omega)) \cap L^1(0,T;L^q(\Omega))$ for all $1 \leq q < \frac{N}{N-2}$ such that $f|_{t=0} = f_0$ in Ω. In addition, $T_R(f^\varepsilon)$ converges to $T_R(f)$ weakly in $L^2(0,T;H^1(\Omega))$ and (E.6), (E.7) hold. In order to conclude the proof of Theorem E.1, we only need to show that (E.9) holds or in other words that we can pass to the limit in the term $\beta''(f^\varepsilon)|\nabla f^\varepsilon|^2$ since all the other terms can be handled easily in view of the convergence of f^ε we just established.

Step 3. Convergence in $L^2(0,T;H^1(\Omega))$ of the truncations. We first observe that it is enough to show that $T_R(f^\varepsilon)$ converges to $T_R(f)$ (as $\varepsilon \to 0_+$) in $L^2(0,T;H^1(\Omega))$ since $\beta''(f^\varepsilon)|\nabla f^\varepsilon|^2 = \beta''(f^\varepsilon) \cdot |\nabla T_R(f^\varepsilon)|^2$ for R large enough and $a^\varepsilon b^\varepsilon \underset{\varepsilon}{\to} ab$ in $L^1(\Omega)$ if $a^\varepsilon \underset{\varepsilon}{\to} a$, $b^\varepsilon \underset{\varepsilon}{\to} b$ in $L^1(\Omega)$ and b^ε is bounded in $L^\infty(\Omega)$.

Next, we remark that we have for all $\delta \in (0,T)$

$$\int_0^\delta dt \int_\Omega dx\,|\nabla T_R(f^\varepsilon)|^2\,dx \;\leq\; \int_0^\delta dt \int_\Omega dx\,R|F^\varepsilon| + \int_\Omega R|f^\varepsilon(\delta) - f_0^\varepsilon|\,dx$$

and thus this quantity is small if δ is small, uniformly in $\varepsilon \in (0,1]$. Hence, we only have to show that $T_R(f^\varepsilon)$ converges in $L^2(\delta,T;H^1(\Omega))$ to $T_R(f)$.

Next, we write $f^\varepsilon = g^\varepsilon + h^\varepsilon$ where $g^\varepsilon, h^\varepsilon$ are respectively solutions of

$$\left.\begin{aligned}\frac{\partial g^\varepsilon}{\partial t} + u^\varepsilon \cdot \nabla g^\varepsilon - \Delta g^\varepsilon &= 0 \quad \text{in} \quad \Omega \times (0,T), \\ \frac{\partial g^\varepsilon}{\partial n} &= 0 \quad \text{on } \partial\Omega \times (0,T), \quad g^\varepsilon|_{t=0} = f_0^\varepsilon \quad \text{in } \Omega,\end{aligned}\right\} \quad \text{(E.18)}$$

$$\left.\begin{aligned}\frac{\partial h^\varepsilon}{\partial t} + u^\varepsilon \cdot \nabla h^\varepsilon - \Delta h^\varepsilon &= F^\varepsilon \quad \text{in} \quad \Omega \times (0,T), \\ \frac{\partial h^\varepsilon}{\partial n} &= 0 \quad \text{on } \partial\Omega \times (0,T), \quad h^\varepsilon|_{t=0} = 0 \quad \text{in } \Omega.\end{aligned}\right\} \quad \text{(E.19)}$$

Notice that $h^\varepsilon \geq 0$ in $\overline{\Omega} \times [0,T]$ and that everything we did in step 2 above applies to h^ε and g^ε: in particular, h^ε and g^ε converge in $C([0,T]; L^1(\Omega))$ and in $L^1(0,T; L^q(\Omega))$ for all $1 \leq q < \frac{N}{N-2}$ to some h and g respectively. In addition, we claim that for each $\delta > 0$, g^ε is bounded in $L^\infty(\Omega \times (\delta,T))$ uniformly in $\varepsilon \in (0,1]$. Indeed, we first observe that for each $m \geq 1$, $g_m^\varepsilon = t^m g^\varepsilon$ solves

$$\left.\begin{aligned}\frac{\partial g_m^\varepsilon}{\partial t} + u^\varepsilon \cdot \nabla g_m^\varepsilon - \Delta g_m^\varepsilon &= m g_{m-1}^\varepsilon \quad \text{in} \quad \Omega \times (0,T), \\ \frac{\partial g_m^\varepsilon}{\partial n} &= 0 \quad \text{on } \partial\Omega \times (0,T), \quad g_m^\varepsilon|_{t=0} = f_0^\varepsilon \quad \text{in } \Omega.\end{aligned}\right\}$$

Of course, $g_0^\varepsilon = g^\varepsilon$ is bounded in $L^{q_0}(\Omega \times (0,T))$ for $1 < q_0 < \frac{N+2}{N}$. Next, if $g_{m-1}^\varepsilon \in L^{q_{m-1}}$ for some $q_{m-1} \geq q_0$ ($m \geq 1$), we multiply the equation for g_m^ε by $|g_m^\varepsilon|^{q_{m-1}-2} g_m^\varepsilon$ and we obtain easily a bound on g_m^ε in $C([0,T]; L^{q_{m-1}}(\Omega))$ and on $\nabla\left[|g_m^\varepsilon|^{\frac{q_{m-1}}{2}-1} g_m^\varepsilon\right]$ in $L^2(\Omega \times (0,T))$, and thus in $L^2(0,T; H^1(\Omega))$ on $|g_m^\varepsilon|^{\frac{q_{m-1}}{2}-1} g_m^\varepsilon$. Using Sobolev (and Gagliardo–Nirenberg if $N = 2$) inequalities and Hölder inequalities, we deduce that g_m^ε is bounded in $L^{q_m}(\Omega \times (0,T))$ with $q_m = \frac{N+2}{N} q_{m-1}$. Therefore, for m large $\left(m > \log\left(1+\frac{N}{2}\right)\left[\log\left(1+\frac{2}{N}\right)^{-1}+1\right]\right)$, $g_{m-1}^\varepsilon \in L^{q_{m-1}}(\Omega \times (0,T))$ where $q_{m-1} > \frac{N+2}{2}$. To simplify notation, from now on, we write g in place of g_m^ε and $\tilde{g}$ in place of g_{m-1}^ε and q in place of q_{m-1}. Then, for $p > 1$, we multiply the equation satisfied by g by $|g|^{p-2}g$ and we obtain easily (where C denotes various constants independent of $g, \varepsilon, \tilde{g}$)

$$\sup_{t \in [0,T]} \|g(t)\|_{L^p(\Omega)}^p + \|\nabla |g|^{p/2}\|_{L^2(\Omega \times (0,T))}^2$$

$$\leq Cp \|\tilde{g}\|_{L^q(\Omega \times (0,T))} \left[\int_0^T dt \int_\Omega dx\, |g|^{\frac{q}{q-1}(p-1)}\right]^{\frac{q-1}{q}}$$

hence, exactly as above, we deduce

$$\|g\|_{L^{p\frac{N+2}{N}}(\Omega \times (0,T))} \leq (Cp)^{1/p} \|g\|_{L^{p\frac{q}{q-1}}(\Omega \times (0,T))}^{\frac{p-1}{p}} \|\tilde{g}\|_{L^q(\Omega \times (0,T))}^{1/p}.$$

We then observe that $\frac{N+2}{2}\frac{q-1}{q} = \lambda > 1$ and we rewrite the preceding inequality as

$$
\left.
\begin{aligned}
&\|g\|_{L^{\lambda p}(\Omega\times(0,T))} \leq (Cp)^{1/\theta p}\, \|g\|_{L^p(\Omega\times(0,T))}^{\frac{\theta p-1}{\theta p}}\, \|\tilde{g}\|_{L^q(\Omega\times(0,T))}^{1/\theta p} \\
&\qquad \text{for all}\quad p > 1/\theta\,,\ \theta = \frac{q-1}{q},
\end{aligned}
\right\}
$$

hence

$$
\left.
\begin{aligned}
&\max\left\{\|g\|_{L^{\lambda p}(\Omega\times(0,T))}, \|\tilde{g}\|_{L^q(\Omega\times(0,T))}\right\} \\
&\leq (Cp)^{1/\theta p}\max\left(\|g\|_{L^p(\Omega\times(0,T))}, \|\tilde{g}\|_{L^q(\Omega\times(0,T))}\right)\quad \text{for all } p > 1/\theta.
\end{aligned}
\right\}
$$

If we choose $p > 1/\theta$ and we iterate this inequality with $p, \lambda p, \ldots, \lambda^{k-1}p$ we find for all $k \geq 1$

$$
\left.
\begin{aligned}
&\|g\|_{L^{\lambda^k p}(\Omega\times(0,T))} \\
&\leq (Cp)^{\frac{1}{\theta p}\frac{\lambda^k-1}{\lambda^k}\frac{\lambda}{\lambda-1}}\,\exp\left\{\frac{1}{\theta p}\log\lambda\,\frac{\lambda}{\lambda^2-1}\,\frac{\lambda^k-k\lambda+k-1}{\lambda^k}\right\} \\
&\qquad\cdot\max\left\{\|g\|_{L^p(\Omega\times(0,T))}, \|\tilde{g}\|_{L^q(\Omega\times(0,T))}\right\}
\end{aligned}
\right\}
$$

and letting k go to $+\infty$, we finally deduce

$$
\|g\|_{L^\infty(\Omega\times(0,T))} \leq C\|\tilde{g}\|_{L^q(\Omega\times(0,T))}\,.
$$

In particular, g_ε is bounded in $L^\infty(\Omega\times(0,T))$ uniformly in $\varepsilon\in(0,1]$ and $g_\varepsilon(\delta)$ converges in $L^2(\Omega)$ to $g(\delta)$. Furthermore, $u^\varepsilon\cdot\nabla g^\varepsilon = \operatorname{div}(u^\varepsilon g^\varepsilon)$ and $u^\varepsilon g^\varepsilon$ converges in $L^2(\Omega\times(0,T))$ to ug. Therefore, g^ε converges to g in view of (E.18) in $L^2(\delta,T;H^1(\Omega))$. We then claim that the proof of Theorem E.1 is complete as soon as we show that $T_R(h^\varepsilon)$ converges in $L^2(0,T;H^1(\Omega))$ to $T_R(h)$ for all $R\in(0,\infty)$. Indeed, we have on $(\delta,T)\times\Omega$ for each $\delta > 0$ fixed

$$
\begin{aligned}
|\nabla T_R(f^\varepsilon)|^2 &= 1_{(|f^\varepsilon|<R)}|\nabla f^\varepsilon|^2 \\
&= 1_{(|f^\varepsilon|<R)}\left\{|\nabla g^\varepsilon|^2 + 2\nabla g^\varepsilon\cdot\nabla h^\varepsilon + |\nabla h^\varepsilon|^2\right\} \\
&= 1_{(|f^\varepsilon|<R)}\left\{|\nabla g^\varepsilon|^2 + 2\nabla g^\varepsilon\cdot\nabla T_{R'}(h^\varepsilon) + |\nabla T_{R'}(h^\varepsilon)|^2\right\}
\end{aligned}
$$

where $R' = R + C_\delta$, $|g_\varepsilon| \leq C_\delta$ on $[\delta,T]\times\overline{\Omega}$. The quantity in brackets converges in $L^1(\Omega\times(\delta,T))$ to $|\nabla g+\nabla T_{R'}(h)|^2$ ($=|\nabla f|^2$ if $|f|\leq R$). Hence

$$
\overline{\lim_{\varepsilon\to 0_+}}\int_0^T\!\!\int_\Omega |\nabla T_R(f^\varepsilon)|^2\,dx\,dt \leq \int_0^T\!\!\int_\Omega 1_{(|f|\leq R)}|\nabla g+\nabla T_{R'}(h)|^2\,dx\,dt
$$

$$
= \int_0^T\!\!\int_\Omega |\nabla T_R(f)|^2\,dx\,dt
$$

using Lebesgue's lemma (and its converse), and we conclude.

Step 4. Convergence in $L^2(0,T;H_0^1(\Omega))$ **of** $T_R(h^\varepsilon)$. It only remains to show that $T_R(h^\varepsilon)$ converges to $T_R(h)$ in $L^2(0,T;H^1(\Omega))$. We observe that $h^\varepsilon \geq 0$ since $h^\varepsilon|_{t=0} \equiv 0$ in $\overline{\Omega}$ and $F^\varepsilon \geq 0$ in $\overline{\Omega} \times (0,T)$. Next, we claim it is enough to show that, for instance, $(1+h^\varepsilon)^{-1/2}$ converges to $(1+h)^{-1/2}$ in $L^2(0,T;H^1(\Omega))$. Indeed, if it is the case, extracting a subsequence if necessary, there exists $C \in L^1(\Omega \times (0,T))$ such that

$$\left.\begin{array}{ll}\dfrac{|\nabla h^\varepsilon|^2}{(1+h^\varepsilon)^3} \leq C & \text{a.e. in } \Omega \times (0,T),\\[2mm] \nabla h^\varepsilon \xrightarrow[\varepsilon]{} \nabla h & \text{a.e. in } \Omega \times (0,T).\end{array}\right\}$$

In particular, $|\nabla T_R(h^\varepsilon)|^2 \leq C(1+R)^3$, $\nabla T_R(h^\varepsilon) \xrightarrow[\varepsilon]{} \nabla T_R(h)$ a.e. in $\Omega \times (0,T)$ and we conclude.

Next, we set $\beta_n(t) = \frac{t}{1+t/n}$ for $t \geq 0$, $n \geq 1$, $\gamma_n(t) = \frac{t}{1-(1-\frac{1}{n})t}$ for $0 \leq t < \left(1-\frac{1}{n}\right)^{-1}$, $n \geq 1$. Let us remark that $\beta_n = \gamma_n \circ \beta_1$, β_n is concave on $[0,\infty)$ while γ_n is convex on $\left[0,\left(1-\frac{1}{n}\right)^{-1}\right)$. Next, we notice that we have for all $k \geq 1$, $\varepsilon \in (0,1]$.

$$\left.\begin{array}{l}\dfrac{\partial}{\partial t}\beta_k(h^\varepsilon) + \operatorname{div}\{u^\varepsilon \beta_k(h^\varepsilon)\} - \Delta\beta_k(h^\varepsilon)\\[2mm] \qquad = \beta_k'(h^\varepsilon)F^\varepsilon + (-\beta_k''(h^\varepsilon)|\nabla h^\varepsilon|^2) \quad \text{in } \Omega \times (0,T)\\[2mm] \dfrac{\partial}{\partial n}\beta_k(h^\varepsilon) = 0 \quad \text{on } \partial\Omega \times (0,T), \quad \beta_k(h^\varepsilon)|_{t=0} = 0 \quad \text{in } \Omega.\end{array}\right\}$$

Letting ε go to 0, we deduce easily that there exists a bounded nonnegative measure μ_k on $\overline{\Omega} \times [0,T]$ ($\forall\, k \geq 1$) such that

$$\left.\begin{array}{l}\dfrac{\partial}{\partial t}\beta_k(h) + \operatorname{div}\{u\,\beta_k(h)\} - \Delta\beta_k(h)\\[2mm] \qquad = \beta_k'(h)F + (-\beta_k''(h)|\nabla h^\varepsilon|^2) + \mu_k \quad \text{in } \Omega \times (0,T).\end{array}\right\} \quad \text{(E.20)}$$

Indeed, $-\beta_k''(h^\varepsilon)|\nabla h^\varepsilon|^2 = \frac{2}{(1+h^\varepsilon)^3}|\nabla h^\varepsilon|^2 = 8|\nabla(1+h^\varepsilon)^{-1/2}|^2$ and $(1+h^\varepsilon)^{-1/2}$ converges weakly in $L^2(0,T;H^1(\Omega))$ to $(1+h)^{-1/2}$ in view of the convergences already shown.

We next claim that $\mu_k \geq \mu_1$ in $\overline{\Omega} \times [0,T]$. Formally, this is straightforward since we deduce from the equation satisfied by $\beta_1(h)$

$$\begin{aligned}&\frac{\partial}{\partial t}\beta_k(h) + \operatorname{div}\{u\beta_k(h)\} - \Delta\beta_k(h)\\[2mm] &= \frac{\partial}{\partial t}\left(\gamma_k \circ \beta_1(h)\right) + \operatorname{div}\{u\gamma_k \circ \beta_1(h)\} - \Delta(\gamma_k \circ \beta_1(h))\\[2mm] &= \gamma_k'(\beta_1(h))\left\{\beta_1'(h)F - \beta_1''(h)|\nabla h|^2 + \mu_1\right\} - \gamma_k''(\beta_1(h))(\beta_1'(h))^2|\nabla h|^2\\[2mm] &= \beta_k'(h)F - \beta_k''(h)|\nabla h|^\varepsilon + \gamma_k'(\beta_1(h))\mu_1\,.\end{aligned}$$

Since $\gamma'_k(\beta_1(h)) = \frac{(1+h)^2}{\left(1+\frac{h}{k}\right)^2} \geq 1$, we deduce that $\mu_k \geq \mu_1$.

To justify these computations and the conclusion, we have only to show that we have

$$\frac{\partial}{\partial t} \beta_k(h) + \text{div}\,\{u\,\beta_k(h)\} - \Delta\beta_k(h)$$
$$\geq \beta'_k(h)F - \beta''_k(h)|\nabla h|^2 + \mu_1 \quad \text{in } \Omega \times (0,T).$$

This is done exactly as in step 1 above, the final step consisting in showing that if

$$-\Delta H = G + m \quad \text{in } \Omega, \quad H \in L^\infty(\Omega) \cap H^1(\Omega), \quad \frac{\partial H}{\partial n} = 0 \quad \text{on } \partial\Omega, \quad (\text{E}.21)$$

where $\|H\|_{L^\infty(\Omega)} < \left(1 - \frac{1}{k}\right)^{-1}$, $H \geq 0$ a.e. in Ω, $G \in L^1(\Omega)$, $m \geq 0$ is a bounded measure on $\overline{\Omega}$, then

$$\left. \begin{array}{ll} -\Delta\gamma_k(H) \geq \gamma'_k(H)G - \gamma''_k(H)|\nabla H|^2 + m & \text{in } \Omega, \\[2mm] \dfrac{\partial}{\partial n}\gamma_k(H) = 0 & \text{on } \partial\Omega. \end{array} \right\} \quad (\text{E}.22)$$

To this end, we use the weak formulation of (E.21) with $\gamma'_k(H_\alpha)\varphi$ where φ is arbitrary in $C^1(\overline{\Omega})$, $H_\alpha \in C^1(\overline{\Omega})$ for $\alpha > 0$, $\sup_\alpha \|H_\alpha\|_{L^\infty(\Omega)} < \left(1 - \frac{1}{k}\right)^{-1}$, $H_\alpha \geq 0$ in Ω, H_α converges to H in $H^1(\Omega)$ and a.e. in Ω as α goes to 0. We then obtain if $\varphi \geq 0$ in $\overline{\Omega}$

$$\int_\Omega \nabla H \cdot \{\gamma''_k(H_\alpha)\nabla H_\alpha\varphi + \gamma'_k(H_\alpha)\nabla\varphi\}\,dx \geq \int_\Omega \gamma'_k(H_\alpha)G + m\varphi\,dx$$

and we recover (E.22) letting α go to 0.

Therefore, we deduce from the weak formulation of (E.20)

$$\int_\Omega \beta_k(h(x,T))\,dx \geq \int_0^T dt \int_\Omega dx\,\beta'_k(h)F + \mu_1(\overline{\Omega} \times [0,T)),$$

and letting k go to $+\infty$, we obtain

$$\int_\Omega h(x,T)\,dx \geq \int_0^T dt \int_\Omega dx\,F + \mu_1(\overline{\Omega} \times [0,T)).$$

On the other hand, (E.19) yields

$$\int_\Omega h^\varepsilon(x,T)\,dx = \int_0^T dt \int_\Omega dx\,F^\varepsilon$$

hence, upon letting ε go to 0_+, $\int_\Omega h(x,T)\,dx = \int_0^T dt \int_\Omega dx\, F$. Therefore, $\mu_1 \equiv 0$ and this implies that $\nabla(1+h^\varepsilon)^{-1/2}$ converges to $\nabla(1+h)^{-1/2}$ in $L^2(\Omega \times (0,T))$, and the proof of Theorem E.1 is complete. $\quad\square$

We conclude this appendix with a simple observation.

Theorem E.2. *Let $p \in [1,\infty)$ and let $\alpha \in (0,p)$. We assume that $f \in L^p(\Omega \times (0,T))$, $\nabla f \in L^1(\Omega \times (0,T))$ (for instance) and that f satisfies for all $R \geq 1$*

$$\int_0^T dt \int_\Omega dx\, |\nabla f|^p\, 1_{(|f|<R)} \ \leq\ C\,R^\alpha \qquad (E.23)$$

for some $C \geq 0$. Then, for all $\beta \in (\alpha,p)$

$$(1+|f|^2)^{\frac{1}{2}-\frac{\beta}{2p}} \ \in\ L^p(0,T;W^{1,p}(\Omega)). \qquad (E.24)$$

Remarks E.2. 1) Many variants and extensions are possible (less restrictive conditions on f, unbounded domains, time-dependent f, etc.); we skip these since the argument given below is extremely simple and can be easily adapted to various situations.

2) Of course, the norm in $L^p(0,T;W^{1,p}(\Omega))$ is estimated in terms of the constant C in (E.24) and a bound on f in $L^p(\Omega \times (0,T))$, say.

3) Using Sobolev inequalities we deduce easily (at least if $p < N$) from (E.24) that $f \in L^{p-\beta}(0,T;L^q(\Omega))$ where $q = \frac{N(p-\beta)}{N-p}$. $\quad\square$

Proof of Theorem E.2. It is enough to show that we have

$$\int_0^T dt \int_\Omega dx\, |\nabla f|^p\, (1+|f|^2)^{-\frac{\beta}{2}} \ \leq\ C. \qquad (E.25)$$

Then, we write

$$\int_0^T dt \int_\Omega dx\, |\nabla f|^p\, (1+|f|^2)^{-\frac{\beta}{2}}$$

$$\leq\ C + \sum_{n\geq 0} \int_0^T dt \int_\Omega dx\, |\nabla f|^p\, (1+|f|^2)^{-\frac{\beta}{2}}\, 1_{(2^n \leq |f| < 2^{n+1})}$$

$$\leq\ C + \sum_{n\geq 0} 2^{-n\beta} \int_0^T dt \int_\Omega dx\, |\nabla f|^p\, 1_{(|f| < 2^{n+1})}$$

$$\leq\ C + C \sum_{n\geq 0} 2^{-n\beta}\, 2^{n\alpha} \ \leq\ C. \quad\square$$

BIBLIOGRAPHY

[1] R. Agemi, *The initial boundary value problem for inviscid barotropic fluid motion.* Hokkaido Math. J., **10** (1981), p. 156–182.

[2] V.I. Agoshkov, D. Ambrosi, V. Pennati, A. Quarteroni and F. Saleri, *Mathematical and numerical modelling of shallow water flow.* To appear in J. Comput. Mech.

[3] V.I. Agoshkov, A. Quarteroni and F. Saleri, *Recent developments in the numerical simulation of shallow water equations. I. Boundary conditions.* Preprint, 1994.

[4] S. Alinhac, *Un phénomène de concentration évanescente pour des flots non-stationnaires incompressibles en dimension deux.* Comm. Math. Phys., **127** (1990), p. 585–596.

[5] S. Alinhac, *Remarques sur l'instabilité du problème des poches de tourbillon.* J. Funct. Anal., **38** (1991), p. 361–379.

[6] G. Allain, *Thèse de 3ème cycle,* Ecole Polytechnique, Palaiseau, 1990.

[7] G. Allain, *Small-time existence for the Navier–Stokes equations with a free surface.* Appl. Math. Optim., **16** (1987), p. 37–50.

[8] A. Alvino, P.L. Lions and G. Trombetti, *Comparison results for elliptic and parabolic equations via Schwarz symmetrization.* Ann. I.H.P. Anal. Nonlin., **7** (1990), p. 37–65.

[9] A. Alvino, P.L. Lions and G. Trombetti, *Comparison results for elliptic and parabolic equations via Schwarz symmetrization: a new approach.* J. Diff. Int. Eq., **4** (1991), p. 25–50.

[10] H. Amann, *Stability of the rest state of a viscous incompressible fluid.* Arch. Rat. Mech. Anal., **126** (1994), p. 231–242.

[11] A.A. Amosov and A.A. Zlotnik, *A family of difference schemes for the equations of one-dimensional magnetogasdynamics: properties and global error estimates.* Sov. Math. Dokl., **37** (1988), p. 545–549.

[12] A.A. Amosov and A.A. Zlotnik, *Global generalized solutions of the equations of the one-dimensional motion of a viscous heat-conducting gas.* Sov. Math. Dokl., **38** (1989), p. 1–5.

[13] A.A. Amosov and A.A. Zlotnik, *Solvability "in the large" of a system of equations of the one-dimensional motion of an*

inhomogeneous viscous heat-conducting gas. Mat. Zametki, **52** (1992), p. 3–16.

[14] J.D. Anderson, Modern compressible flow: with historical perspective. McGraw-Hill, New York, 1982.

[15] J.D. Anderson, Fundamentals of aerodynamics. McGraw-Hill, New York, 1984.

[16] S.N. Antontsev and A.V. Kazhikov, Mathematical study of flows of nonhomogeneous fluids. Lecture Notes, Novosibirsk State University, 1973 (in Russian).

[17] S.N. Antontsev, A.V. Kazhikov and V.N. Monakhov, Boundary values problems in mechanics of nonhomogeneous fluids. North-Holland, Amsterdam, 1990.

[18] P. Azerad, *Analyse et approximation du problème de Stokes dans un bassin peu profond.* C.R. Acad. Sci. Paris, **318** ser. I (1994), p. 53–58.

[19] H. Bahouri and J.Y. Chemin, *Equations de transport relatives à des champs de vecteurs non lipschitziens et mécanique des fluides.* Preprint, 1995.

[20] C. Bandle, Isoperimetric inequalities and applications. Pitman, London, 1980.

[21] C. Bardos, *Existence et unicité de l'équation d'Euler en dimension deux.* J. Math. Anal. Appl., **40** (1972), p. 769–790.

[22] A.J.C. Barré de Saint-Venant, *Mémoire sur la dynamique des fluides.* C.R.A.S. Sc. Paris, **17** (1843), p. 1240–1243.

[23] B. Barthes-Biesel, M. Jaffrin, F. Jouaillee and H. Viviand, Ecoulements de fluides réels. Cours de l'Ecole Polytechnique, Palaiseau, 1989.

[24] G.K. Batchelor, The theory of homogeneous turbulence. Cambridge University Press, Cambridge, 1953.

[25] G.K. Batchelor, An introduction to fluid dynamics. Cambridge University Press, Cambridge, 1967.

[26] J.T. Beale, *The initial value problem for the Navier–Stokes equations with a free surface.* Comm. Pure Appl. Math., **34** (1981), p. 359–392.

[27] J.T. Beale, *Large-time regularity of viscous surface waves.* Arch. Rat. Mech. Anal., **84** (1984), p. 307–352.

[28] J.T. Beale, T. Kato and A. Majda, *Remarks on the breakdown of smooth solutions for the 3D Euler equations.* Comm. Math. Phys., **94** (1984), p. 61–66.

[29] J.T. Beale and A. Majda, *Rates of convergences for viscous splitting of the Navier–Stokes equations.* Math. Comp., **37** (1981), p. 243–259.

[30] H. Beirão da Veiga, *Un théorème d'existence dans la dynamique des fluides compressibles.* C.R. Acad. Sci. Paris, **289** (1979), p. 297–299.

[31] H. Beirão da Veiga, *On an Euler type equation in hydrodynamics.* Ann. Mat. Pura Appl., **125** (1980), p. 279–285.

[32] H. Beirão da Veiga, *On the barotropic motion of compressible perfect fluids.* Ann. Scuola Norm. Sup. Pisa, **8** (1981), p. 371–351.

[33] H. Beirão da Veiga, *Homogeneous and non-homogeneous boundary value problems for first-order linear hyperbolic systems arising in fluid mechanics.* Part I, Comm. P.D.E., **7** (1982), Part II, Comm. P.D.E., **8** (1983), p. 407–432.

[34] H. Beirão da Veiga, *Diffusion on viscous fluids. Existence and asymptotic properties of solutions.* Ann. Scuola Norm. Sup. Pisa, **10** (1983), p. 341–355.

[35] H. Beirão da Veiga, *On the solutions in the large of the two-dimensional flow of a non-viscous incompressible fluid.* J. Diff. Eq., **54** (1984), p. 373–389.

[36] H. Beirão da Veiga, *An L^p-theory for the n-dimensional, stationary, compressible Navier–Stokes equations and the incompressible limit for compressible fluids. The equilibrium solution.* Comm. Math. Phys., (1987), p. 229-248.

[37] H. Beirão da Veiga, *Stationary motions and the incompressible limit for compressible viscous fluids.* Houston J. Math., **13** (1987), p. 527–544.

[38] H. Beirão da Veiga, *The stability of one-dimensional stationary flows of compressible viscous fluids.* Ann. Inst. Henri Poincaré, Anal. non Lin., **7** (1990), p. 259–268.

[39] H. Beirão da Veiga, *Data dependence in the mathematical theory of compressible inviscid fluids.* Arch. Rat. Mech. Anal., **119** (1992), p. 109-127.

[40] H. Beirão da Veiga, *Attracting properties for one-dimensional flows of a general barotropic viscous fluid. Periodic flows.* Ann. Mat. Pura Appl., **CLXI** (1992), p. 156–165.

[41] H. Beirão da Veiga, *Perturbation theorems for linear hyperbolic mixed problems and applications to the compressible Euler equations.* Comm. Pure Appl. Math., **XLVI** (1993), p. 221–259.

[42] H. Beirão da Veiga, *On the singular limits for slightly compressible fluids.* Preprint, 1994.

[43] H. Beirão da Veiga, *Singular limits in Fluid dynamics. (II).* Preprint, 1994.

[44] H. Beirão da Veiga, *On the solutions in the large of the two-dimensional flow of a non-viscous incompressible fluid.* Preprint, 1994.

[45] H. Beirão da Veiga, *An extension of the classical Prodi–Serrin's sufficient condition concerning the Navier–Stokes equations.* Preprint, 1994.

[46] H. Beirão da Veiga and A. Valli, *Existence of C^∞ solutions of the Euler equations for non-homogeneous fluids.* Comm. P.D.E., **5** (1980), p. 95–107.

[47] H. Beirão da Veiga, R. Serapioni and A. Valli, *On the motion of non-homogeneous fluids in the presence of diffusion.* J. Math. Anal. Appl., **85** (1982), p. 179–191.

[48] N.P. Belov, <u>Numerical methods of weather prediction.</u> Gidrometeoizdat, Leningrad, 1975 (in Russian).

[49] A. Benabdallah-Lagha, *Limite des équations d'un fluide compressible lorsque la compressibilité tend vers 0.* Preprint, 1995.

[50] M. Ben-Artzi, *Global solutions of $2 - D$ Navier–Stokes and Euler equations.* Arch. Rat. Mech. Anal., **128** (1994), p. 329–358.

[51] D. Benedetto, C. Marchioro and M. Pulvirenti, *On the Euler flow in $\mathbb{R}^2$.* Arch. Rat. Mech. Anal., **123** (1993), p. 377–386.

[52] A.F. Bennett and P.E. Kloeden, *The simplified quasigeostrophic equations: existence and uniqueness of strong solutions.* Mathematika, **27** (1980), p. 287–311.

[53] A.F. Bennett and P.E. Kloeden, *The quasigeostrophic equations: approximation, predictability and equilibrium spectra of solutions.* Quart. J. Roy. Mat. Soc., **107** (1981), p. 121–136.

200   *Bibliography*

[54] A.F. Bennett and P.E. Kloeden, *The periodic quasigeostrophic equations: existence and uniqueness of strong solutions.* Proc. Roy. Soc. Edin., **91 A** (1982), p. 185–203.

[55] C. Bernardi, M. O. Bristeau, O. Pironneau and M.G. Vallet, *Numerical analysis for compressible viscous isothermal stationary flows.* In <u>Applied and industrial mathematics</u>, R. Spigler (ed.), p. 231–243; Kluwer, Amsterdam, 1991.

[56] C. Bernardi and O. Pironneau, *On the shallow water equations at low Reynolds number.* Comm. P.D.E., **16** (1991), p. 59–104.

[57] A.L. Bertozzi, *Existence, uniqueness and a characterization of solutions to the contour dynamics equation.* PhD Thesis, Princeton University, 1991.

[58] A.L. Bertozzi and P. Constantin, *Global regularity for vortex patches.* Comm. in Math. Phys., **152** (1993), p. 19–26.

[59] H. Bessaih, *Limite de modèles de fluides compressibles.* To appear in Portugaliae Mathematica.

[60] H. Bessaih, *Stability for solutions of Navier–Stokes equations when the Mach number goes to zero.* Preprint, 1995.

[61] O. Besson and M.R. Laydi, *Some estimates for the anisotropic Navier–Stokes equations and the hydrostatic approximation.* RAIRO M^2AN, **26** (1992), p. 855–866.

[62] O. Besson, M.R. Laydi and R. Touzani, *Un modèle asymptotique en océanographie.* C.R. Acad. Sci. Paris, **310** ser. I (1990), p. 661–665.

[63] F. Bethuel and J.M. Ghidaglia, *Weak limits of solutions to the steady incompressible two-dimensional Euler equation in a bounded domain.* Asymptotic Anal., **8** (1994), p. 277–291.

[64] R.B. Bird, W.E. Stewart and E.N. Lightfoot, <u>Transport phenomena</u>. Wiley, New York, 1960.

[65] D. Blanchard, *Truncations and monotonicity methods for parabolic equations.* Nonlinear Analysis T.M.A., **21** (1993), p. 725–743.

[66] D. Blanchard and F. Murat, *Renormalized solutions of nonlinear parabolic problems with L^1 data: Existence and uniqueness.* Preprint, 1994.

[67] L. Boccardo, J.I. Diaz, D. Giachetti and F. Murat, *Existence of a solution for a weaker form for a nonlinear elliptic equation.* Preprint, 1994.

[68] W. Borchers and T. Miyakawa, *Algebraic L^2 decay for Navier–Stokes flows in exterior domains.* Acta Math., **165** (1990), p. 189–227.

[69] W. Borchers and T. Miyakawa, L^2 *decay for Navier–Stokes flows in unbounded domains, with applications to exterior stationary flows.* Arch. Rat. Mech. Anal., **118** (1992), p. 273–295.

[70] W. Borchers and T. Miyakawa, *On the stability of exterior stationary Navier–Stokes flows.* Preprint, 1994.

[71] W. Borchers and H. Sohr, *On the semigroup of the Stokes operator for exterior domains in L^q spaces.* Math. Z., **196** (1987), p. 415–425.

[72] W. Borchers and W. Varnhorn, *On the boundedness of the Stokes semigroup in two-dimensional exterior domains.* Math. Z., **213** (1993), p. 275–299.

[73] A.J. Bourgeois and J.T. Beale, *Validity of the quasigeostrophic model for large-scale flow in the atmosphere and ocean.* Preprint, 1994.

[74] J. Bouttes, <u>Mécanique des fluides</u>. Ellipses. Ecole Polytechnique, Palaiseau, 1988.

[75] G.L. Browning, W.R. Holland, H.O. Kreiss and S.J. Worley, *An accurate hyperbolic system for approximately hydrostatis and incompressible oceanographic flows.* Dyn. Atmos. Oceans, **14** (1980), p. 303–332.

[76] G.L. Browning, A. Kasahara and H.O. Kreiss, *Initialization of the primitive equations by the bounded derivative method.* J. Atmos. Sci., **37** (1980), p. 1424–1436.

[77] L. Caffarelli, R.V. Kohn and L. Nirenberg, *On the regularity of the solutions of Navier–Stokes equations.* Comm. Pure Appl. Math., **35** (1982), p. 771–831.

[78] A.P. Calderón and A. Zygmund, *On the existence of certain singular integrals.* Acta Math., **88** (1952), p. 85–139.

[79] S. Candel, <u>Mécanique des fluides</u>. Dunod, Paris, 1990.

[80] M. Cannone, *Ondelettes, paraproduits et Navier–Stokes.* Thesis, Univ. Paris-Dauphine, Paris, 1994.

[81] M. Cannone, Y. Meyer and F. Planchon, *Solutions auto-similaires des équations de Navier–Stokes.* Exposé n. VIII, In <u>Séminaire</u>

Equations aux Dérivées Partielles, Ecole Polytechnique, Palaiseau, 1994.

[82] M. Cannone, Y. Meyer and F. Planchon, work in preparation.

[83] L. Cattabriga, *Su un problema al contorno relativo al sistema di equazioni di Stokes.* Rend. Sem. Mat. Univ. Padova, **31** (1961), p. 308–340.

[84] D. Chae, *Remarks on the regularity of weak solutions of the Navier–Stokes equations.* Comm. P.D.E., **17** (1992), p. 1267–1286.

[85] D. Chae, *Weak solutions of 2D incompressible Euler equations.* Nonlinear Analysis T.M.A., **23** (1994), p. 629–638.

[86] J.G. Charney, *Geostrophic turbulence.* J. Atmos. Sci., **28** (1971), p. 1087–1095.

[87] J.Y. Chemin, *Sur le mouvement des particules d'un fluide parfait incompressible bidimensionnel.* Invent., **103** (1991), p. 599–629.

[88] J.Y. Chemin, *Persistance de structures géométriques dans les fluides incompressibles bidimensionnels.* In Séminaire EDP, 1990-1991, Ecole Polytechnique, Palaiseau.

[89] J.Y. Chemin, *Remarques sur l'existence globale pour le système de Navier–Stokes incompressible.* SIAM J. Math. Anal., **23** (1992), p. 20–28.

[90] J.Y. Chemin, *Une facette mathématique de la mécanique des fluides. I.* Preprint, Ecole Polytechnique, 1993.

[91] J.Y. Chemin and N. Lerner, *Flot de champs de vecteurs non lipschitziens et équations de Navier–Stokes.* Preprint, 1993.

[92] G.Q. Chen, *The theory of compensated compactness and the system of isentropic gas dynamics.* Preprint MCS-P154-0590, Univ. of Chicago, 1990.

[93] X. Chen and W. Xie, *Discontinuous solutions of steady state, viscous compressible Navier–Stokes equations.* IMA Preprint ♯ 1027, 1992.

[94] A.J. Chorin, Vorticity and turbulence. Appl. Math. Sciences ♯ 103, Springer, 1991.

[95] R. Coifman, P.L. Lions, Y. Meyer and S. Semmes, *Compensated-compactness and Hardy spaces.* J. Math. Pures Appl., **72** (1993), p. 247–286.

[96] R. Coifman and G. Weiss, *Extensions of Hardy spaces and their use in analysis*. Bull. A.M.S., **83** (1977), p. 569–645.

[97] C. Conca, F. Murat and O. Pironneau, *The Stokes and Navier–Stokes equations with boundary conditions involving the pressure*. Japan J. Math. vol. **20** n. 2, (1994), p. 279–318.

[98] P. Constantin, *Remarks on the Navier–Stokes equations*. Comm. Math. Phys., **129** (1990), p. 241.

[99] P. Constantin, *Geometric statistics in turbulence*. SIAM Review, **36** (1994), p. 73.

[100] P. Constantin and C. Fefferman, *Direction of vorticity and the problem of global regularity for the Navier–Stokes equations*. Ind. Univ. Math. J., **42** (1993), p. 775.

[101] P. Constantin and C. Fefferman, *Scaling exponents in fluid turbulence: some analytic results*. Preprint, 1994.

[102] P. Constantin and C. Foias, Navier–Stokes equations. University of Chicago Press, Chicago, 1988.

[103] P. Constantin, A. Majda and E.G. Tabak, *Singular front formation in a model for quasi-geostrophic flow*. Preprint, 1995.

[104] P. Constantin and E. Titi, *On the evolution of nearly circular vortex patches*. Comm. Math. Phys., **119** (1988), p. 177–198.

[105] V. Coscia and M. Padula, *Nonlinear energy stability in a compressible atmosphere*. Geophys. Astrophys. Fluid Dynamics, **54** (1990), p. 49–83.

[106] G. Cottet, *Equations de Navier–Stokes dans le plan avec tourbillon initial mesure*. C.R. Acad. Sci. Paris, **303** (1986), p. 105–108.

[107] G. Cottet, *Two dimensional incompressible fluid flow with singular initial data*. Mathematical topics in fluid mechanics (J.F. Rodrigues and A. Sequeira, eds.), Longman Scientific, (1992), p. 32–49.

[108] R. Courant and K.O. Friedrichs, Supersonic flow and shock waves. Springer, New York, 1948 and 1976.

[109] A.D.D. Craik, Wave interactions and fluid flows. Cambridge University Press, Cambridge, 1985.

[110] A.D.D. Craik, *The stability of unbounded two and three-dimensional flows subject to body forces: some exact solutions*. J. Fluid Mech., **198** (1989), p. 275–292.

[111] A.D.D. Craik, *Time-dependent solutions of the Navier–Stokes equations with spatially uniform velocity gradients*. Proc. Roy. Soc. Edin., **124** (1994), p. 127–136.

[112] A.D.D. Craik and H.R. Allen, *The stability of three-dimensional time-periodic flows with spatially uniform strain rates*. J. Fluid Mech., **234** (1992), p. 613–627.

[113] A.D.D. Craik and W.O. Criminale, *Evolution of wavelike disturbances in shear flows: a class of exact solutions of the Navier–Stokes equations*. Proc. Roy. Soc. London, **406** (1986), p. 13–26.

[114] J.S. Darrozes and C. François, Mécanique des fluides incompressibles. Springer, Berlin, 1982.

[115] R. Dautray and J.L. Lions, Analyse mathématique et calcul numérique pour les sciences et les techniques. Tome 3. Masson, Paris, 1985.

[116] K. Deckelnick, *Decay estimates for the compressible Navier–Stokes equations in unbounded domains*. Math. Z., **209** (1992), p. 115–130.

[117] K. Deckelnick, *L^2 decay for the compressible Navier–Stokes equations in unbounded domains*. Comm. P.D.E., **18** (1993), p. 1445–1476.

[118] J.M. Delort, *Existence de nappes de tourbillon en dimension deux*. J. Amer Math. Soc., **4** (1991), p. 553–586.

[119] J.M. Delort, *Une remarque sur le problème des nappes de tourbillon axisymétriques sur $\mathbb{R}^3$*. J. Funct. Anal., **108** (1992), p. 274–295.

[120] I.V. Denisova, *A priori estimates of the solution of a linear time-dependent problem connected with the motion of a drop in a fluid medium*. Trudy. Matem. Inst. Steklov, **188** (1990), p. 1–24.

[121] B. Desjardins, *Linear transport equations with initial values in Sobolev spaces and application to the Navier–Stokes equations*. Preprint, 1995.

[122] P. Deuring, *The Stokes system in exterior domains: existence, uniqueness, and regularity of solutions in L^p-spaces*. Comm. Part. Diff. Eq., **16** (1991), p. 1513–1528.

[123] P. Deuring, W. von Wahl and Weidemaier, *Das lineare Stokes system in $\mathbb{R}^3$, II. Das Außenraumproblem*, Bayreuth. Math. Schr., **28** (1988), p. 1–109.

[124] R.J. DiPerna, *Convergence of the viscosity method for isentropic gas dynamics.* Comm. Math. Phys., **91** (1983), p. 1-30.

[125] R.J. DiPerna and P.L. Lions, *On the Fokker–Planck–Boltzmann equations.* Comm. Math. Phys., **120** (1988), p. 1–23.

[126] R.J. DiPerna and P.L. Lions, *Equations différentielles ordinaires et équations de transport avec des coefficients irréguliers.* In Séminaire EDP 1988-1989, Ecole Polytechnique, Palaiseau, 1989.

[127] R.J. DiPerna and P.L. Lions, *On the global existence for Boltzmann equations: global existence and weak stability.* Ann. Math., **130** (1989), p. 321–366.

[128] R.J. DiPerna and P.L. Lions, *Ordinary differential equations, Sobolev spaces and transport theory.* Invent. Math., **98** (1989), p. 511-547.

[129] R.J. DiPerna and A. Majda, *Concentrations in regularizations for 2D incompressible flow.* Comm. Pure Appl. Math., **40** (1987), p. 301-345.

[130] R.J. DiPerna and A. Majda, *Reduced Hausdorff dimension and concentration-cancellation for two dimensional incompressible flow.* J. Amer. Math. Soc., **1** (1988), p. 59–95.

[131] F.V. Dolzhanskii, V.I. Klyatskin, A.M. Obukhov and M.A. Chusov, Nonlinear systems of hydrodynamic type, Nauka, Moscow, 1973 (in Russian)

[132] P.G. Drazin and W.H. Reid, Hydrodynamic stability. Cambridge University Press, Cambridge, 1981.

[133] G.F.D. Duff, *Derivative estimates for the Navier–Stokes equations in a three dimensional region.* Acta Math., **164** (1990), p. 145–210.

[134] J.A. Dutton, *The nonlinear quasigeostrophic equations: Existence and uniqueness of solutions on a bounded domain.* J. Atmos. Sci., **31** (1974), p. 422–433.

[135] J.A. Dutton, Ceaseless wind, an introduction to the theory of atmosphere motion, McGraw-Hill, New York, 1976.

[136] W.E, *Dynamics of vortices in Ginzburg–Landau theories with applications to superconductivity.* Physica D., **77** (1994), p. 383–404.

[137] W.E, *Dynamics of vortex liquids in Ginzburg–Landau theories with applications to superconductivity.* Phys. Rev. B., **50** (1994), p. 1126–1135.

[138] W.E, *Propagation of oscillations in the solutions of 1 − d compressible fluid equations.* Preprint.

[139] D.G. Ebin, *The initial boundary value problem for subsonic fluid motion.* Comm. Pure Appl. Math., **32** (1979), p. 1–19.

[140] L. Euler, *Opera Omnia.* Series Secunda, **12** (1755), p. 274–361.

[141] L.C. Evans, <u>Weak convergence methods for nonlinear partial differential equations</u>. CBMS ♯ 74, AMS, Providence, 1990.

[142] L.C. Evans and S. Müller, *Hardy spaces and the two-dimensional Euler equations with nonnegative vorticity.* J. Amer. Math. Soc., **7** (1994), p. 199–219.

[143] E.B. Fabes, B.F. Jones and N.M. Rivière, *The initial value problem for the Navier–Stokes equations with data in L^p.* Arch. Rat. Mech. Anal., **45** (1972), p. 222–240.

[144] E. Fabes, J. Lewis and N. Rivière, *Singular integrals and hydrodynamic potentials.* Amer. J. Math., **99** (1977), p. 601–625.

[145] E. Fabes, J. Lewis and N. Rivière, *Boundary-value problems for the Navier–Stokes equations.* Amer. J. Math., **99** (1977), p. 626–688.

[146] M. Farge, *Ondelettes continues : application à la turbulence.* J. Ann. Soc. Math. France, 1990, p. 17–62.

[147] M. Farge, *Wavelet transforms and their applications to turbulence.* Annal. Review Fluid Mech., **24** (1992), p. 395–457.

[148] P. Federbush, *Navier and Stokes meet the wavelet.* Comm. Math. Phys., **155** (1993), p. 219–248.

[149] C. Fefferman and E. Stein, *H^p spaces of several variables.* Acta Math., **129** (1972), p. 137–193.

[150] E. Fernández-Cara and F. Guillén, *The existence of nonhomogeneous, viscous and incompressible flow in unbounded domains.* Comm. P.D.E., **17** (1992), p. 1253–1265.

[151] R. Finn, *On the exterior stationary problem for the Navier–Stokes equations, and associated perturbation problems.* Arch. Rat. Mech. Anal., **19** (1965), p. 363–406.

[152] C. Foias, C. Guillopé and R. Temam, *Lagrangian representation of a flow.* J. Diff. Eq., **57** (1985), p. 440–449.

[153] C. Foias and R. Temam, *Some analytic and geometric properties of the solutions of the evolution Navier–Stokes equations.* J. Math. Pures Appl., **58** (1979), p. 339–368.

[154] C. Foias and R. Temam, *Self-similar universal homogeneous statistical solutions of the Navier–Stokes equations.* Comm. Math. Phys., **90** (1983), p. 187–206.

[155] C. Foias, O.P. Manley and R. Temam, *New representation of Navier–Stokes equations governing self-similar homogeneous turbulence.* Phys. Rev. Lett., **51** (1983), p. 617–620.

[156] J. Frehse and M. Ružička, *Regularity for the stationary Navier–Stokes equations in bounded domains.* Arch. Rat. Mech. Anal., **128** (1994), p. 361–380.

[157] J. Frehse and M. Ružička, *On the regularity of the stationary Navier–Stokes equations.* Ann. Sci. Norm. Sup. Pisa, **21** (1994), p. 63–95.

[158] J. Frehse and M. Ružička, *Weighted estimates for the stationary Navier–Stokes equations.* To appear in Acta Appl. Math.

[159] J. Frehse and M. Ružička, *Existence of regular solutions to the stationary Navier–Stokes equations.* To appear in Math. Ann.

[160] H. Fujita and T. Kato, *On the Navier–Stokes initial value problem I.* Arch. Rat. Mech. Anal., **16** (1964), p. 269–315.

[161] G.P. Galdi, *On the energy equation and on the uniqueness for D-solutions to steady Navier–Stokes equations in exterior domains.* In Mathematical problems related to the Navier–Stokes equation. (G.P. Galdi, Ed.), Advances in Mathematics for Applied Science, **11** (1992), World Sci., p. 36–80.

[162] G.P. Galdi, *Existence and uniqueness at low Reynolds number of stationary plane flow of a viscous fluid in exterior domains.* In Recent developments in theoretical fluid mechanics. (G.P. Galdi and J. Nečas, eds.), Pitman Research Notes in Mathematics, London, Vol. **291** (1993), p. 1–33.

[163] G.P. Galdi, An introduction to the mathematical theory of Navier–Stokes equations. Volume 1, Springer Tracts in Natural Philosophy, vol. **38** (1994).

[164] G. Galdi and P. Maremonti, *Monotonic decreasing and asymptotic behavior of the kinematic energy for weak solutions of the Navier–Stokes equations in exterior domains.* Arch. Rat. Mech. Anal., **94** (1986), p. 253–266.

[165] G.P. Galdi and C.G. Simader, *Existence, uniqueness and L^q-estimates for the Stokes problem in an exterior domain.* Arch. Rat. Mech. Anal., **112** (1990), p. 291–318.

[166] G.P. Galdi and C.G. Simader, *New estimates for the steady-state Stokes problem in exterior domains with applications to the Navier–Stokes problem.* Diff. Int. Eq., **7** (1994), p. 847–861.

[167] P. Gamblin and X. Saint Raymond, *On three dimensional vortex patches.* Preprint, 1994.

[168] R.S. Gellrich, *Free boundary value problems for the stationary Navier–Stokes equations in domains with noncompact boundaries.* J. Anal. Appl., **12** (1993), p. 425–456.

[169] P. Gérard, *Résultats sur les fluides parfaits incompressibles bidimensionnels (d'après J.Y. Chemin et J.M. Delort).* Sém. Bourbaki, n. 757, in Astérisque **206** (1992), SMF, Paris, 1993.

[170] P. Germain, Mécanique des milieux continus. T. 1. Masson, Paris, 1973.

[171] P. Germain, Mécanique. Ellipses. Ecole Polytechnique, Palaiseau, 1986.

[172] M. Giga, Y. Giga and H. Sohr, L^p *estimates for the Stokes system.* In Functional analysis and related topics, 1991, Ed. H. Komatsu, Springer, Berlin, 1992.

[173] Y. Giga, *Analyticity of the semi-group generated by the Stokes operator in L_r spaces.* Math. Z., **178** (1981), p. 297–329.

[174] Y. Giga, *The nonstationary Navier–Stokes system with some first-order boundary conditions.* Proc. Jap. Acad. **58** (1982), p. 101–104.

[175] Y. Giga, *Domains of fractional powers of the Stokes operators in L_r spaces.* Arch. Rat. Mech. Anal., **89** (1985), p. 251–265.

[176] Y. Giga, *Solutions for semilinear parabolic equations in L^p and regularity of weak solutions of the Navier–Stokes system.* J. Diff. Eq., **61** (1986), p. 186–212.

[177] Y. Giga, *Review of "The equations of Navier–Stokes and abstract parabolic equations" by W. von Wahl.* Bull. AMS **19** (1988), p. 337–340.

[178] Y. Giga and T. Kambe, *Large time behavior of the viscosity of two dimensional viscous flow and its applications.* Comm. Math. Phys., **117** (1988), p. 549–568.

[179] Y. Giga and T. Miyakawa, *Solutions in L_r to the Navier–Stokes initial value problem.* Arch. Rat. Mech. Anal., **89** (1985), p. 267–281.

[180] Y. Giga and T. Miyakawa, *Navier–Stokes flows in $\mathbb{R}^3$ and Morrey spaces*. Comm. P.D.E., **14** (1989), p. 577–618.

[181] Y. Giga, T. Miyakawa and H. Osada, *Two-dimensional Navier–Stokes flow with measures as initial vorticity.* Arch. Rat. Mech. Anal., **104** (1988), p. 223–250.

[182] Y. Giga and H. Sohr, *On the Stokes operator in exterior domains.* J. Fac. Sci. Univ., Tokyo, Sec **I.A.36** (1989), p. 103–130.

[183] Y. Giga and H. Sohr, *Abstract L^p estimates for the Cauchy problem with applications to the Navier–Stokes equations in exterior domains.* J. Funct. Anal., **102** (1991), p. 72–94.

[184] Y. Giga and S. Takahashi, *On global weak solutions of the non-stationary two-phase Stokes flow.* Hokkaido Univ. Preprint ♯ 149, 1992.

[185] P. Goldreich and S. Tremaine, *The excitation of density waves at the Lindblad and corotation resonances by an external potential.* Astrophysics J., **233** (1979), p. 857–871.

[186] M.E. Goldstein, Aeroacoustics. McGraw-Hill, New York, 1976.

[187] K.K. Golovkin, *Potential theory of nonstationary linear Navier–Stokes equations in the case of three space variables.* Trudy Mat. Inst. Steklov, **59** (1960), p. 87–99 (in Russian).

[188] K.K. Golovkin, *The plane motion of a viscous incompressible fluid.* Amer. Math. Soc. Transl., **35** (1964), p. 297–350.

[189] K.K. Golovkin, *Vanishing viscosity in Cauchy's problem for hydromechanics equations.* In Proceedings of the Steklov Institute of Mathematics, vol. 92 (O.A. Ladyzenskaya ed.), 1966 (AMS translations, 1968).

[190] K.K. Golovkin and O.A. Ladyzenskaya, *Solutions of non-stationary boundary value problems for the Navier–Stokes equations.* Trudy Mat. Inst. Steklov, **59** (1960), p. 100–114 (in Russian).

[191] K.K. Golovkin and V.A. Solonnikov, *On the first boundary-value problem for nonstationary Navier–Stokes equations.* Soviet Math. Dokl., **2** (1961), p. 1188–1191.

[192] J. Goodman and T. Hou, *New stability estimates for the 2D vortex method.* Comm. Pure Appl. Math., **44** (1991), p. 1015–1031.

[193] R. Grauer and T. Sideris, *Numerical computation of three dimensional incompressible ideal fluids with swirl.* Phys. Rev. Lett., **67** (1991), p. 3511.

[194] C. Greengard and E. Thomann, *On DiPerna-Majda concentration sets for two-dimensional incompressible flow.* Comm. Pure Appl. Math., **41** (1988), p. 295–303.

[195] G. Grubb, *Solution dans les espaces de Sobolev L_p anisotropes des problèmes aux limites pseudo-différentiels paraboliques et des problèmes de Stokes.* C.R. Acad. Sci. Paris, **312, I** (1991), p. 89–92.

[196] G. Grubb, *Initial-value problems for the Navier–Stokes equations with Neumann conditions.* In The Navier–Stokes equations II, J.G. Heywood, K. Masuda, R. Rautmann and S.A. Solonnikov Eds., Proc. Conf. Oberwolfach, 1991. Lectures Notes in Math., 1530, Springer, Berlin, 1992, p. 262–283.

[197] G. Grubb, *Nonhomogeneous Navier–Stokes problems in L_p Sobolev spaces over interior and exterior domains.* Proceedings of the Navier–Stokes Equations III, Oberwolfach 1994.

[198] G. Grubb and N.J. Kokholm, *A global calculus of parameter-dependent pseudo-differential boundary problems in L_p Sobolev spaces.* Acata Mathematica **171** (1993), p. 165–229.

[199] G. Grubb and V.A. Solonnikov, *Reduction of basic initial-boundary value problems for the Stokes equation to initial-boundary value problems for systems of pseudo-differential equations.* Zapiski Nauchn. Sem. L.O.M.I. **163** (1987), p. 37–48; J. Soviet. Math. **49** (1990), p. 1140–1147.

[200] G. Grubb and V.A. Solonnikov, *Solution of parabolic pseudo-differential initial-value boundary value problems.* J. Diff. Eq. **87** (1990), p. 256–304.

[201] G. Grubb and V.A. Solonnikov, *Reduction of basic initial-boundary value problems for the Navier–Stokes equations to nonlinear parabolic systems of pseudo-differential equations.* Zap. Nauchn. Sem. L.O.M.I. **171** (1989), p. 36–52; J. Soviet Math. **56** (1991), p. 2300-2308.

[202] G. Grubb and V.A. Solonnikov, *Boundary value problems for the nonstationary Navier–Stokes equations treated by pseudo-differential methods.* Math. Scand. **69** (1991), 217–290.

[203] C. Guillopé, C. Foias and R. Temam, *Lagrangian representation of a flow.* J. Diff. Eq., **57** (1985), p. 440–449.

[204] N. Gunter, *On the motion of fluid in the moving container.* Izv. Akad. Nauk. SSSR. Ser. Fiz.-Mat., **20** (1926), p. 1323–1348, p.

1503–1532; **21** (1927), p. 621–656, p. 735–756, p. 1139–1162; **22** (1928), p. 9–30.

[205] O.V. Guseva, *On a nonstationary boundary-value problem of the hydrodynamics of a viscous incompressible fluid*. Vestnik Leningrad Univ. Ser. Mat. Meh. Astronim., **19** (1961), in Russian.

[206] F. Habergel, Thèse de l'Université Paris-Sud Orsay, Orsay, 1992.

[207] W.D. Hayes and R.F. Probstein, Hypersonic flow theory. 2nd edition, Academic Press, New York, 1962.

[208] W.D. Henshaw, H.O. Kreiss and L.G. Reyna, *Smallest scale estimates for the Navier–Stokes equations for incompressible fluids*. Arch. Rat. Mech. Anal., **112** (1990), p. 21–44.

[209] J.G. Heywood, *On uniqueness questions in the theory of viscous flow*. Acta Math., **136** (1976), p. 61–102.

[210] J.G. Heywood, *A uniqueness theorem for non-stationary Navier–Stokes flow past an obstacle*. Ann. Sc. Norm. Sup. Pisa, **6** ser. 6 (1979), p. 427–445.

[211] J.G. Heywood, *Remarks on the possible global regularity of solutions of the three-dimensional Navier–Stokes equations*. In Progress in theoretical and computational fluid mechanics. Eds. G.P. Galdi, J. Málek and J. Nečas, Longman, Harlow, 1994.

[212] J.O. Hinze, Turbulence. 2nd edition, McGraw-Hill, New York, 1975.

[213] D. Hoff, *Construction of solutions for compressible, isentropic Navier–Stokes equations in one space dimension with nonsmooth initial data*. Proc. Royal Soc. Edinburgh, **A 103** (1986), p. 301–315.

[214] D. Hoff, *Global existence for 1D compressible, isentropic Navier–Stokes equations with large initial data*. Trans. Amer. Math. Soc., **303** (1987), p. 169–181.

[215] D. Hoff, *Discontinuous solutions of the Navier–Stokes equations for compressible flow*. Arch. Rat. Mech. Anal., **114** (1991), p. 15–46.

[216] D. Hoff, *Global well-posedness of the Cauchy problem for nonisentropic gas dynamics with discontinuous initial data*. J. Diff. Eq., **95** (1992), p. 33–73.

[217] D. Hoff, *Spherically symmetric solutions of the Navier–Stokes equations for compressible, isothermal flow with large, discontinuous initial data*. Ind. Univ. Math. J., **41** (1992), p. 1225–1302.

[218] D. Hoff, *Global solutions of the Navier–Stokes equations for multi-dimensional compressible flow with discontinuous initial data.* Preprint, 1994.

[219] D. Hoff, *Continuous dependence on initial data for discontinuous solutions of the Navier–Stokes equations for one-dimensional, compressible flow.* Preprint, 1994.

[220] D. Hoff and R. Zarnowski, *Continuous dependence in L^2 for discontinuous solutions of the viscous p-system.* Preprint, 1994.

[221] E. Hopf, *Uber die Anfengswertaufgabe für die hydrodynamischen Grundgleichangen.* Math. Nacht., **4** (1951), p. 213–231.

[222] L. Hörmander, *Estimates for translation invariant operators in L^p spaces.* Acta Math., **104** (1960), p. 93–140.

[223] P. Huerre, <u>Mécanique des fluides</u>. Cours de l'Ecole Polytechnique, Palaiseau, 1991.

[224] N. Itaya, *On the Cauchy problem for the system of fundamental equations describing the movement of compressible viscous fluid.* Ködai Math. Sem. Rep., **23** (1971), p. 60–120.

[225] N. Itaya, *Some results on the piston problem related with fluid mechanics.* J. Math. Kyoto Univ., **23** (1983), p. 631–641.

[226] H. Iwashita, *$L^q - L^r$ estimates for solutions of non-stationary Stokes equations in exterior domains and the Navier–Stokes initial value problems in L_q spaces.* Math. Ann., **285** (1989), p. 265–288.

[227] S. Jiang, *On initial boundary value problems for a viscous, heat-conducting, one-dimensional real gas.* J. Diff. Eq., **110** (1994), p. 157–181.

[228] S. Jiang, *On the asymptotic behavior of the motion of a viscous, heat-conducting, one-dimensional real gas.* Math. Z., **216** (1994), p. 317–336.

[229] S. Jiang, *Remarks on the global existence in the dynamics of a viscous, heat-conducting, one-dimensional gas.* <u>Proc. of the Workshop on Qualitative Aspects and Appl. of Nonlinear Evol. Eqns</u> (H. Beirão da Veiga and Ta-Tsien Li, eds.), World Scientific Publ., Singapore, 1994, p. 156–162.

[230] S. Jiang, *Global smooth solutions to the equations of a viscous, heat-conducting, one-dimensional gas with density-dependent viscosity.* Preprint, 1995.

[231] S. Jiang, *Global spherical symmetric solutions to the equations of a viscous polytropic gas in an exterior domain.* Preprint, 1995.

[232] J.I. Kanel', *On a model system of equations for one-dimensional gas motion.* Differencial'nye Uravnenija, **4** (1968), p. 721–734 (in Russian).

[233] J.I. Kanel', *Cauchy problem for the dynamic equations for a viscous gas.* Sibirsk Mat. Zh., **20** (1979), p. 293–306.

[234] S. Kaniel and M. Shinbrot, *Smoothness of weak solutions of the Navier–Stokes equations.* Arch. Rat. Mech. Anal., **24** (1967), p. 302–324.

[235] K. Karamcheti, Principles of ideal-fluid aerodynamics. Wiley, New York, 1966.

[236] T. Kato, *On classical solutions of the two-dimensional non-stationary Euler equation.* Arch. Rat. Mech. Anal., **25** (1967), p. 188–200.

[237] T. Kato, *Non-stationary flows of viscous and ideal fluids in* $\mathbb{R}^3$. J. Funct. Anal., **9** (1972), p. 296–305.

[238] T. Kato, *Remarks on zero viscosity limit for nonstationary Navier–Stokes flows with boundary.* In Seminar on Nonlinear PDE, ed. S.S. Chern, MSRI, 1984.

[239] T. Kato, *Strong* L^p-*solutions of the Navier–Stokes equations in* $\mathbb{R}^m$, *with applications to weak solutions.* Math. Z., **187** (1984), p. 471–480.

[240] T. Kato, *Remarks on the Euler and Navier–Stokes equations in* $\mathbb{R}^2$. Proc. Symp. Pure Math., **45** (1986), Part 2, p. 1–17.

[241] T. Kato, *Strong solutions of the Navier–Stokes equations in Morrey spaces.* Boll. Soc. Brasil Mat. (N.S.), **22-2** (1992), p. 127–155.

[242] T. Kato, *The Navier–Stokes equation for an incompressible fluid in* $\mathbb{R}^2$ *with a measure as the initial vorticity.* Diff. Int. Eq., **7** (1994), p. 949–966.

[243] T. Kato, *A remark on a theorem of C. Bardos on the 2D Euler equation.* Preprint.

[244] T. Kato and H. Fujita, *On the non-stationary Navier–Stokes system.* Rend. Sem. Mat. Univ. Padova, **32** (1962), p. 243–260.

[245] T. Kato and G. Ponce, *Well-posedness of the Euler and Navier–Stokes equations in the Lebesgue spaces* $L_s^p(\mathbb{R}^2)$. Revista Mat. Iberoamer., **2** (1986), p. 73–88.

[246] T. Kato and G. Ponce, *On nonstationary flows of viscous and ideal fluids in $L_s^p(\mathbb{R}^2)$*. Duke Math. J., **55** (1987), p. 487–499.

[247] T. Kato and G. Ponce, *Commutator estimates and the Euler and Navier–Stokes equations*. Comm. Pure Appl. Math., **41** (1988), p. 891–907.

[248] S. Kawashima, *Asymptotic behavior of solutions to the equations of a viscous gas*. Preprint, 1995.

[249] S. Kawashima and A. Matsumura, *Asymptotic stability of travelling wave solutions of systems for one-dimensional gas motion*. Comm. Math. Phys., **101** (1985), p. 97–127.

[250] S. Kawashima, A. Matsumura and T. Nishida, *On the fluid dynamical approximation to the Boltzmann equation at the level of the Navier–Stokes equation*. Comm. Math. Phys., **70** (1979), p. 97–124.

[251] S. Kawashima and T. Nishida, *Global solutions to the initial value problem for the equations of one-dimensional motion of viscous polytropic gases*. J. Math. Kyoto Univ., **21** (1981), p. 825–837.

[252] S. Kawashima and M. Okada, *On the equations of one-dimensional motion of compressible viscous fluids*. J. Math. Kyoto Univ., **23** (1983), p. 55–71.

[253] B. Kawohl, *Global existence of large solutions to initial boundary value problems for a viscous, heat-conducting, one-dimensional real gas*. J. Diff. Eq., **58** (1985), p. 76–103.

[254] A.V. Kazhikov, *Resolution of boundary value problems for nonhomogeneous viscous fluids*. Dokl. Akad. Nauh., **216** (1974), p. 1008–1010 (in Russian).

[255] A.V. Kazhikov, *To a theory of boundary value problems for equation of one-dimensional nonstationary motion of viscous heat-conduction gases, boundary value problems for hydrodynamical equations*. Inst. Hydrodynamics, Siberian Branch Akad. USSR. N. **50** (1981), p. 37–62 (in Russian).

[256] A.V. Kazhikov, *Cauchy problem for viscous gas equations*. Sibirskii Mat. Zh., **23** (1982), p. 60–64.

[257] A.V. Kazhikov, *The equation of potential flows of compressible viscous fluid at low Reynolds number: existence, uniqueness and stabilization of solution*. Sib. Mat. J., **34** (1993), p. 70–80 (in Russian).

[258] A.V. Kazhikov, *Some new statements for initial boundary value problems for Navier–Stokes equations of viscous gas.* In <u>Progress in theoretical and computational fluid mechanics</u>. Eds. G.P. Galdi, J. Málek and J. Nečas, Longman, Harlow, 1994.

[259] A.V. Kazhikov and V.V. Shelukhin, *Unique global solution with respect to time of the initial boundary value problems for one-dimensional equations of a viscous gas.* J. Appl. Math. Mech., **41** (1977), p. 273–282.

[260] A.V. Kazhikov and S.H. Smagulov, *The correctness of boundary-value problems in a diffusion model of an inhomogeneous liquid.* Sov. Phys. Dokl., **22** (1977), p. 249–259.

[261] A.V. Kazhikov and S.H. Smagulov, *The correctness of boundary-value problems in a certain diffusion model of an inhomogeneous fluid.* Čisl. Metody Meh. Splošn. Sredy, **7** (1978), p. 75–92 (in Russian).

[262] R. Kerr, *Evidence for a singularity of the three-dimensional, incompressible Euler equations.* Phys. of Fluids A., vol. 5, **7** (1993), p. 1725–1746.

[263] B.A. Khesin and Yu.V. Chekanov, *Invariants of the Euler equations for ideal and barotropic hydrodynamics and superconductivity in D dimensions.* Physica D, **40** (1989), p. 119–131.

[264] U.J. Kim, *Weak solutions of an initial boundary value problem for an incompressible viscous fluid with nonnegative density.* SIAM J. Math. Anal., **18** (1987), p. 89-96.

[265] S. Klainerman and A. Majda, *Singular limits of quasilinear hyperbolic systems with large parameters and the incompressible limit of compressible fluids.* Comm. Pure Appl. Math., **34** (1981), p. 481–524.

[266] S. Klainerman and A. Majda, *Compressible and incompressible fluids.* Comm. Pure Appl. Math., **35** (1982), p. 629–653.

[267] T. Kobayashi and T. Muramatu, *Abstract Besov space approach to the non-stationary Navier–Stokes equations.* Math. Methods Appl. Sci., **15** (1992), p. 599–620.

[268] H. Kozono and T. Ogawa, *Two-dimensional Navier–Stokes flow in unbounded domains.* Math. Ann., **297** (1993), p. 1–31.

[269] H. Kozono and T. Ogawa, *Decay properties of strong solutions for the Navier–Stokes equations in two-dimensional unbounded domains.* Arch. Rat. Mech. Anal., **122** (1993), p. 1–17.

[270] H. Kozono and H. Sohr, *New a priori estimates for the Stokes equations in exterior domains.* Indiana Univ. Math. J., **40** (1991), p. 1–28.

[271] H. Kozono and H. Sohr, *On a new class of generalized solutions for the Stokes equations in exterior domains.* Ann. Scuola Norm. Pisa, **4** (1992).

[272] H. Kozono and H. Sohr, *On stationary Navier–Stokes equations in unbounded domains.* To appear in Ricerche Mat.

[273] H. Kozono and M. Yamasaki, *Semilinear heat equations and the Navier–Stokes equation with distributions in new function spaces as initial data.* Comm.P.D.E., **19** (1994), p. 959–1014.

[274] H. Kozono and M. Yamasaki, *The exterior problem for the non-stationary Navier–Stokes equation with data in the space $L^{n,\infty}$.* C.R. Acad. Sci. Paris, **320** (1995), p. 685–690.

[275] H. Kozono and M. Yamasaki, *The stability of small stationary solutions in Morrey spaces of the Navier–Stokes equation.* Preprint, 1995.

[276] G.A. Kuz'min, *Ideal incompressible hydrodynamics in terms of the vortex momemtum density.* Phys. Lett. **96 A** (1983), p. 88–90.

[277] O.A. Ladyzhenskaya, The mathematical theory of viscous incompressible flow. Gordon and Breach, London, 1969.

[278] O.A. Ladyzhenskaya, *Unique solvability in the large of three-dimensional Cauchy problems for the Navier–Stokes equations in the presence of axial symmetry.* In Boundary value problems of mathematical physics and related aspects of function theory, II, Steklov Inst. Seminars in Mathematics, Leningrad, vol. 7, translated by Consultants Bureau, New York, 1970.

[279] O.A. Ladyzhenskaya and V.A. Solonnikov, *Unique solvability of an initial and boundary-value problem for viscous incompressible nonhomogeneous fluids.* J. Soviet. Math., **9** (1978), p. 697–749.

[280] H. Lamb, Hydrodynamics. Dover, New York, 1945.

[281] L. Landau and E. Lifschitz, Mécanique des fluides. 2nd edition, Editions MIR, Moscow, 1989.

[282] P.D. Lax, *Hyperbolic systems of conservation laws. II.* Comm. Pure Appl. Math., **10** (1957), p. 537–566.

[283] J. Leray, *Etude de diverses équations intégrales nonlinéaires et de quelques problèmes que pose l'hydrodynamique.* J. Math. Pures Appl., **12** (1933), p. 1–82.

[284] J. Leray, *Essai sur les mouvements plans d'un liquide visqueux que limitent des parois.* J. Math. Pures Appl., **13** (1934), p. 331-418.

[285] J. Leray, *Essai sur le mouvement d'un liquide visqueux emplissant l'espace.* Acta Math., **63** (1934), p. 193-248.

[286] M. Lesieur, Turbulence in fluids. Martinus Nijhoff, Dordrecht, 1987.

[287] M. Lesieur, Turbulence in fluids. 2nd edition, Kluwer, Dordrecht, 1990.

[288] L. Lichtenstein, *Über einige Existenzprobleme der Hydrodynamik homogener unzusammendrückbarer, reibunglosser Flüssikeiten und die Helmholtzschen Wirbelsalitze.* Mat. Zeit., **23** (1925), p. 89-154; **26** (1927), p. 193-323, p. 387-415, p. 725; **32** (1930), p. 608.

[289] J. Lighthill, Waves in fluids. Cambridge University Press, Cambridge, 1978.

[290] C.C. Lin, The theory of hydrodynamic stability. Cambridge University Press, Cambridge, 1955.

[291] C.K. Lin, *On the incompressible limit of compressible Navier–Stokes equations.* Comm. P.D.E., **20** (1995), p. 677-707.

[292] J.L. Lions, *Sur la régularité et l'unicité des solutions turbulentes des équations de Navier–Stokes.* Rend. Sem. Mat. Padova, **30** (1960), p. 16-23.

[293] J.L. Lions, Quelques méthodes de résolution des problèmes aux limites nonlinéaires. Dunod, Paris, 1969.

[294] J.L. Lions, *Perturbations singulières dans les problèmes aux limites et en contrôle optimal.* L. N. in Mathematics **323**, Springer, Berlin (1973).

[295] J.L. Lions and G. Prodi, *Un théorème d'existence et d'unicité dans les équations de Navier–Stokes en dimension 2.* C.R. Acad. Sci. Paris, **248** (1959), p. 3519-3521.

[296] J.L. Lions, R. Temam and S. Wang, *On the equations of the large-scale ocean.* Nonlinearity, **5** (1992), p. 1007-1053.

[297] J.L. Lions, R. Temam and S. Wang, *New formulations of the primitive equations of atmosphere and applications.* Nonlinearity, **5** (1992), p. 237-288.

[298] J.L. Lions, R. Temam and S. Wang, *Modèles et analyse mathématiques du système Océan/Atmosphère. I. Structure des sous-systèmes.* C.R. Acad. Sci. Paris, **316** (1993), p. 113-119.

[299] J.L. Lions, R. Temam and S. Wang, *Modèles et analyse mathéma-tiques du système Océn/Atmosphère. II. Couplage.* C.R. Acad. Sci. Paris, **316** (1993), p. 211–215.

[300] J.L. Lions, R. Temam and S. Wang, *Models for the coupled atmos-phere and ocean.* In Comput. Mech. Advances, vol. 1, Ed. J.T. Oden, Elsevier, Amsterdam, 1993, p. 1–112.

[301] J.L. Lions, R. Temam and S. Wang, *Numerical analysis of coupled atmosphere and ocean models.* Preprint, 1994.

[302] J.L. Lions, R. Temam and S. Wang, *Mathematical theory for coupled atmosphere and ocean models.* Preprint, 1994.

[303] P.L. Lions, *Existence globale de solutions pour les équations de Navier–Stokes compressibles isentropiques.* C.R. Acad. Sci. Paris, **316** (1993), p. 1335–1340.

[304] P.L. Lions, *Compacité des solutions des équations de Navier–Stokes compressibles isentropiques.* C.R. Acad. Sci. Paris, **317** (1993), p. 115–120.

[305] P.L. Lions, *Limites incompressibles et acoustique pour des fluides visqueux compressibles isentropiques.* C.R. Acad. Sci. Paris, **317** (1993), p. 1197–1202.

[306] P.L. Lions, *Compactness in Boltzmann's equation via Fourier integral operators.* J. Math. Kyoto Univ., I, **34** (1994), p. 391–428; II, **34** (1994), p. 429–462; III, **34** (1994), p. 539–584.

[307] P.L. Lions, *Remarks on incompressible models in fluid mechanics.* Preprint, 1993.

[308] P.L. Lions and F. Murat, work in preparation.

[309] P.L. Lions and P. Orenga, work in preparation.

[310] P.L. Lions, B. Perthame and P.E. Souganidis, *Existence and compactness of entropy solutions for the one-dimensional isentropic gas dynamics systems.* To appear in Comm. Pure Appl. Math.

[311] P.L. Lions, B. Perthame and E. Tadmor, *Kinetic formulation of the isentropic gas dynamics and p-systems.* Comm. Math. Phys., **163** (1994), p. 415–431.

[312] P.L. Lions, B. Perthame and E. Tadmor, *A kinetic formulation of multi-dimensional scalar conservation laws and related equations.* J. Amer. Math. Soc., **7** (1994), p. 169–191.

[313] T.P. Liu, *Nonlinear stability of shock waves for viscous conservation laws.* Preprint, 1994.

[314] D. Lyndon-Bell and J.A. Kalnajs, *On the generating mechanisms of spiral structure.* Mon. Not. R. Astron. Soc., **157** (1974), p. 1–30.

[315] A. Majda, Compressible fluid flow and systems of conservation laws in several space dimensions. Appl. Math. Sci. ♯ 53, Springer, Berlin, 1984.

[316] A. Majda, *Vorticity and the mathematical theory of incompressible fluid flow.* Comm. Pure Appl. Math., **39** (1986), p. 187–220.

[317] A. Majda, *The interaction of nonlinear analysis and modern applied mathematics.* Proc. ICM Kyoto 1990, vol. 1, Springer, Tokyo, (1991), p. 175–191.

[318] A. Majda, *Remarks on weak solutions for vortex sheets with a distinguished sign.* Ind. Univ. Math. J., **42** (1993), p. 921–939.

[319] T. Makino, *On a local existence theorem for the evolution equation of gaseous stars.* In Patterns and waves. Ed. T. Nishida, M. Mimura and H. Fujii. Kinokuniya/North Holland, Amsterdam, 1991.

[320] T. Makino and N. Okada, *Free boundary problem for the equation of spherically symmetric motion of viscous gas.* Preprint, 1995.

[321] P. Manneville, Dissipative structures and weak turbulence. Academic Press, Boston, 1990.

[322] C. Marchioro and M. Pulvirenti, *Hydrodynamics in two dimensions and vortex theory.* Comm. Math. Phys., **84** (1982), p. 483–503.

[323] C. Marchioro and M. Pulvirenti, Vortex methods in two-dimensional fluid dynamics. Springer Lecture Notes in Physics ♯ 203, Berlin, 1984.

[324] C. Marchioro and M. Pulvirenti, Mathematical theory of incompressible non-viscous fluids. Appl. Math. Sci. ♯ 96, Springer, New York, 1994.

[325] G.I. Marchuk, Numerical solution of problems of atmospheric and ocean dynamics. Gidrometeoizdat, Leningrad, 1974 (in Russian).

[326] P. Maremonti, *On the asymptotic behaviour of the L^2-norm of suitable weak solutions to the Navier–Stokes equations in three-dimensional exterior domains.* Comm. Math. Phys., **118** (1988), p. 335–400.

[327] J.E. Marsden, *Well-posedness of the equations of a non-homogeneous perfect fluid.* Comm. P.D.E., **1** (1976), p. 215–230.

[328] V.N. Maslennikova and M.A. Timoshin, L^p theory for the Stokes system in exterior domains. Proc. Symposium on new methods in analysis and differential equations, University of Voronez Press, (1989), p. 63–77 (in Russian).

[329] V.N. Maslennikova and M.A. Timoshin, L^p theory for the Stokes equations in unbounded domains with a compact boundary. Dokl. Akad. Nauk SSSR, **313** (1990), p. 1341–1346 (in Russian).

[330] K. Masuda, *Weak solutions of Navier–Stokes equations.* Tôhoku Math. J., **36** (1984), p. 623–646.

[331] A. Matsumura and T. Nishida, *The initial value problem for the equations of motion of compressible and heat-conductive fluids.* Proc. Japan Acad., **A 55** (1979), p. 337–342.

[332] A. Matsumura and T. Nishida, *The initial value problem for the equations of motion of viscous and heat-conductive gases.* J. Math. Kyoto Univ., **20** (1980), p. 67–104.

[333] A. Matsumura and T. Nishida, *Initial boundary value problems for the equations of motion of general fluids.* Computing Meth. in Appl. Sci. and Engin. V (R. Glowinski and J.L. Lions, eds.), North-Holland, Amsterdam, 1982, p. 389–406.

[334] A. Matsumura and T. Nishida, *Initial boundary value problems for the equations of motion of compressible and heat-conductive fluids.* Comm. Math. Phys., **89** (1983), p. 445–464.

[335] A. Matsumura and T. Nishida, *Initial boundary value problems for the equations of motion of compressible viscous fluids.* Contemporary Mathematics, vol. 17, Nonlinear Partial Differential Equations, ed. J. Smoller, Amer. Math. Soc., Providence, 1983.

[336] A. Matsumura and T. Nishida, *On the stability of travelling wave solutions of a one-dimensional model system for compressible viscous gas.* Japan J. Appl. Math., 1985.

[337] A. Matsumura and T. Nishida, *Asymptotics towards the rarefaction waves of the solutions of a one-dimensional model system for compressible viscous gas.* Preprint, 1994.

[338] M. McCracken, *The resolvent problem for the Stokes equations on half spaces in L_p.* SIAM J. Math. Anal., **12** (1981), p. 201–228.

[339] F.J. McGrath, *Nonstationary plane flow of viscous and ideal fluids.* Arch. Rat. Mech. Anal., **27** (1968), p. 328–348.

[340] J.C. McWilliams and B.R. Gent, *Intermediate models of planetary circulations in the atmosphere and ocean.* J. Atmos. Sci., **37** (1980), p. 1657–1678.

[341] A.B. Mergulis, *On existence of two dimensional nonstationary flows of an ideal incompressible fluid admitting a curl nonsummable to any power greater than 1.* Siberian J. Math., **33** (1992), p. 934–937.

[342] N. Meyer-Vernet and B. Sicardy, *On the physics of resonant disk-satellite interaction.* Icarus, **69** (1987), p. 157–175.

[343] M. Michaux and J.M. Rakotoson, *Remarks on Navier–Stokes equation with measures as data.* Appl. Math. Lett., **6-6** (1993), p. 75–77.

[344] L.M. Milne-Thomson, Theoretical hydrodynamics. Macmillan, New York, 1960.

[345] T. Miyakawa, *On nonstationary solutions of the Navier–Stokes equations in exterior domains.* Hiroshima Math. J., **12** (1982), p. 119–140.

[346] T. Miyakawa and H. Sohr, *On energy inequality, smoothness and large time behavior in L^2 for weak solutions of the Navier–Stokes equations in exterior domains.* Math. Z., **199** (1988), p. 455–478.

[347] T. Miyakawa and M. Yamada, *Planar Navier–Stokes flows in a bounded domain with measures as initial vorticities.* Hiroshima Math. J., **22** (1992), p. 401–420.

[348] A.S. Monin and A.M. Yaglom, Statistical fluid mechanics. M.I.T. Press, Cambridge, 1975.

[349] F. Murat, *Compacité par compensation.* Ann. Sc. Norm. Sup. Pisa, **5** (1978), p. 489–507.

[350] F. Murat, *Compacité par compensation, II.* In Proceedings of the International Meeting on Recent Methods on Nonlinear Analysis, E. De Giorgi, E. Magenes, U. Mosco eds., Pitagora, Bologna, 1979.

[351] F. Murat, *Compacité par compensation, III.* Ann. Sc. Norm. Sup. Pisa, **8** (1981), p. 69–102.

[352] T. Nagasawa, *On the one-dimensional motion of the polytropic ideal gas non-fixed on the boundary.* J. Diff. Eq., **65** (1986), p. 49–67.

[353] T. Nagasawa, *On the outer pressure problem of the one-dimensional polytropic ideal gas.* Japan J. Appl. Math., **5** (1988), p. 53–85.

[354] T. Nagasawa, *On the one-dimensional free boundary problem for the heat-conductive compressible viscous gas.* Lecture Notes in

Num. Appl. Anal. (M. Mimura and T. Nishida, eds.), vol. 10,
Kinokuniya/ North-Holland, Tokyo, 1989, p. 83–99.

[355] J. Nash, *Le problème de Cauchy pour les équations différentielles
d'un fluide général.* Bull. Soc. Math. France, **90** (1962), p.
487–497.

[356] C.L.M.H. Navier, *Mémoire sur les lois du mouvement des fluides.*
Mém. Acad. Sci. Inst. France, **6** (1822), p. 375–394.

[357] V.B. Nikolaev, *On the solvability of mixed problem for one-dimen-
sional axisymmetrical viscous gas flow.* Dinamicheskie zadachi
Mekhaniki sploshnoj sredy, **63** Sibirsk. Otd. Acad. Nauk SSSR,
Inst. Gidrodinamiki, 1983 (Russian).

[358] T. Nishida, *Equations of motion of compressible viscous fluids.* In
Patterns and waves. Ed. T. Nishida, M. Mimura and H. Fujii.
Kinokuniya/North-Holland, Amsterdam, 1991.

[359] T. Nishida and M. Okada, *Free boundary problems for the equation
of a one-dimensional motion of viscous gas.* Preprint, 1994.

[360] A. Nouri and F. Poupaud, *An existence theorem for the multifluid
Navier–Stokes problem.* Preprint, 1994.

[361] A. Nouri, F. Poupaud and Y. Demay, *An existence theorem for the
multifluid Stokes problem.* Preprint, 1994.

[362] A. Novótny, *Steady flows of viscous compressible fluids in exterior
domains under small perturbations of great potential forces.* M3AS,
3 (1993), p. 725–757.

[363] A. Novótny, *Steady flows of viscous compressible fluids.* L^2
approach. Preprint, 1994.

[364] A. Novótny and M. Padula, L^p-approach to steady flows of viscous
compressible fluids in exterior domains. Arch. Rat. Mech. Anal.,
126 (1994), p. 243–297.

[365] A. Novótny and M. Padula, *On the asymptotic decay of gradient of
velocity for steady-state Navier–Stokes equations.* Preprint, 1994.

[366] A. Novótny and M. Padula, *Physically reasonable solutions to
steady compressible Navier–Stokes equations in 3D-exterior domains
I ($v_\infty = 0$).* Preprint, 1994.

[367] A. Novótny and P. Penel, L^p-approach to steady flows of viscous
compressible heat conductive fluids in exterior domains. I. Preprint,
1995.

[368] F.K.G. Odqvist, *Beiträge zur Theorie der nichtstationären zähen Flüssigkeitsbewegungen. I.* Ark. Mat. Astr. Pys., **28** (1932), p. 1–22.

[369] T. Ohyama, *Interior regularity of weak solutions of the time-dependent Navier–Stokes equation.* Proc. Japan Acad., **36** (1960), p. 273–277.

[370] M. Okada, *The free boundary value problem for the equations of one-dimensional motion of compressible viscous fluids.* Preprint, 1994.

[371] P. Orenga, *work in preparation.*

[372] C.W. Oseen, Hydrodynamik. Leipzig, 1927.

[373] V.I. Oseledets, *New form of the Navier–Stokes equation. Hamiltonian formalism.* (In Russian). Moskov. Matemat. Obshch. **44** n.3 (267) (1989), p. 169–170.

[374] M. Padula, *Existence and continuous dependence for solutions to the equations of a one-dimensional model in gas dynamics.* Meccanica, (1981), p. 128–135.

[375] M. Padula, *On the uniqueness of viscous, compressible, steady flows.* In Trends in applications of pure mathematics in mechanics. vol. IV, Pitman, London, 1982.

[376] M. Padula, *Uniqueness theorems for steady, compressible, heat-conducting fluids.* I, Acc. Naz. Lincei, **LXXIV** (1983), p. 380–387; II, Acc. Naz. Lincei, **LXXV** (1983), p. 56–60.

[377] M. Padula, *Nonlinear energy stability for the compressible Bénard problem.* Bull. U.M.I., **6** (1986), p. 581–602.

[378] M. Padula, *Existence and uniqueness for viscous, steady compressible motions.* Arch. Rat. Anal., **77** (1987), p. 89–102.

[379] M. Padula, *An existence theorem for non-homogeneous motions in exterior domains.* Math. Z., **203** (1990), p. 581–604.

[380] M. Padula, *On the existence and uniqueness of non-homogeneous motions in exterior domains.* Math. Z., **203** (1990), p. 581–604.

[381] M. Padula, *A representation formula for the steady solutions of a compressible fluid moving at low speed.* Transport Theory Sta. Phys., **21** (1992), p. 593–613.

[382] M. Padula, *Stability properties of regular flows of heat-conducting compressible fluids.* J. Math. Kyoto Univ., **32** (1992), p. 401–442.

[383] M. Padula, *On the exterior steady problem for the equations of a viscous isothermal gas.* Orw. Carol. Praha.

[384] M. Padula, *Mathematical properties of motions of viscous compressible fluids.* In Progress in theoretical and computational fluid mechanics. Eds. G.P. Galdi, J. Málek and J. Nečas, Longman, Harlow, 1994.

[385] M. Padula and C. Pileckas, *Steady flows of a viscous ideal gas in domains with non-compact boundaries: existence and asymptotic behavior in a pipe.* Preprint, 1994.

[386] J. Pedlosky, Geophysical fluid dynamics. Springer, Berlin, 1979.

[387] R. Peyret and T.D. Taylor, Computational methods for fluid flow. Springer, Berlin, 1983.

[388] O. Pironneau, Finite element methods for fluids. Wiley, New York, 1989.

[389] F. Planchon, *Solutions globales pour les équations de Navier–Stokes à valeurs dans $H^s(\mathbb{R}^3)$ ou $L^p(\mathbb{R}^3)$.* Preprint, 1994.

[390] S.D. Poisson, *Mémoire sur les équations générales de l'équilibre et du mouvement des corps solides élastiques et des fluides.* Jour. de l'Ecole Polytechnique, **13** (1831), p. 1–74.

[391] G. Ponce, *Remarks on a paper by J.T. Beale, T. Kato and A. Majda.* Comm. Math. Phys., **98** (1985), p. 349–353.

[392] G. Ponce, *On two dimensional incompressible fluids.* Comm. P.D.E., **11** (1986), p. 483–511.

[393] L. Prandtl and O.E. Tietjens, Applied hydro- and aeromechanics. Dover, New York, 1957.

[394] G. Prodi, *Un teorema di unicità per le equazioni di Navier–Stokes.* Ann. di Mat., **48** (1959), p. 173–182.

[395] G. Prodi, *Qualche risultato riguardo alle equazioni di Navier–Stokes nel caso bidimensionale.* Rem. Sem. Mat. Padova, **30** (1960), p. 1–15.

[396] A. Pumir and E.D. Siggia, *Finite time singularities in the axisymmetric three dimensional Euler equations.* Phys. Rev. Lett., **68** (1992), p. 1511.

[397] A. Pumir and E.D. Siggia, *Developemnt of singular solutions to the axisymmetric Euler equations.* Phys. Fluids A, **4** (1992), p. 1472.

[398] J.L. Rhyming, Dynamique des fluides. Presses polytechniques romandes, Lausanne, 1985.

[399] L.F. Richardson, Weather prediction by numerical process. Cambridge Univ. Press, Cambridge, 1922.

[400] B.L. Rozhdestvenskii and N.N. Yanenko, Systems of quasilinear equations and their applications to gas dynamics. 2nd ed., Nauka, Moscow, 1978 (in Russian); English transl., Amer. Math. Soc., Providence, R.I., 1983.

[401] X. Saint Raymond, *Régularité stratifiée et équation d'Euler 3D à temps grand.* Preprint, 1995.

[402] J. Salençon, Mécanique des milieux continus. Ellipses. Ecole Polytechnique, Palaiseau, 1988.

[403] R. Salvi, *On the existence of weak solutions of a nonlinear mixed problem for the Navier–Stokes equations in a time dependent domain.* J. Fac. Sci. Univ. Tokyo, **22** (1985), p. 213–221.

[404] V. Scheffer, *Turbulence and Hausdorff dimension.* In Turbulence and the Navier–Stokes equations. Lecture Notes in Math. ♮565, Springer (1976), p. 94–112.

[405] V. Scheffer, *Hausdorff measure and the Navier–Stokes equations.* Comm. Math. Phys., **55** (1977), p. 97–112.

[406] V. Scheffer, *The Navier–Stokes equations in space dimension four.* Comm. Math. Phys., **61** (1978), p. 41–68.

[407] V. Scheffer, *The Navier–Stokes equations in a bounded domain.* Comm. Math. Phys., **73** (1980), p. 1–42.

[408] H. Schlichting, Boundary-layer theory. 7th edition, McGraw-Hill, New York, 1987.

[409] S. Schochet, *The compressible Euler equations in a bounded domain. Existence of solutions and the incompressible limit.* Comm. Math. Phys., **104** (1986), p. 49–75.

[410] S. Schochet, *The weak vorticity formulation of the 2D Euler equations and concentration-cancellation.* Preprint, 1995.

[411] S. Schochet, *Point-vortex methods for weak solutions of the 2D Euler equations.* Preprint, 1995.

[412] M.E. Schonbek, L^2 *decay for weak solution of the Navier–Stokes equations.* Arch. Rat. Mech. Anal., **88** (1985), p. 209–222.

[413] M.E. Schonbek, *Large time behavior of solutions to the Navier–Stokes equations.* Comm. P.D.E., **11** (1986), p. 733–763.

[414] P. Secchi, *On a stationary problem for the compressible Navier–Stokes equations; the self-gravitating equilibrium solutions.* J. Diff. Int. Eq., **7** (1994), p. 463–482.

[415] L. Sedov, <u>Mécanique des milieux continus.</u> vol. I. MIR, Moscou, 1975.

[416] Ph. Serfati, *Etude mathématique de flammes infiniment minces en combustion. Résultats de structure et de régularité pour l'équation d'Euler incompressible.* Thèse de Doctorat, Univ. Paris VI, 1992.

[417] Ph. Serfati, *Structures holomorphes à faible régularité spatiale en mécanique des fluides.* Preprint, 1993.

[418] Ph. Serfati, *Vortex patches et régularité stratifiée pour le Laplacien.* Preprint, 1993.

[419] Ph. Serfati, *Une preuve directe d'existence globale des vortex patches 2D.* Preprint, 1993.

[420] Ph. Serfati, *Régularité stratifiée et équation d'Euler 3D à temps grand.* Preprint, 1993.

[421] Ph. Serfati, *Régularité C^∞ en temps de solutions sous-lipschitz de l'équation d'Euler.* Preprint, 1993.

[422] Ph. Serfati, *Pertes de régularité pour le laplacien et l'équation d'Euler sur $\mathbb{R}^n$.* Preprint, 1994.

[423] D. Serre, *Invariants et dégénérescence symplectique de l'équation d'Euler des fluides parfaits incompressibles.* C.R. Acad. Sc. Paris, **298** (1984), p. 349–352.

[424] D. Serre, *Solutions faibles globales des équations de Navier–Stokes pour un fluide compressible.* C.R. Acad. Sci. Paris, **303** (1986), p. 629–642.

[425] D. Serre, *Sur l'équation monodimensionnelle d'un fluide visqueux, compressible et conducteur de chaleur.* C.R. Acad. Sci. Paris, **303** (1986), p. 703–706.

[426] D. Serre, *Variations de grande amplitude pour la densité d'un fluide visqueux compressible.* Physica D, **48** (1991), p. 113–128.

[427] J. Serrin, *On the interior regularity of weak solutions of the Navier–Stokes equations.* Arch. Rat. Mech. Anal., **9** (1962), p. 187–195.

[428] J. Serrin, *The initial value problem for the Navier–Stokes equations.* In Nonlinear problems. Ed. R.E. Langer, Univ. of Wisconsin Press, p. 69–80, 1963.

[429] J. Serrin, *Mathematical principles of classical fluid mechanics.* In Handbuch der Physik VIII/1, Springer, Berlin, 1972, p. 125–262.

[430] A.H. Shapiro, The dynamics and thermodynamics of compressible flow. Ronald Press, New York, 1953.

[431] V.V. Shelukhin, *On the structure of generalized solutions of the one-dimensional equations of a polytropic viscous gas.* Prikl. Matem. Mekhan., **48** (1984), p. 912–920.

[432] V.V. Shelukhin, *The problem of predicting the temperature of the ocean from average data over a preceding period of time.* Russian Acad. Sci. Dokl. Math., **45** (1992), p. 644–648.

[433] C.G. Simader, *Mean value formulas, Weyl's lemma and Liouville theorem for* Δ^2 *and Stokes' system.* Resultate der Mathematik, **22** (1992), p. 761–780.

[434] C.G. Simader and H. Sohr, The weak and strong dirichlet problem for Δ in L^q in bounded and exterior domains. Pitman Research Notes in Mathematics. To appear.

[435] J. Simon, *Ecoulement d'un fluide non-homogène avec une densité initiale s'annulant.* C.R. Acad. Sci. Paris, **15** (1978), p. 1009–1012.

[436] J. Simon, *Sur les fluides visqueux incompressibles et non homogènes.* C.R. Acad. Sci. Paris, **309** (1989), p. 447–452.

[437] J. Simon, *Non-homogeneous viscous incompressible fluids; existence of velocity, density and pressure.* SIAM J. Math. Anal., **21** (1990).

[438] H. Sohr, *Zur Regularitätstheorie der instationären Gleichungen von Navier–Stokes.* Math. Z., **184** (1983), p. 359–375.

[439] H. Sohr and W. Varnhorn, *On decay properties of the Stokes equations in exterior domains.* In Navier–Stokes equations: theory and numerical methods. (J.G. Heywood, K. Masuda, R. Rautmann and V.A. Solonnikov, eds.) Springer Lecture Notes in Mathematics, Vol. **1431** (1990), p. 134–151.

[440] H. Sohr and W. Von Wahl, *On the singular set and the uniqueness of weak solutions of the Navier–Stokes equations.* Manus. Math., **49** (1984), p. 27–59.

[441] H. Sohr and W. Von Wahl, *A new proof of Leray's structure theorem and the smoothness of weak solutions of Navier–Stokes*

equations for large $|x|$. Bayreuth Math. Schr., **20** (1985), p. 153–204.

[442] H. Sohr and W. Von Wahl, *On the regularity of the pressure of weak solutions of Navier–Stokes equations.* Arch. Math., **46** (1986), p. 428–439.

[443] H. Sohr, W. Von Wahl and M. Wiegner, *Zur Asymptotik der Gleichungen von Navier–Stokes.* Nachr. Akad. Wiss. Göttingen, **3** (1986), p. 1–15.

[444] V.A. Solonnikov, *Estimates of the solutions of a nonstationary linearized system of Navier–Stokes equations.* Trudy Mat. Inst. Steklov, vol. 70, 1964, In Amer. Math. Soc. Transl., Series 2, vol. 75, p. 1–17.

[445] V.A. Solonnikov, *A priori estimates for second-order parabolic equations.* Trudy Mat. Inst. Steklov, vol. 70, 1964, In Amer. Math. Soc. Transl., Series 2, vol. 65, p. 51–137.

[446] V.A. Solonnikov, *Estimates for solutions of nonstationary Navier–Stokes systems.* Zap. Nauchn. Sem. L.O.M.I. **38** (1973), p. 153–231; J. Soviet Math. **8** (1977), p. 467–529.

[447] V.A. Solonnikov, *Estimates of solutions of an initial- and boundary-value problem for the linear nonstationary Navier–Stokes system.* Zap. Nauchn. Sem. L.O.M.I. **59** (1976), p. 172–254; J. Soviet Math. **10** (1978), p. 336–393.

[448] V.A. Solonnikov, *Solvability of the problem of evolution of an isolated volume of viscous incompressible capillary fluid.* J. Soviet Math., **32** (1986), p. 223–228.

[449] V.A. Solonnikov, *On the evolution of an isolated volume of viscous incompressible capillary fluid for large values of time.* Vestnik Leningrad Univ. Math., **15** (1987), p. 52–58.

[450] V.A. Solonnikov, *On the transient motion of an isolated volume of viscous incompressible fluid.* Math. USSR Izv., **31** (1988), p. 381–405.

[451] V.A. Solonnikov, *Unsteady motion of a finite mass of fluid, bounded by a free surface.* J. Soviet Math., **40** (1988), p. 672–686.

[452] V.A. Solonnikov, *Solvability of a problem on the motion of a viscous incompressible fluid bounded by a free surface.* Math. USSR Izv., **31** (1988), p. 381–405.

[453] V.A. Solonnikov, *On a nonstationary motion of a finite isolated mass of self gravitating fluid*. Algebra and Anal., **1** (1989), p. 207–246.

[454] E. Stein, *On the functions of Littlewood–Paley, Lusin and Marcinkiewicz*. Trans. Amer. Math. Soc., **88** (1958), p. 430–466.

[455] E. Stein and G. Weiss, *On the theory of H^p spaces*. Acta Math., **103** (1960), p. 25–62.

[456] G.G. Stokes, *On the theories of internal friction of fluids in motion and of the equilibrium and motion of elastic solids*. Trans. Camb. Phil. Soc., **8** (1849), p. 207–319.

[457] I. Straškraba and A. Valli, *Asymptotic behaviour of the density for one-dimensional Navier–Stokes equations*. Manuscripta Math., **62** (1988), p. 401–416.

[458] M. Struwe, *Regular solutions of the stationary Navier–Stokes equations on $\mathbb{R}^5$*. Preprint, 1995.

[459] H. Swann, *The convergence with vanishing viscosity of nonstationary Navier–Stokes flow to ideal flow in $\mathbb{R}_3$*. Trans. Amer. Math. Soc., **157** (1971), p. 373–397.

[460] H.L. Swinney and J.P. Gollub, Hydrodynamic instabilities and the transition to turbulence. Springer, Berlin, 1981.

[461] G. Sylvester, *Large-time existence of small viscous surface waves without surface tension*. Comm. P.D.E., **15** (1990), p. 823–903.

[462] G. Talenti, *Elliptic equations and rearrangements*. Ann. Sc. Norm. Sup. Pisa, **3**(1976), p. 697–718.

[463] N. Tanaka, *Global existence of two-phase nonhomogeneous viscous incompressible fluid flow*. Preprint.

[464] A. Tani, *On the first initial boundary value problem of compressible viscous fluid motion*. Publ. Res. Math. Sci. Kyoto Univ., **13** (1977), p. 193–253.

[465] A. Tani, *Two-phase free boundary problem for compressible viscous fluid motion*. J. Math. Kyoto Univ., **24** (1984), p. 243–267.

[466] A. Tani, *Multiphase free boundary problem for the equation of motion of general fluids*. Comm. Math. Univ. Carolinae, **26** (1985), p. 201–208.

[467] L. Tartar, Topics in nonlinear analysis. Publications Mathématiques d'Orsay, Université Paris-Sud Orsay, 1978.

[468] L. Tartar, *Compensated compactness and applications to partial differential equations.* In Nonlinear analysis and mechanics, Heriot-Watt Symposium, IV. Pitman, London, 1979.

[469] L. Tartar, In Systems of nonlinear partial differential equations. Reidel, Dordrecht, 1983.

[470] L. Tartar, In Macroscopic modelling of turbulent flows. Lecture Notes in Physics ♯ 230, Springer, Berlin, 1985.

[471] M.E. Taylor, *Analysis on Morrey spaces and applications to Navier-Stokes and other evolution equations.* Comm. in P.D.E., **17** (1992), p. 1407–1456.

[472] R. Temam, Navier–Stokes equations. North-Holland, Amsterdam, 1977.

[473] H. Tennekes and J.L. Lumley, A first course in turbulence. M.I.T. Press, Cambridge, 1973.

[474] A.A. Townsend, The structure of turbulent shear flow. Cambridge University Press, Cambridge, 1976.

[475] D.J. Tritton, Physical fluid dynamics. 2nd edition, Van Nostrand Reinhold, New York, 1988.

[476] B. Turkington, *Vortex rings with swirl.* SIAM J. Math. Anal., **20** (1989), p. 57–70.

[477] S. Ukaï, *The incompressible limit and the initial layer of the compressible Euler equation.* J. Math. Kyoto Univ., **26** (1988), p. 323–331.

[478] M. Van Dyke, Perturbation methods in fluid mechanics. The Parabolic Press, Stanford, 1975.

[479] M. Van Dyke, An album of fluid motion. The Parabolic Press, Stanford, 1982.

[480] A. Valli, *On the existence of stationary solutions to compressible Navier–Stokes equations.* Ann. I.H.P., Anal. Non Lin., **4** (1987), p. 99–113.

[481] A. Valli, *Mathematical results for compressible flows.* Mathematical topics in fluid mechanics (J.F. Rodrigues and A. Sequeira, eds.), Pitman Research Notes in Math. Ser. 274, Wiley, New York, 1992, p. 193–229.

[482] A. Valli and W.M. Zajaczkowski, *Navier–Stokes equations for compressible fluids: Global existence and qualitative properties of*

the solutions in the general case. Comm. Math. Phys., **103** (1986), p. 259–296.

[483] T. von Karman, <u>Aerodynamics</u>. McGraw-Hill, New York, 1954.

[484] W. von Wahl, <u>The equations of Navier–Stokes and abstract parabolic equations</u>. Braunschweig, Vieweg, 1985.

[485] V.A. Weigant, *Example of non-existence in the large for the problem of the existence of solutions of Navier–Stokes equations for compressible viscous barotropic fluids*. (In Russian). Dokl. Akad. Na., **339** (1994), p. 155–156.

[486] V.A. Weigant and A.V. Kazhikov, *The global solvability of initial boundary value problem for potential flows of compressible viscous fluid at low Reynolds numbers*. (In Russian). Dokl. Akad. Na., **340** (1995), p. 460–462.

[487] V.A. Weigant and A.V. Kazhikov, *Global solutions to equations of potential flows of compressible viscous fluid at low Reynolds number*. To appear in Diff. Eq.

[488] F.B. Weissler, *The Navier–Stokes initial value problem in L^p*. Arch. Rat. Mech. Anal., **74** (1981), p. 219–230.

[489] M. Wiegner, *Decay results for weak solutions of the Navier–Stokes equations in $\mathbb{R}^n$*. J. London Math. Soc., **35** (1987), p. 303–313.

[490] W. Wolibner, *Un théorème sur l'existence du mouvement plan d'un fluide parfait, homogène, incompressible, pendant un temps infiniment long*. Math. Z., **37** (1933), p. 698–726.

[491] H.F. Yashima and R. Benabidallah, *Unicité de la solution de l'équation monodimensionnelle ou à symétrie sphérique d'un gaz visqueux et calorifère*. Rendi. del Circolo Mat. di Palermo, Ser. II, **XLII** (1993), p. 195–218.

[492] H.F. Yashima and R. Benabidallah, *Equation à symétrie sphérique d'un gaz visqueux et calorifère avec la surface libre*. To appear in Annal Mat. Pura Applicata.

[493] V.I. Yudovich, *Non-stationary flow of an ideal incompressible liquid*. Zh. Vych. Mat., **3** (1963), p. 1032–1066 (in Russian).

[494] V.I. Yudovich, *Uniqueness theorem for the basic nonstationary problem in dynamics of ideal compressible fluid*. Math. Rev. Lett., **2** (1995), p. 27–38.

[495] N. Zabusky, M. Hugues and K. Roberts, *Contour dynamics for the Euler equations in two dimensions*. J. Comput. Phys., **30** (1979), p. 96–106.

[496] Y. Zheng, *Concentration-cancellation for the velocity fields in two dimensional incompressible fluid flows.* Comm. Math. Phys., **135** (1991), p. 581–594.

[497] A. Zygmund, *On a theorem of Marcinkiewicz concerning interpolation of operators.* J. Math. Pures Appl., **35** (1956), p. 223–248.

INDEX